Arithmetic and Calculators

Arithmetic and Calculators

HOW TO DEAL WITH ARITHMETIC IN THE CALCULATOR AGE

William G. Chinn
SAN FRANCISCO CITY COLLEGE

Richard A. Dean
CALIFORNIA INSTITUTE OF TECHNOLOGY

Theodore N. Tracewell
CALIFORNIA STATE UNIVERSITY, HAYWARD

W. H. FREEMAN AND COMPANY
San Francisco

Library of Congress Cataloging in Publication Data

Chinn, William G.
  Arithmetic and calculators.

  Includes index.
  1. Arithmetic—1961–   2. Calculating-machines.
I. Dean, Richard A., joint author. II. Tracewell,
Theodore N., joint author. III. Title.
QA107.C62   513'.028'5   77-11111
ISBN 0-7167-0016-6
ISBN 0-7167-0015-8 pbk.

Printed in the United States of America

AMS 1970 subject classifications: 98A05, 98B99

2 3 4 5 6 7 8 9

Contents

Contents

Contents

Prelude

To our dear readers:

This is a book about arithmetic—about the arithmetic of the calculator age.

We are not going to overwhelm you with rules about carrying and borrowing or with a list of terms like *"trial divisors,"* although we shall surely try some divisors. We shall not make a fuss about addends, subtrahends, and minuends. Instead, we shall concentrate on what makes arithmetic tick and what arithmetic can do for you—the kinds of problems arithmetic can solve and how to solve them.

The arithmetic itself is to be done in large part by a calculator. You will need an inexpensive electronic calculator to enjoy this book. A calculator cannot tell you how to do a problem. It *can help* you learn. It *can help* you to apply what is important in arithmetic. We think your calculator may change your attitude toward arithmetic. It should make arithmetic more useful to you. It will free you from the drudgery of calculating. You will be free to concentrate on the applications of arithmetic. It will give you a new tool to solve everyday problems. And we hope that it will give you some fun in playing and experimenting with numbers.

Let's begin with a psychological experiment. How do you respond to these questions?

What do you think of when you think of numbers?

How do you picture the counting numbers,

$$1, 2, 3, 4, 5, 6, 7, 8, 9, 10, 11, 12, \ldots ?$$

Express your ideas in a few words, or if you like, draw a picture! Let your feeling toward numbers go wild! If you can, get some other persons to answer these questions.

Compare your answers. You will discover that people give different and often personal answers to these questions. We are sure you will find these answers, interesting and informative. Afterward, but please not before, you may want to look at Appendix A, where we have recorded some of the responses given by an adult class in arithmetic.

There is no unique image of counting numbers that all people carry around. Indeed, there are many different impressions of numbers and arithmetic. This means that, to perform arithmetic operations, you may want to call upon your own personal view of how operations work.

Many of our likes and dislikes are shaped by our earliest contacts. In the case of arithmetic, this contact often comes from unhappy experiences in early school learning situations. Too often our natural desires to quantify our experiences are snubbed by demands for a rote recall of number facts for which no need has been established. As we grow older, our feelings and frustrations toward arithmetic grow into a helpless confusion in a sea of numbers. Now with the aid of these relatively inexpensive calculators, we have the capability of dealing swiftly and correctly with the processes of arithmetic and can concentrate on their use. Incidentally, the development of these electronic calculators is the first technological breakthrough in computation to become available to most people since the abacus.

Another reward for having a calculator at hand is that you will be able to play and to experiment with numbers. You can use a calculator to discover things about numbers not easily revealed through hand computation. In this way, you will see things about numbers that you would not be able to appreciate by hand. In this way, a calculator is to arithmetic what a microscope is to nature. With a microscope, you see things that you cannot with the unaided eye.

In arithmetic, observations are often expressed through the recognition of patterns. Here are some elementary examples.

Example 0.1:   Counting by 2's.

2,	4,	6,	8,	10,
12,	14,	16,	18,	20,
22,	24,	26,	28,	30,

. . .

These are the so-called "*even*" numbers. An even number is one that is a multiple of 2. Numbers that end in a 2, 4, 6, 8, or 0 are multiples of 2. This fact is suggested by the table above for we can see how the rows repeat. Thus, for example, 48 is an even number because it ends in an 8. Indeed,

$$48 = 24 \times 2.$$

Example 0.2: Counting by 5's.

$$5, \quad 10,$$
$$15, \quad 20,$$
$$25, \quad 30,$$
$$\cdots$$

These numbers are the multiples of 5. The pattern of the table suggests that any number ending in a 5 or a 0 is a multiple of 5. Thus, for example, 75 is a multiple of 5. Indeed,

$$75 = 15 \times 5$$

Example 0.3: Some special products. Explain what is happening here:

$$8 \times 9 \qquad = 72$$
$$8 \times 99 \qquad = 792$$
$$8 \times 999 \qquad = 7992$$
$$8 \times 9999 \qquad = 79992$$
$$8 \times 99999 \qquad = 7\underline{\hspace{1cm}}2 \quad \text{(Fill in the blank)}$$
$$8 \times 999999 \qquad = \underline{\hspace{1.5cm}} \quad \text{(Write in the answer)}$$
$$8 \times \underline{\hspace{1.5cm}} = 79999992 \quad \text{(Fill in the blank)}$$

If you see the pattern, then you can make a good guess even without checking the answer.

Other interesting patterns arise in division.

Example 0.4: Division by 3; the decimal part. (It will help if you use the calculator. If its operation is unfamiliar, go on to Chapter 1 and then return to this example when you are ready.) Begin systematically. We recorded the answers as they will appear on an 8-digit calculator.

$$1 \div 3 = 0.3333333$$

$$2 \div 3 = 0.6666666$$

$$3 \div 3 = 1. \qquad \text{(or 1.0000000 on some calculators)}$$

$$4 \div 3 = 1.3333333$$

$$5 \div 3 = 1.6666666$$

$$6 \div 3 = 2. \qquad \text{(or 2.0000000 on some calculators)}$$

$$7 \div 3 = 2.3333333$$

$$8 \div 3 = 2.6666666$$

$$9 \div 3 = 3. \qquad \text{(or 3.0000000 on some calculators)}$$

and so on.

Do you see a pattern? Can you predict the decimal part of $11 \div 3$? of $13 \div 3$?

Example 0.5: Division by 7; the decimal part. What is the decimal form of the remainder when a whole number is divided by 7?

$$1 \div 7 = 0.1428571$$

$$2 \div 7 = 0.2857142$$

$$3 \div 7 = 0.4285714$$

$$4 \div 7 = 0.5714285$$

$$5 \div 7 = 0.7142857$$

$$6 \div 7 = 0.8571428$$

$$7 \div 7 = 1. \qquad \text{(or 1.0000000 on some calculators)}$$

$$8 \div 7 = 1.1428571$$

$$9 \div 7 = 1. \underline{\qquad} \qquad \text{(Complete the answer!)}$$

Do you see a pattern—or perhaps several patterns? Can you predict the decimal part of $15 \div 7$?

Here are some things you might notice about Example 0.5. The decimal part always begins and ends with the same digit. The sequence of the digits is always the same, but the starting place differs.

> 1 is followed by 4
>
> 4 is followed by 2
>
> 2 is followed by 8
>
> 8 is followed by 5
>
> 5 is followed by 7
>
> 7 is followed by 1
>
> and now the cycle repeats.

If you like these patterns, try divisions by 13 and look for similar patterns.

Example 0.6:   Magic squares. The following arrangement of the whole numbers from 1 to 9 has been known for a long time. We can imagine that arrangements of this type were first found by doodling—very much as we are suggesting playing around with a calculator. To find what is magic about the arrangement below, add the numbers in each row and column.

8	1	6
4	9	2
3	5	7

row sum = _____

row sum = _____

row sum = _____

Perhaps that's not so impressive. Now add the sums of each column. Each column sums to 15 also! Does this seem surprising?

Now consider this arrangement of the same numbers.

2	7	6
9	5	1
4	3	8

Again add the numbers in each row and column. The result is 15. But there is more magic! Add the numbers on each of the diagonals as indicated by the dotted lines in the square above. Again 15! (More magic and more magic squares will be discussed in Chapter 11.)

There may seem to be much that is magic about numbers and arithmetic. There is! Sometimes solutions to difficult problems call for insight and a stroke of cleverness born of intuition and experience. But more often, systematic methods for the analysis of problems are available. Indeed, it is the power of such systems that make mathematics a useful discipline. These systems are based on universal mathematical concepts and constructions. Together they compose the world of ordered beauty that is mathematics. So, before we can fly on wings of magic we must walk along the roads leading to these basic mathematical principles. Your calculator will help you gain the experience and technique so necessary to the development of insight and intuition. Join us on this journey.

$$\boxed{1}$$

Getting Started
on Your Calculator

1.1   A Look at Your Calculator

Electronic calculators come in many sizes and shapes. They differ in the kinds of mathematical functions they perform and in the complexity of problems they solve with these functions. In this book we shall suppose that you have a calculator that—

1. accepts eight-digit numbers for computation (an *"8-place"* calculator);

2. performs the four basic arithmetic operations, addition (+), subtraction (−), multiplication (×), and division (÷);

3. uses *"floating-point"* arithmetic; and

4. uses *"algebraic logic."*

Many calculators have features to supplement these basic features. Some have special keys for square roots, percents, reciprocals, and powers. Others will compute logarithms and trigonometric functions. Some calculators have one or more memory registers, activated by pressing memory keys (M), which permit partial computations to be stored for later recall while still other calculations are performed. Some machines have *"fixed-point"* arithmetic. For these, the results of computations are displayed to a fixed number of decimals—say, to hundredths. For others, the position of the decimal point may vary from problem to problem according to the computations. These have a *"floating-point"* mode. Then again, some machines can use either mode. The floating-point mode is the more flexible and will have the greatest use for us in studying arithmetic. We shall illustrate some of these extra features in special sections as we need them.

Look at your calculator and become familiar with its operation. Because of the wide variation in calculators, it is impossible to write detailed instructions to fit every machine. Read the operation manual accompanying your machine and practice with it.

Still, we must agree on a language and symbolism to convey the arithmetical procedures we want to study in this book. So let us have a look at the parts of a calculator and do some easy exercises to gain familiarity and to fix our notation for its operation.

1.2 The Physical Set-Up

Look at your calculator. Find the "off–on" switch. It won't work unless you turn it on! To save batteries, turn it off when you have completed a calculation and recorded the final answer.

Prominently displayed on your calculator is a "window." The numbers you wish to manipulate will appear here as you enter them into the calculator. The answer to your computations will appear in this window also.

The rest of the face of your calculator is occupied by its keys. Each key is labeled to show its function. Some sophisticated machines give two or more different functions for the same key. We shall assume that you have a fairly simple machine with one function per key.

Most calculators have a central cluster of *numerical* keys, which look like this:

7	8	9
4	5	6
1	2	3
0		

In machines using *"algebraic logic,"* these numerical keys are surrounded by operational or command keys, including decimal-point (.), equal ($=$), addition ($+$), subtraction ($-$), multiplication ($\times$), and division ($\div$) keys. In this book, use of these keys will be indicated by the following symbols.

7	8	9	÷
4	5	6	×
1	2	3	−
0	·	=	+

There are other keys as well. The clear (C) key permits you to begin a new problem by erasing all numbers associated with any previous calculation. Turning the machine off and then on also has that same effect, because electrical impulses are necessary to keep a number in a register. Another key that most calculators have is a *"clear entry"* (CE) key, which permits you to correct a mistake in the number you are entering without disturbing other numbers or calculations already stored elsewhere in the machine. On some machines, this function is combined with the (C) key. Read your operating manual. If your machine has a memory, there are special keys to control it.

Some machines have a *constant* key or switch (K). When you activate this switch, you may make several multiplications (or divisions) with the same factor (or divisor). For example, in calculating these products.

$$8 \times 2, \quad 4 \times 2, \quad 36 \times 2,$$

the constant switch causes the machine to remember the constant factor, 2, and so this factor need not be re-entered for each multiplication. Many machines have this capability as an automatic feature of their logic, so that a separate switch is unnecessary.

1.3   Entering Numbers into Your Calculator

Here are some easy exercises to get started with your calculator.

Example 1.1:   Enter 325.

Put the number, 325, into the calculator so that, for example, something could be added to it.

Action		*Display*
Turn your calculator **ON**	Read the display:	*0.*
Press numerical key, **3**	Read the display:	*3*
Press numerical key, **2**	Read the display:	*32*
Press numerical key, **5**	Read the display:	*325*

You are not done yet! The calculator does not know that there are only three digits in your number. For all it knows, you might have been trying to enter

3257895674893.234562831853942477780125663!

You must tell the calculator that you have finished entering a number.

Press operation key **(=)**   Read the display: *325.*

Pressing the (=) key tells the machine that a number has been entered. Some machines will, at this point, add the decimal point as we have indicated to acknowledge that the number is complete. Other machines "blink" as you press the (=) key. As a matter of fact, you can end the entry of a number by pressing any operation key. The key you choose should depend on what you want to do next. However, because any other operation key may follow an (=) command, it is a good one for us to use here.

In the future, we shall describe the actions in Example 1.1 simply as follows.

325 ▣

In this book, numerals printed as

0 1 2 3 4 5 6 7 8 9

designate numbers that either have been entered into a calculator or are the results of calculator computations.

Example 1.2: Enter 3.25.

This shows how to use the decimal point. We would use it in working with $3.25.

Enter 3.25	*Action*	*Display*	*Remarks*
	Press **(C)**	*0.*	Clear all registers.
	Press **3**	*3*	
Key in 3.25	Press **(.)**	*3.*	
	Press **2**	*3.2*	
	Press **5**	*3.25*	
	Press **(=)**	*3.25*	Complete entry.

or, more briefly,

3.25 ▣

Example 1.3: Enter .25.

This would be useful in working with 25 cents, or $0.25.

Enter .25	*Action*	*Display*
	Clear	*0.*
	Press (.)	*0.*
Key in .25	Press 2	*0.2*
	Press 5	*0.25*
	Press (=)	*0.25*

or, briefly,

$$0.25 \; \boxed{=}$$

Exercises 1.3

1. Enter these numbers in your calculator. Be sure that you do so correctly.
   - •a. 26
   - •b. 306
   - •c. 780
   - •d. .9341
   - e. 8.763
   - •f. 6.032
   - g. 89760031
   - h. 7321653
   - i. 655.3293

•2. Find out what happens to your calculator when you try to key in a nine-digit number such as 123456789. (Every machine has some symbol or action to let you know that you have tried to exceed the capacity of the machine.)

3. For fun, enter 0.07734 and turn your calculator so that you are reading the display upside-down. Does your calculator speak to you? What other words or messages can you form in this way?

4. What numbers should be entered in your calculator to record the following amounts?
   - •a. one dollar and a half
   - •b. fifty-five cents
   - c. eleven dollars and a quarter
   - d. one hundred sixty-five dollars and fifty-six cents
   - •e. one thousand dollars and one cent

5. What is the largest amount of money (in dollars and cents—you must include the cents) that you can record on your calculator—
   - a. if you may use the same digit more than once?
   - •b. if you must use a different digit in each place?

1.4   Elementary Operations with Your Calculator

Here are some examples to show how the four main functions are used. We indicate entering of numbers in abbreviated style.

Example 1.4:   Add 2 plus 3. In symbols, $2 + 3$. The answer is 5. In symbols, $2 + 3 = 5$.

2 + 3	Action	Display	Remarks
	Clear	0.	
	Key in **2**	2	
	Press **(+)**	2.	Pressing (+) completes the entry of 2 into the calculator.
	Key in **3**	3	
	Press **(=)**	5.	Answer!

In the future, we shall indicate this sequence of machine operations by

$$2 \boxplus 3 \boxminus 5.$$

Example 1.5:   Multiply 2 by 3. In symbols, 2×3. The answer is 6. In symbols, $2 \times 3 = 6$.

2 × 3	Action	Display	Remarks
	Clear	0.	
	Key in **2**	2	
	Press **(×)**	2.	This completes the entry of 2.
	Key in **3**	3	
	Press **(=)**	6.	Answer!

In the future, we shall indicate this by

$$2 \boxtimes 3 \boxminus 6.$$

Example 1.6:   Subtract 2 from 3. In symbols, $3 - 2$. The answer is 1. In symbols, $3 - 2 = 1$.

3 − 2	Action	Display	Remarks
	Clear	0.	
	Key in **3**	3	
	Press **(−)**	3.	
	Key in **2**	2	
	Press **(=)**	1.	Answer!

Briefly,

$$3 \boxminus 2 \boxminus 1.$$

Example 1.7: Divide 6 by 3. In symbols, $6 \div 3$. The answer is 2. In symbols, $6 \div 3 = 2$.

6 ÷ 3	Action	Display	Remarks
	Clear	*0.*	
	Key in **6**	*6*	
	Press (÷)	*6.*	Completes entry of 6.
	Key in **3**	*3*	
	Press (=)	*2.*	Answer!

Briefly,

$$6 \;\boxed{\div}\; 3 \;\boxed{=}\; 2.$$

Next, we show some examples found in common business situations.

Example 1.8: Add 1.25 and 2.50 (as in \$1.25 + \$2.50). For example, Sandy bought a can of coffee for \$1.25 and a case of soft drinks for \$2.50; what was the total bill? The answer is \$3.75.

1.25 + 2.50	Action	Display	Remarks
	Clear	*0.*	
	Key in **1.25**	*1.25*	
	Press (+)	*1.25*	
	Key in **2.5**	*2.5*	On machines using floating arithmetic, the final zeros after the decimal point need not be entered although no harm occurs if you do so.
	Press (=)	*3.75*	

Briefly,

$$1.25 \;\boxed{+}\; 2.5 \;\boxed{=}\; 3.75$$

Example 1.9: Multiply .25 by 6. For example, Peter bought 6 candy bars. Each costs 25¢ or \$0.25. How much is the total cost?

.25 × 6	Action	Display	Remarks
	Clear	*0.*	
	Key in **.25**	*0.25*	
	Press (×)	*0.25*	
	Key in **6**	*6*	
	Press (=)	*1.5*	Answer! Some machines may show 1.50. Interpret our answer as \$1.50.

Briefly,

$$.25 \;\boxed{\times}\; 6 \;\boxed{=}\; 1.5$$

Example 1.10: Divide 3 by 4. For example, Cindy bought 4 burgers. The total bill was $3. How much was each burger?

3 ÷ 4	Action	Display	Remarks
	Clear	*0.*	
	Key in **3**	*3*	Even if you were thinking of 3 as $3.00, in floating-point it is unnecessary to key in the additional zeroes.
	Press (÷)	*3.*	
	Key in **4**	*4*	
	Press (=)	*0.75*	Interpret this answer for the burger problem as $0.75 or 75 cents.

Briefly,

$$3 \; ÷ \; 4 \; = \; 0.75$$

Example 1.11: Subtract 2.75 from 5. For example, you pay for a purchase of $2.75 with a $5 bill. What is your change?

5 — 2.75	Action	Display
	Clear	*0.*
	Key in **5**	*5*
	Press (—)	*5.*
	Key in **2.75**	*2.75*
	Press (=)	*2.25*

Briefly,

$$5 \; - \; 2.75 \; = \; 2.25$$

Exercises 1.4

Perform the following operations with your calculator and compare each result with the answer provided in parentheses.

1. 1234 + 567	(1801.)		7. 11 ÷ 11	(1.)
2. 1234 × 567	(699678.)		• 8. 11 × 11	(121.)
3. 1234 − 567	(667.)		• 9. 11 × 121	(1331.)
•4. 1234 ÷ 567	(2.1763668)		•10. 11 × 1331	(14641.)
5. 11 + 11	(22.)		•11. 101 × 10201	(1030301.)
6. 11 − 11	(0.)		•12. 101 × 1030301	(guess!)

In each of the examples we have just examined (Examples 1.4 through 1.11), we have performed an operation on *two* numbers. Your calculator can only perform an operation on a pair of numbers at one time. This is not a limitation due to the small size of your calculator; *all* computers operate that way. However, because your calculator keeps the result of one such operation (on display or hidden inside,) it is possible to perform an operation combining that number with a new one. Here are some examples.

Example 1.12:   Add 2, 3, and 6 together. In symbols, $2 + 3 + 6$. The strategy is to add 2 and 3 (obtain 5), and then add 6. In symbols, we write $(2 + 3) + 6$, enclosing the numbers 2 and 3 in parentheses to show that they have been combined first.

> $2 + 3 + 6 = (2 + 3) + 6$

On your calculator,

$$(2 + 3) + 6 = ?$$
$$(2\ \boxplus\ 3) + 6 \to 5\ \boxplus\ 6\ \boxed{=}\ 11$$

Example 1.13:   Add 12 and 13, and then multiply by 3. In symbols, $(12 + 13) \times 3$. Here again, the first operation, $(12 + 13)$, is enclosed in parentheses to show that it is performed before the multiplication by 3.

   This computation is involved in answering: "Betty bought 12 pieces of gum and 13 gumdrops. If each item cost 3 cents, what was the cost of the purchase?"

Analysis: Betty bought $(12 + 13)$ items at 3 cents each. The total cost was $(12 + 13) \times 3$ cents.

$$(12 + 13) \times 3 = ?$$
$$(12\ \boxplus\ 13) \times 3 \to 25\ \boxed{\times}\ 3\ \boxed{=}\ 75$$

The answer is interpreted as 75 cents.

Perhaps you have noticed that we have made no mention of *fractions*. We have not given any examples with fractions like $\frac{1}{2}$. This is because calculators work only with decimals, not with common fractions. On the calculator, $\frac{1}{2}$ is processed through its basic interpretation, $1 \div 2$. More of this in Chapter 8.

   Later, we shall become proficient at converting fractions to decimals (it's easy on a calculator) and shall work entirely with decimals. We are already familiar with these conversions when working with dollars and cents. The name of the coin may be a "quarter," but we write checks for $3.25, not "three and one quarter dollars."

Exercises 1.5

1. Perform these operations on your calculator. Check your work against the answers given in parentheses.

   a. $6 + 3$ (9)   e. $121 \div 11$ (11)

   b. 6×3 (18)   •f. $1975 - 1924$ (51)

   c. $6 - 3$ (3)   •g. 2806×26 (72956)

   •d. $16 + 101$ (117)

•2. Compute the following sums. Is there a pattern?   [HINT: Don't clear your calculator between parts!]

   a. $1 + 2$ c. $1 + 2 + 3 + 4$

   b. $1 + 2 + 3$ d. $1 + 2 + 3 + 4 + 5$

•3. Perform the following operations. Is there a pattern? Compare your answer with that for Exercise 2.   [HINT: For part a, the procedure can be written as $(2 \times 3) \div 2 = 3$. In other words, key in **2**, press **(×)**, key in **3**, press **(÷)**, key in **2**, press **(=)**; the answer is 3.]

   a. $(2 \times 3) \div 2$ c. $(4 \times 5) \div 2$

   b. $(3 \times 4) \div 2$ d. $(5 \times 6) \div 2$

4. A food cooperative bought the following supplies. What was the total expense?

   > 8 dozen eggs at $0.75 per dozen;
   > 8 steaks at $2.50 each; and
   > 8 gallons of milk at $1.25 per gallon.

5. Make up five problems of your own invention to use in practicing your skills with the calculator.

2

Counting and
Decimal Notation

2.1   Once Upon a Time . . .

In this chapter, we deal with the basic concepts of numbers and ways
to record numbers. These concepts are absolutely necessary to fully
understand your calculator and to understand more advanced mathematical processes.

Much of this chapter will undoubtedly be familiar to you.

- •Do you know what the digits in the number 1339.25 signify?

- •Do you know whether it is larger or less than 1339.2499?

- •Do you know why decimal notation makes it easy to compare numbers?

If you know these things, you may want to skip this chapter!

On the other hand, you may be entertained by the fable we tell in
this chapter to explain how counting and the decimal system were
invented. You may also find some of the exercises interesting. We
invite even the experts among our readers to browse through this
chapter and its exercises.

2.2 The Old "Bean-in-the-Pot" Routine

No one can be sure just when people first began to develop a procedure for counting. No one can be sure why someone made a start along this direction. Perhaps it was curiosity; perhaps it was need. Whatever the motivation, through however many countless years, people did develop a routine and a powerful notation for counting. We want to tell you a tale to describe how counting might have begun. Although our tale is fictitious, its ideas cannot be too far from the thinking that must have gone on, even if those thoughts were not symbolized and written down.

Imagine a land of long ago, when the hills were young and life was simple. In that land there lives a shepherd who tends a flock of sheep. He is worried that not all his sheep will return from pasture. To keep track of his flock, he lets out his sheep one by one. As each sheep leaves the fold, he puts a bean in a pot:

one sheep → one bean.

When the sheep return to the fold from grazing, he reverses the procedure. He removes a bean from the pot as a sheep returns:

one bean → one sheep.

Figure 2.1
The connecting arrows show that each sheep is matched with one bean, and each bean with one sheep. This is a one-to-one correspondence between sheep and beans.

Our story makes it clear that, by pairing one bean with one sheep, there must be the same number of sheep as beans. Such a pairing is called a *one-to-one correspondence,* or *1–1 correspondence.*

Example 2.1: Set up a 1–1 correspondence between these games and numbers:

        Games: basketball, football, hockey, baseball

and

        Numbers: 5, 11, 6, 9

Solution: A 1–1 correspondence is set up as indicated by the arrows in the table below. Here, a game is associated with the number designating the number of players on a team.

Games	Numbers
basketball ⟵———————⟶	5
football ⟵———————⟶	11
hockey ⟵———————⟶	6
baseball ⟵———————⟶	9

Correspondences need not be one-to-one. Many practical examples involve less regular types of correspondence.

Example 2.2: Show a correspondence between flowers and colors by drawing an arrow from a *flower* to a *color* if there is a variety of the flower having the color.

Solution:

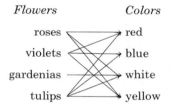

For two sets to have a 1–1 correspondence, they must have the same number of elements.

Example 2.3: Can these two sets be placed in a 1–1 correspondence?

B: Three candy bars

C: Four children

Solution: No. There are fewer candy bars than persons.

The very idea of 1–1 correspondence gives us a method of determining whether one number is larger than another.

Example 2.4: A teacher wonders if she has enough crayons for her class. Which is greater, the number of crayons or the number of pupils?

Solution: Start passing out crayons, one to each pupil. If each child does not get a crayon, there are more children. If there are some crayons left, there are more crayons. If each child has a crayon and there are no crayons left over, then the number of crayons is the same as the number of pupils.

Example 2.5: Suppose the teacher's class has seven members and the teacher has ten crayons. Is 10 greater than 7?

Solution: We can picture the attempt at a 1–1 correspondence as illustrated in the following diagram. We see that when we have associated one crayon with each child, there will be crayons left over. Hence, 10 is greater than 7.

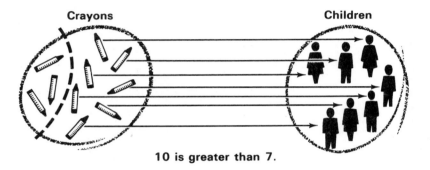

10 is greater than 7.

Viewing the same situation the other way around, we see that each child is associated with one crayon, and that there are crayons with which no child is associated. Hence, 7 is less than 10.

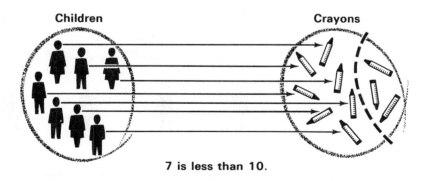

7 is less than 10.

1. For each part, pair the members of the sets according to the given directions. That is, set up a correspondence.

   •a. Pair the states listed in S with their capital cities listed in C.

      S = {Arizona, California, Florida, New Jersey, New York, Ohio, Washington}

      C = {Albany, Columbus, Olympia, Phoenix, Sacramento, Trenton, Tallahassee}

   b. Pair the countries listed in C with the names of their basic units for money in M.

      C = {Australia, England, France, Italy, United States}

      M = {dollar, franc, lira, pound}

   c. Six persons are seated around a circle as shown. Set up a correspondence between each person in P and his or her neighbors in the circle from N.

      P = {Charity, Dick, Faith, Harry, Hope, Tom}

      N = {Charity, Dick, Faith, Harry, Hope, Tom}

   •d. Pair the upper-case letters listed in U with their lower-case letters listed in L.

      U = {A, R, T} L = {r, a, t}

2. For each part of Exercise 1, state whether or not the correspondence is one-to-one.

•3. A class has 25 students. Three correspondences are given below. Explain why each correspondence need not be one-to-one.

   a. Each student is paired with his grade.

   b. Each student is paired with his major subject area.

   c. Each student is paired with his birthdate.

4. Tell which parts of Exercise 3 *cannot* give a 1–1 correspondence. Tell which parts of Exercise 3 *may* give a 1–1 correspondence.

5. Is there a 1–1 correspondence between a set of Social Security numbers and the set of persons having these Social Security numbers? Explain.

•6. Suggest two sets between which there is a 1–1 correspondence.

7. Without counting objects, determine which set has more members.

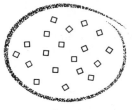

Set 1   **Set 2**

2.3 Place-Value Notation

We continue our story of the shepherd and his flock.

Times were good and the flock prospered. In no time the flock increased in size. In counting his flock, the shepherd filled one pot after another with beans. He just had to find a way to keep the count without using so many beans! One day, as he was dropping beans in the first pot, he had an inspiration!

When he got as many beans in the pot as he had fingers on both hands, he dumped out all the beans from this pot and put one (1) bean in a second pot at his left to show that he now had two handfuls (10) of sheep. Then he continued to put beans in the first pot for each sheep he counted.

Before the idea   **After the idea**

Each bean stands for   Each bean in the   Each bean in the
one sheep: 13 beans,   left pot stands   right pot stands
13 sheep.   for 10 sheep.   for 1 sheep.

THIRTEEN   EQUALS   TEN   PLUS   THREE

Figure 2.2
How the shepherd recorded 13 sheep using his new idea.

Because each bean in the pot on the left counts for two handfuls (ten) of sheep, while each bean in the pot on the right counts for one sheep, the shepherd has counted 10 and 3 sheep. Today, of course, we say 10 + 3 or 13 (thirteen) sheep.

He continued to count and somewhat later he found that he had three beans in the left pot and one bean in the right pot. How many sheep had he counted by then?

Left pot   **Right pot**

Each bean in the left   Each bean in this pot
pot stands for 10 sheep.   stands for 1 sheep.

Three tens = Thirty-   One

Figure 2.3
How the shepherd recorded 31 sheep: 30 + 1 = 31.

The shepherd noticed that his system used the same number of beans (four in all) to record 13 sheep as to record 31 sheep. The difference lies in the placement of those four beans! In our decimal notation, 13 and 31 use the same digits (1 and 3), but the places in which the "1" and the "3" occur is all important.

Things seem to be going well. The counting continued until the shepherd had nine beans in the left pot. As the beans collected in the right pot, the shepherd knew he had to do something or he would soon have a build-up of too many beans in the left pot.

Left pot   **Right pot**

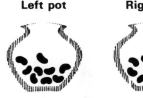

Nine tens = 90   Nine ones = 9   More sheep to come.
How shall they be counted?

Figure 2.4
The shepherd has counted 99 sheep. One more sheep makes a big change in his system. It leads to the invention of a new pot farther to the left.

Our shepherd was a logical fellow, so he did precisely as he had done before. He got a third pot, placed it to the left of the other two pots he already had, and emptied the two pots. Then, he placed a single bean into the new pot to show that he had ten tens of sheep.

His reasoning went like this. One more bean in the "ones" pot causes a dumping out of the beans in this pot in exchange for putting one bean in the "tens" pot. This brings the "tens" pot to an overload. So there is a dumping out of the ten beans in the "tens" pot in exchange for one bean in the next pot. Thus, he continued.

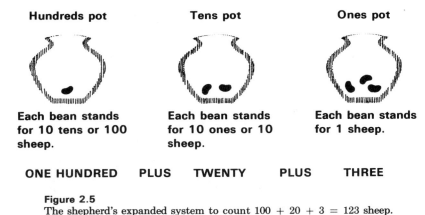

Hundreds pot	**Tens pot**	**Ones pot**
Each bean stands for 10 tens or 100 sheep.	Each bean stands for 10 ones or 10 sheep.	Each bean stands for 1 sheep.
ONE HUNDRED PLUS	**TWENTY** PLUS	**THREE**

Figure 2.5
The shepherd's expanded system to count $100 + 20 + 3 = 123$ sheep.

Once this giant step was taken, the shepherd was prepared to continue this process repeatedly. He knew that he had enough ideas to count all the stars he could see at night—if only he had enough beans! At least he didn't have to have one bean for each star! He wondered how many stars there were . . . .

The next job was to invent symbols to represent the number of beans in each pot. These we now use were derived mostly from Arabic script

١	٢	٣	٤	٥	٦	٧	٨	٩
1	2	3	4	5	6	7	8	9

and from early Hindu characters

1	2	4	6	7

The numeral 0, a space occupier, comes from the Hindus. It is used to show a zero count of the ones, tens, hundreds, or other bunches of tens. For these historical reasons, our system of writing numerals is known as the Hindu-Arabic system. Because the groupings are in bunches of ten, the system is also referred to as the *decimal numeration system,* or *base-ten system.* Incidentally, the word, *decimal* comes from the Latin word *decem,* meaning "ten."

A decimal numeral, then, is built by combining the symbols

0, 1, 2, 3, 4, 5, 6, 7, 8, or 9

occupying particular decimal "places." The placement of these digits is exactly like that of the shepherd's pots. Thus, for example, the symbol

3 5 4 3 8

represents, proceeding from right to left,

8 ones

3 tens

4 hundreds

5 thousands

3 ten-thousands

Displayed horizontally,

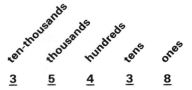

Figure 2.6
35438 is thirty-five thousand,
four hundred thirty-eight,
or 35 thousand, 438.

The legend of the shepherd is a story about the invention of a system to *record* numbers—the result of a counting process. The idea of counting, of a 1–1 correspondence, must precede any notation for recording the result of the count. A symbol such as **12** or **XII,** is called a *numeral,* and designates a number. Some designations follow for the number of objects that may be placed in a 1–1 correspondence with this set of dots.

12	**(base-ten numeral)**
XII	**(Roman numeral)**
⊘))	**(Egyptian numeral)**
⊥⊥⊥⊥ ⊥⊥⊥⊥ //	**(Robinson Crusoe's tally marks)**
twelve	**(English word)**
dozen	**(English word)**
10 + 2	**(arithmetic fact of addition)**
6 × 2	**(arithmetic fact of multiplication)**

The word *number* is used to express many ideas, concepts, and constructs in mathematics. The numbers, 1, 2, 3, . . . , that occur in counting are often called, informally, "*counting numbers.*" In arithmetic we consider several kinds of numbers. We want to tune our vocabulary to the current usage in mathematics. We shall introduce some special terms as we go along. For now, we introduce the term

integer

to mean one of the numbers 0, 1, 2, 3, . . . , or the negative of one of these. (See Chapter 6.) The word *integer* comes from a Latin word meaning "untouched or whole." Informally, the numbers 0, 1, 2, 3, . . . are also called *"whole numbers"* to distinguish them from fractions like $\frac{1}{2}$. More of that later.

A system that goes hand-in-hand with the decimal system for numeration is the *metric system* for measuring things. The decimal places designate

ones, tens, hundreds, thousands, and so on

on the one hand, and

ones, tenths, hundredths, thousandths, and so on

on the other.

We shall discuss tenths, hundredths, . . . in greater detail in Chapter 8. But we are sure that you have worked enough with these before so that you can see how the decimal system is like the *system* in the metric system.

In the metric system, special prefixes are attached to basic units to label ten of the units, a hundred of the units, and so on. Also, special prefixes are used to label tenths, hundredths, etc. of the basic units. Thus,

a *kilo*gram means *1000* grams;

a *kilo*metre means *1000* metres.

The metric system gives rise to the *International System* (*SI*, for short, to stand for *Système International*). A table of some prefixes for the SI units is given below. The column in the table headed *"factor"* states how many times the unit is multiplied by. The ones that are starred (*) are not used very often.

Prefix	Factor	Prefix	Factor
*tera-	trillion	deci-	tenth
*giga-	billion	centi-	hundredth
mega-	million	milli-	thousandth
kilo-	thousand	micro-	millionth
*hecto-	hundred	*nano-	billionth
*deka-	ten	*pico-	trillionth

1. For each display of pots, tell how many sheep have been counted. Enter the number in your calculator.

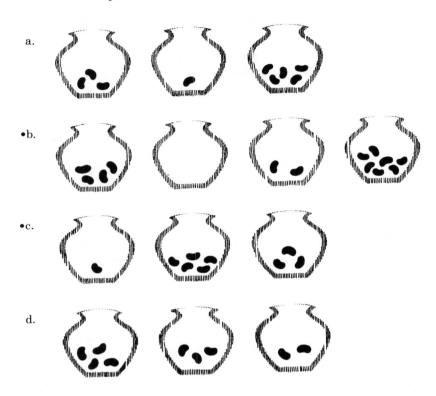

a.

•b.

•c.

d.

2. For each number below, draw a set of three pots and show the number of beans in the appropriate pots to represent the number.

•a. 32   d. 312   •g. 500

b. 132   •e. 405   •h. 114

c. 231   f. 100   i. 411

•3. You are given four beans and three pots. Find all possible numbers that can be represented using all four beans. For example, 112 and 301 are two possibilities. [REMARK: There are altogether *15* possibilities. If you get *10*, you are rated *good;* if you get *13* or more, you are rated *excellent.*]

4. Consider the place farthest to the right as the "*ones*" place. See how many decimal places you can name in order from right to left. [HINT: Look up the word *numeral* in a dictionary. Watch out! The English way of naming numbers is slightly different from that used in the United States.]

5. Using any of the ten digits as a possibility for each of three places, how many different 3-digit license plates can there be? (Do not try to list all of them.)

•6. You are given two pots and three beans. Find all possible numbers that can be represented using three *or fewer* beans. [HINT: There are altogether 10 possibilities.]

2.4 Ordering of Counting Numbers as Base–Ten Numerals

Numbers are used to describe quantities of things. As such, numbers are then related, one to another, in order to *compare* quantities. We may do this loosely.

> A bird has fewer legs than a spider.
>
> There are more eggs in the refrigerator than there are cans of beer.
>
> More people attended one *Rose Bowl* football game than one *World Series* baseball game.

Or we may do this precisely with numbers. Arithmetic has symbols to record the numbers and the result of the comparison.

Two *is less than* four.	In symbols: $2 < 4$
Twelve *is greater than* six.	In symbols: $12 > 6$
One-hundred-two thousand, eight hundred twelve *is greater than* fifty thousand, one hundred six.	In symbols: $102812 > 50106$

In this section, we shall learn how our base–ten system for writing numbers makes it easy to determine which of two numbers is greater, as well as ways for our calculator to make this comparison.

Notation. Suppose you have two numbers, one called A, another called B. If number A is less than number B, write

$$A < B \qquad \text{and read} \qquad \text{"}A \text{ is less than } B.\text{"}$$

Of course, to say that number A is less than number B is the same as saying that number B is greater than number A. So we can equally well write

$$B > A \qquad \text{and read} \qquad \text{"}B \text{ is greater than } A.\text{"}$$

To remember how the notation is used, think of the symbols $<$ or $>$ as arrowheads that always point to the *lesser* number. Thus

$2 < 4$	or, equivalently,	$4 > 2$
$6 < 12$	or, equivalently,	$12 > 6$

$$50106 < 102812 \qquad \text{or, equivalently,} \qquad 102812 > 50106$$
$$25 < 26 \qquad \text{or, equivalently,} \qquad 26 > 25$$

The 1–1 correspondence scheme behind the shepherd's accounting system allows him to determine whether or not two sets (the flock of sheep and the pile of beans) have the same number of members. Also, if they don't, he can determine which set has more members. However, the place-value scheme developed by the shepherd is even more efficient in deciding which of two numbers is the larger. Using the full knowledge of this scheme, we do not have to go through the trial of making a 1–1 correspondence. This trial would be unthinkable for large numbers: Is the national debt in dollars larger than the population of the world? It may come as a surprise that a task we can perform so easily is the result of a series of involved decisions.

Example 2.6: Consider the situations in which the shepherd has three pots as illustrated below.

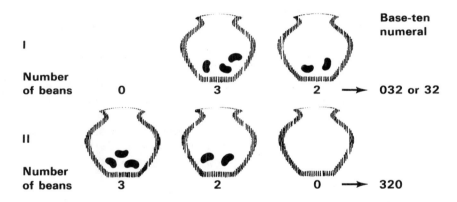

Figure 2.7
Comparing numbers using place value. The representation using more pots is the greater number: 320 > 32.

In both instances he has two beans in one pot and three in another. How can the shepherd decide that illustration II represents more sheep than illustration I?

The decision in Example 2.6 is made by comparing the pots from *left* to *right*. Each bean in a pot to the left represents ten of each bean in the pot immediately to its right.

The first principle used in making comparisons is—

1. If more pots are needed in one representation than another, the number requiring more pots is greater.

Example 2.7:

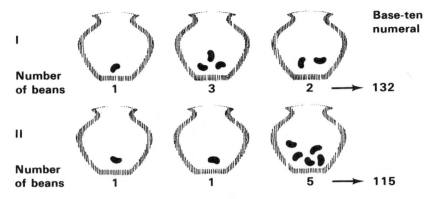

Base-ten
numeral

I

Number
of beans   1   3   2 ⟶ 132

II

Number
of beans   1   1   5 ⟶ 115

Figure 2.8
Comparing numbers using place value. Start at the left. Find the
leftmost pot containing a different number of beans. It is the
second pot: 3 > 1. Conclude: 132 > 115.

In this instance, the shepherd used more beans (7 in all) in illustration II than in illustration I. How can he decide that I represents more
sheep than II?

The decision is again made by beginning at the left. Because each
pot farthest to the left has one (1) bean, the shepherd knows that each
display represents at least 100 sheep, but less than 200. Thus, for
comparison purposes, he can ignore the hundreds pot and concentrate
on the pots to the right of the dotted line shown in Figure 2.9.

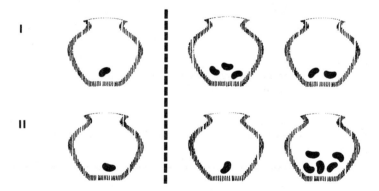

I

II

Figure 2.9
More details in comparing numbers using place value.

By comparing the first pot from the left at which a different number
of beans are used (in this case, the second pot from the left,) the
shepherd concludes that illustration I represents the greater number.

The second principle in making comparisons is—

2. If the same number of pots are used in two representations of numbers, compare individual pots, starting at the left. Find the first pot from the left containing a different number of beans. The representation with the greater number of beans in *this* pot is the greater number.

These rules with pots translate immediately into numbers in the base–ten numeral system.

Example 2.8: Note that 2464 is greater than 985 because 2464 requires four places while 985 requires only three: 2464 > 985. Now note that 2464 is greater than 2439 because the first place (the tens place) at which the numerals differ (circled below) shows that 6 > 3 (6 is greater than 3):

$$2 \; 4 \; ⑥ \; 4 \quad \text{compared with} \quad 2 \; 4 \; ③ \; 9$$

So it follows that

$$2464 > 2439$$

Now, we restate the two principles in terms of numerals and places.

Principle 1. If two whole numbers in decimal form have different numbers of *digits,* the one with the greater number of digits is the greater.

Principle 2. If two whole numbers in decimal form have the same number of digits, compare the numbers digit-by-digit starting at the left. At the first place where the digits differ, the number with the larger digit at that place is the greater.

Example 2.9:

(1) Compare 1234567 with 123456.
1234567 is greater than 123456 because "1234567" has more digits than "123456."

(2) Compare 1234567 with 1235467.
These two numbers have the same number of digits. So we compare digit-by-digit from left to right.

$$1 \; 2 \; 3 \; 4 \; 5 \; 6 \; 7$$
$$\updownarrow$$
$$1 \; 2 \; 3 \; \mathbf{5} \; 4 \; 6 \; 7$$

The digits first differ at the fourth place from the left (shown by the double arrow.) Because

$$4 \text{ is less than } 5,$$

we have

$$1234567 \text{ is less than } 1235467$$

As efficient and simple as this system, or algorithm, is for the human eye and brain, our calculators are not set up to utilize this algorithm. Calculators could be made to do so, and some may actually make comparisons in this way. We can, however, use our simple calculators to make comparisons of numbers by means of subtraction or division. Because we have not yet discussed the theory and practice of subtraction or division, we shall just indicate with examples how the process might be carried out.

A comparison of two numbers based on subtraction uses a property that you may recall from earlier school days. For example, "3 can be subtracted from 7" because 7 is greater than 3:

$$7 - 3 = 4$$

On the other hand, "7 cannot be subtracted from 3 because 3 is less than 7." Actually, $3 - 7$ *is* possible as we shall see in Chapter 6:

$$3 - 7 = -4$$

The minus sign before the "4" shows that 3 is less than 7. If you try this on your calculator, it will either show the $(-)$ sign or give some other indication that this "improper" operation has been performed.

Example 2.10: $3 - 7 = ?$

$$3 \; \boxed{-} \; 7 \; \boxed{=} \; -4.$$

NOTE: Some machines will show some "overflow" signal (such as a blinking) instead of a minus sign $(-)$ to show that the calculator cannot perform this operation.

The different behavior of the calculator for the problem reminds us that we "can't subtract more than we have." We shall discuss such calculator responses in greater detail in Chapter 6. At this time, we notice that the calculator always behaves in this peculiar way whenever the first number is less than the second in a subtraction.

Incidentally, we shall also discuss subtraction on the calculator later. For now, all we need to do is to press the $\boxed{-}$ key at the right time. (Please refer to your own operator's manual for your particular machine.)

Example 2.11: Discover how your calculator responds to these different subtraction problems. (The exact sequence of steps for each operation is the same as in Example 2.10.)

Operation	Display	Remarks
12 − 8	$12 \boxminus 8 \boxminus 4.$	Safe! 12 is greater than 8. (12 > 8)
15 − 23	$15 \boxminus 23 \boxminus \text{-}8.$	Tilt! 15 is less than 23. (15 < 23)
15 − 15	$15 \boxminus 15 \boxminus 0.$	Difference is 0; numbers equal. (15 = 15)
1573 − 1563	$1573 \boxminus$ $1563 \boxminus 10.$	Safe! 1573 is greater than 1563. (1573 > 1563)
1563 − 1573	$1563 \boxminus$ $1573 \boxminus \text{-}10.$	Tilt! 1563 is less than 1573. (1563 < 1573)
1234567 − 1235467	$1234567 \boxminus$ 1235467 $\boxminus \text{-}900.$	1234567 is less than 1235467. (1234567 < 1235467)

In the above, we used subtraction to determine which of two numbers is greater. We can also use a property of *division* in comparing numbers.

If 7 is divided by 3, a number greater than 1 is obtained (3 *"goes into"* 7 more than once):

$$7 \div 3 = 2.3333333; \quad \text{Conclude:} \quad 7 > 3$$

On the other hand, if 3 is divided by 7, a number smaller than 1 is obtained:

$$3 \div 7 = 0.2857142; \quad \text{Conclude:} \quad 3 < 7$$

Exercises 2.4

1. For each of the following numbers, write on the left a number less than the given number but with the same number of digits. On the right, write a number larger than the given number, again with the same number of digits. [The first problem is completed as an example. You should give an alternate answer.]

 a. 29 < 37 < 73   d.   < 199 <

 •b.   < 79 <   e.   < 2946 <

 •c.   < 101 <   •f.   < 3099 <

• 2. Use a digit-by-digit argument to conclude that 1563 is less than 1673.

• 3. Suppose you were to use a digit-by-digit argument to compare 1563 with 94.

    a. Why isn't the "1" compared with the "9"?

    b. What digit is compared with the "1"?

4. Insert the proper symbol ($<$) or ($>$) between these pairs of numbers:

    a. 16   61   •d. 1087   1091

    •b. 161   116   e. 99   1000

    c. 100   99   •f. 104657   104567

• 5. List as many 3-digit numbers as possible using the digits 4, 7, and 9, but repeating no digit. Arrange these numbers in increasing order. [6 numbers are possible.]

6. Repeat Exercise 5, but now allowing digits to be repeated. [27 are possible.]

• 7. You are given a number, say 7401, and you are asked for the sum of its digits. What problem would this correspond to using beans and pots?

8. Find a number (different from 0) that is twice the sum of its digits. [HINT: This can be done by trying various numbers on the calculator systematically. Can this be a one-digit number? Can it be an odd number?]

9. Find a number less than 100 that is four times the sum of its digits. [Aside from 0, there are four such numbers.]

10. Fill two pots, *Left* (*L*) and *Right* (*R*), with four beans in all possible ways. There are five ways.

L	*R*		*L*	*R*		*L*	*R*		*L*	*R*		*L*	*R*
4	0		3	1		2	2		1	3		0	4

Add one bean to each pot. Now how many sheep are counted in each case? Does this give all ways that six beans can be put into two pots? If not, give the missing ones.

11. If the shepherd has 3 pots and 10 beans, what is the largest number that he can represent?

•12. Find the smallest number the shepherd can represent—

    a. using 10 beans in 2 pots;

    b. using 10 beans in 3 pots if the farthest pot to the left cannot be empty.

2.5   The Number Line

We have found that it is possible to compare two numbers. Thus, we can think of arranging them from smaller to larger or from larger to smaller. To keep our discussion simple, in the following, we shall stay with the arrangement (or order) from smaller to larger. As we can *order* any two given numbers, we can apply this process repeatedly to order more than two numbers.

Suppose we have a bucket of numbers. We reach in and take out two of them, say 758 and 23. How do they compare? We find that 23 is less than 758 ($23 < 758$); so we place 23 somewhere to the left of 758:

$$23 \ldots 758$$

Pick another number from the bucket, say, 549. Compare it with 23 and 758.

How does it compare with 23?   549 is greater than 23. $549 > 23$
How does it compare with 758?   549 is less than 758. $549 < 758$

Thus we conclude that 549 is somewhere to the right of 23 and somewhere to the left of 758:

$$23 \ldots 549 \ldots 758$$

Clearly, we can keep going. Pull out another number. Compare it with each of the previous ones. Eventually, we might be able to order all the *counting numbers* (positive integers) 1, 2, 3, . . . provided all these numbers are in the bucket.

But there is a better way to order the counting numbers—by building up systematically.

Start with 1.	1	*1.*
Next, add 1.	$1 + 1 = 2$	*1* ⊞ *1* ▣ *2.*
Then add 1 more.	$2 + 1 = 3$	*2* ⊞ *1* ▣ *3.*
Then add 1 more. and so on.	$3 + 1 = 4$	*3* ⊞ *1* ▣ *4.*

(That is, for any given counting number, the next counting number is greater by 1.) By this routine, we can be sure none of the numbers between are omitted.

Begin with the number, 0. Successively, add 1. We obtain the whole numbers: 0, 1, 2, 3, 4, ... Notice that this is a way to *count*.

It is useful to do this in a geometric setting. Think of a straight line.

Choose a starting point; call it 0. Then choose a unit of length. Now mark off successive intervals of this length. Because we add a unit of length to make the next mark, it is natural to put these marks in 1–1 correspondence with the integers. In this way, we can build a ruler, or we can imagine the counting numbers as points on a long, long, line that has no end.

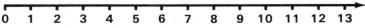

Figure 2.10
Construction of the number line. The choice of the starting point, 0, and the unit of length is arbitrary. Here, we have chosen about $\frac{1}{2}$ centimetre.

The number line lets us visualize certain relationships between numbers. One such relationship is that of order. Observe that *smaller* numbers appear to the *left* of larger numbers on the number line. For instance, 4 is less than 7, and 4 is located to the left of 7.

> *Rule for locating numbers on the number line:*
> If $A < B$ (A is less than B), then A is to the left of B.

> Conversely,
> If A is to the left of B, then $A < B$.

This rule tells us, for example, that $4 < 13$.

Even when numbers are beyond the scope of our diagram, we can continue to visualize what must happen. Thus, consider 102 and 5981. The two points on the number line are "off our map," but we know that we would get to 102 sooner than to 5981. So, $102 < 5981$ (or equivalently, $5981 > 102$; "5981 is greater than 102").

Consider now, another relationship. Start with 0, skip a number, connect, skip, as in Figure 2.11.

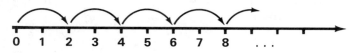

Figure 2.11
Generating the even numbers by skipping.

We touch down on the numbers, 0, 2, 4, 6, 8, ... What do we call these numbers? The *even numbers,* of course! To produce the even numbers on a calculator, we act on the hint from the number line that each even number is followed by another two spaces away. So we start with 0 and add 2 repeatedly.

Suppose we start with 1 instead and add 2 repeatedly. What kinds of numbers would we get? (Try it!) Thus, except for the starting point, we generate the odd numbers in the same manner we generated the even numbers. (See Figure 2.12.)

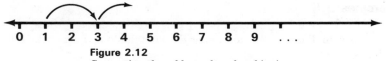

Figure 2.12
Generating the odd numbers by skipping.

We can use the number line further to help us visualize the results of addition.

Example 2.12: On the number line, $2 + 3 = ?$

Solution:

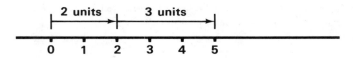

To find the sum, $2 + 3$, on the number line, start at 2. Go 3 units to the right. Better yet, start at 0. Move 2 units to the right to get to 2. Next, go (advance) 3 units to the right. We see:

$$2 + 3 = 5$$

Notice that, starting at 0, and advancing 3 units first, then another 2 units, we reach the same point, 5. In other words,

$$2 + 3 = 3 + 2$$

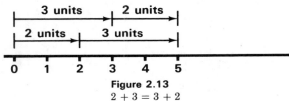

Figure 2.13
$2 + 3 = 3 + 2$

Of course, we can add several numbers on the number line. Simply continue with the next number from the last completed sum.

Example 2.13: Show the addition, $(2 + 3) + 4$ on the number line.

Solution:

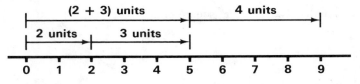

Exercises 2.5

1. Using a small unit, make a number line 40 units long. Locate the following numbers:

   23, 17, 31, 2, 5, 0, 7, 37, 11, 29, 3, 19, 1, 13.

• 2. For each of the following numbers, write the next whole number below it.

   37   79   101   199   2946   3099

   For each, confirm your answer on your calculator. For example:

   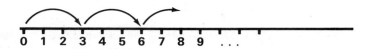

• 3. Start with 0 on the number line.

   a. Skip two numbers, then stop; skip the next two numbers, then stop; and so on. Generate the first eight numbers this way. What do you call such numbers you have produced?

   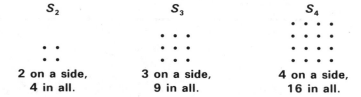

   b. How would you go about producing the above set of numbers using addition on the calculator? Confirm on your calculator. For example,

   $0 \boxplus 3 \boxminus 3.$ $3 \boxplus 3 \boxminus 6.$ . . .

   c. Beginning with 5, generate the first 8 multiples of 5 on your calculator and record the results.

4. You may have heard of square numbers (or *squares*). Here is one way they originate. Line up some pennies in a square array so that each row has the same number of pennies, each column has the same number of pennies, and there are the same number of rows as columns. For example—

   S_2 S_3 S_4

   • •   • • •   • • • •
   • •   • • •   • • • •
                             • • •   • • • •
                                                       • • • •

   2 on a side,   **3 on a side,**   **4 on a side,**
   4 in all.   **9 in all.**   **16 in all.**

   We have labeled each square S with a subscript numeral to indicate the number on a side. We may even think of a single • as a "square" with 1 penny on each side. (Call it S_1.) To find the numbers of pennies in each array is easy. In S_2 there are 2 rows with 2 in each row $\rightarrow 2 \times 2 = 4$; so the square of 2 is 4. Use your calculator to find the following squares:

   a. square of 3   c. square of 1

   b. square of 4   d. square of 23

5. Start with 0 on the number line. Advance (to the right) one step; then 2 steps; next, 3 steps; and so on. Generate the first eight such numbers: 1, 3, 6, ...

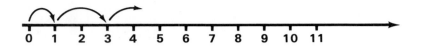

6. The following patterns give rise to triangular numbers. Here are the first four triangular numbers.

T_1 T_2 T_3 T_4

a. Find the first 8 triangular numbers: 1, 3, ...

b. Describe a way of getting triangular numbers using the calculator.

• 7. From the diagrams in Exercise 4, show pictorially that any square number is the sum of two triangular numbers. [HINT: Rearrange the pennies in the triangular arrays so that they look like the display below.]

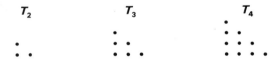

T_2 T_3 T_4

8. Express each of the following in terms of S's and T's.

a. $T_1 + T_2 = ?$

b. $S_4 = T_? + T_?$

c. $S_8 = ? + ?$ (Check your guess on the calculator.)

• 9. Using the calculator, start with a series of odd numbers and add successively as indicated below:

$$1, \qquad 1 + 3, \qquad 1 + 3 + 5, \qquad \ldots$$

What kind of numbers is being generated? (See diagram.)

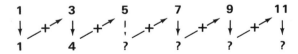

10. Look at the pattern below. To each block of squares, explain why you are adding an *odd* number of points to get the next square number.

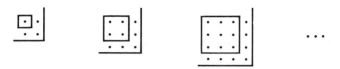

2.6 How Many Counting Numbers?

How many counting numbers are there? To find out, we might start by entering a 1 in the calculator, add 1 for the next counting number, add another 1 for the next, and so on. If your calculator has an 8-digit capacity, eventually you will get to 99999999. (Suppose you can complete one addition per second continuously for 24 hours a day. This task would take a bit over 1157 days, or about 3 years and 2 months of your time.) What happens if you add 1 to this number? Try this by entering as many 9's as your calculator can hold, then add 1.

Does the fact that your calculator "overflows" after reaching capacity mean that there is no larger counting number? What about calculators that have a 12-digit capacity? or a 16-digit capacity? Of course, the number of counting numbers is not limited to the capacity of a calculator. But watching the repeated additions on the calculator confirms an important notion:

> Given a particular counting number, adding 1 to it gives a still larger counting number.

This observation tells us that there is no largest counting number. The answer to the question, "What is the largest number?" is: "There is none."

Let's put a big number, say 10000000, into our calculator and then add 1.

$$1\,0000000\ \boxplus\ 1\ \boxminus\ 1\,000000\,1.$$

Now think of a larger integer—one too large for our calculator. Say, 1 followed by one hundred zeros:

$$G = \underbrace{1000\ldots00}_{100\ \text{zeros}}$$

Is this the largest integer? No, we may add 1 to it.

$$G + 1 = \underbrace{1000\ldots01}_{99 \text{ zeros}}$$

G + 1 is an integer that is larger than G.

Incidentally, the number we labeled G, which is 1 followed by 100 zeros, was named a *"googol"* by a nephew of the mathematician, Edward Kasner, at the time the nephew was nine years old.

The argument that there is no largest integer is thus perfectly general:

If n is an integer, then $(n + 1)$ is a larger integer.

Hence there can be no largest integer. We say that the set of integers is *infinite,* which is simply a fancy way of saying that there is no end to the sequence of integers.

The situation for numbers in our calculators is quite different. Because there are only eight digits, there *IS* a largest integer that our calculator will recognize. It is

$$\underbrace{99999999}_{8 \text{ nines}}$$

Let's see what happens when we try to add 1.

$$\boxed{99999999} \boxplus 1 \boxminus [\text{OVERFLOW}]$$

Your machine will display some symbol to indicate the overflow. It may also display a number. In some cases, this number gives information about the correct answer—or, at least, information about the number of decimal places in the correct answer. From your instruction manual, or by experimentation, see if you can discover what information your machine provides about overflow answers.

If you think about it, you will see that the calculator recognizes only a finite number of numbers of all kinds. This will have some strange consequences for our computations which we will examine as they arise.

Exercises 2.6

•1. Use your calculator to find each of the following:

    a. Enter 9; add 1.   c. Enter 9999; add 1.

    b. Enter 99; add 1.

•2. Enter as many 9's as your calculator can take; add 1.

    a. Describe the result.

    b. What is the correct answer to the above?

3. Enter 10000000 on your calculator; add 1. Give the result, then describe the result of adding 1 to a number made up of 1 followed by 5000 zeros.

4. How many even counting numbers are there? Explain logically how you arrived at your answer.

5. In Section 2.1 we discussed 1–1 correspondences. Here is a 1–1 correspondence between counting numbers and even counting numbers.

counting numbers	1	2	3	4	⋯
	↕	↕	↕	↕	
even counting numbers	2	4	6	8	⋯

Can this correspondence be used to support your answer to Exercise 4? Explain.

•6. Suppose the shepherd has eight pots, each filled with nine beans. He wants to add one more bean to the count. What problem does this correspond to in an 8-digit calculator? What problem does the additional bean create for the shepherd?

2.7 Numeration Schemes for Numbers between Whole Numbers

So far so good. We can count! And we can go on counting forever; there is no largest counting number. But we use numbers for more than counting. For example, the price of a calculator may be $19.75; not a whole number of dollars. In the number, 19.75, the "19" is the whole number of dollars and the ".75" is part of a dollar; 75 cents.

The number, .75, arises in other situations. Suppose a package of four toothbrushes costs $3. How much does each cost? We want to divide $3 into 4 equal parts. Let us do this on a calculator:

$$3 \div 4 = ?$$

The calculator has just divided 3 by 4 and given us the answer. 0.75. (Without any interpretation, it is permissible to read the number, 0.75, as *"point seven five."*) We may interpret this as $3 divided by 4 is $0.75, or *"point seven five dollars,"* or 75 cents.

The notations 0.75 and .75 describe the same number. The initial zero is not really needed. The zero often appears in print because the decimal point is easy to overlook. Some machines automatically insert the initial zero in displays; other machines do not. In any case, you need not enter the initial zero when keying a number into your machine, but it won't change anything if you do.

Calculators handle numbers by digits in *"decimal notation."* For whole numbers, the base–10 system is decimal notation. Our task now is to understand decimal notation for numbers other than whole numbers.

We begin with a situation comedy!

Let us suppose that José Greenthumb has raised a crop of green beans of which he is quite proud. To boast to his friends, he decides to measure the lengths of a few of them. Being a thoroughly modern gardener, he decides to do it with a metric ruler—and for this purpose, he uses an inexpensive (25¢ or $.25, or "point two five" dollars) plastic ruler that has both a scale in inches and one in centimetres. He measured his first bean in this way:

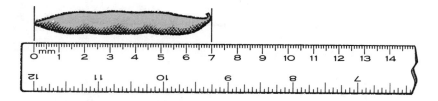

So José recorded 7 centimetres.

The next one he measured looked like this:

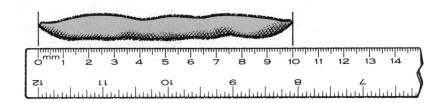

So José recorded 10 centimetres. So far, no problems; the lengths could be recorded in whole number of centimetres.

But the next looked like this:

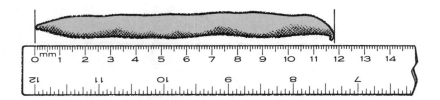

Now José had a bit of a problem. He could simply record its length as 12 centimetres, but that would be wrong. He decided instead to use the smaller marks on the scale. Note that there are 10 equal spaces between each centimetre. José recorded the length in the form

$$11.8$$

with the understanding that ".8" stands for 8 of the smaller marks; 8 marks beyond 11. (Incidentally, the length of each of these small intervals is one millimetre (abbreviated mm); each small mark denotes a millimetre. 10 millimetres = 1 centimetre.)

With this scheme, José has no further trouble recording the measurements of his beans. How did he record the length of this bean?

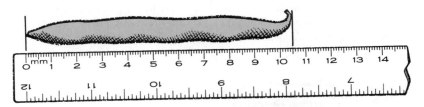

(Note that the halfway point between two successive integers is at a .5 mark.)

Situations that call for this sort of precision arise in almost all measurement problems—measuring lengths, weights, or time (most races are timed in tenths of a second.) Furthermore, they arise in computations involving such numbers. Just as we have seen, they may arise in dividing one integer by another.

The adaptation of the decimal number system for whole numbers to other numbers is easily explained on the number line. Suppose you need to find the number indicated by the arrow in the following number line.

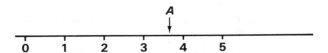

Simply split the interval from 3 to 4 into 10 equal parts as indicated in the next figure, which is a blow-up of part of the previous one.

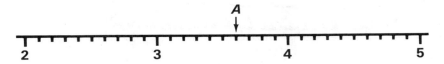

Now record the number as 3.6. This can be read schematically as "three point six," or more correctly as "three and six-tenths."

If the arrow had indicated a point still not on one of these finer marks, we could again split the interval between 3.6 and 3.7 into 10 equal parts. In the next figure we show this in a blow-up of the previous figure. Here we suppose we have located an arrow at B.

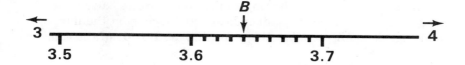

Because the arrow is 4 of the smaller marks to the right of 3.6, we record the number pointed to by the arrow as "3.64". It can be read as "three point six four."

Now why didn't we write the number as "3.6.4"? Instead of a second decimal point, we use *place-value* notation. In the number

$$3.64$$

the decimal point (.) tells us that what follows is less than 1 unit. The "6" tells us that the unit has been split into 10 equal parts, and the number is at or beyond the sixth division mark. The "4" tells us that the interval from 3.6 to 3.7 has been split again into 10 equal parts, and that the number is at (or beyond) the fourth mark. Because there are no other digits following "4," we conclude that the number is at the fourth mark.

But there is still a better explanation and use of place-value that is consistent with the previous use of place-value notation. Notice that when we split the unit interval "3-to-4" into 10 equal parts, we really switched over to a new unit that is a tenth as long as the first unit. This is indicated in the following figure, showing arrow *A* again.

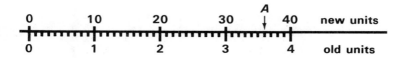

Thus *A* is 36 of the new units. (Count 'em!) In writing the number as 3.6 in the old larger units, the decimal point simply indicates that change.

Now consider the arrow *B*. The intervals have been split into still finer parts. The old unit has been split twice: there are now 100 tiny intervals in each of the old large units. The stage here is exactly the same as splitting the monetary unit, one dollar, first into 10 small units (dimes,) next into 100 tiny units (cents.)

It is very difficult to show this in one single scale, and so we shall just indicate it schematically below.

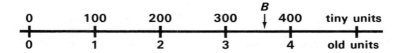

The point *B* is 364 of the tiny units. So in writing 3.64 in the old larger units, the decimal point indicates the change in the unit size, and the "64" indicates that the point marked B is 64 of the smaller units beyond 3.

In terms of the small units (cents),

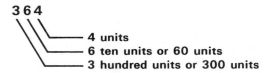

In terms of the large units (dollars),

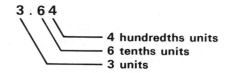

The number 3.64 may be read schematically as "three point six four" or, more correctly, as "3 and 64 hundredths."

We shall go into much greater detail about decimal numbers in Chapter 8, but for now it will be useful to gain some familiarity with the system by completing the exercises at the end of this section.

It is important to note that numbers between 0 and 1 are expressed in decimal notation with a zero (0) in the units place. Thus, for example,

$$0.5 \quad \text{denotes} \quad \text{one-half unit,}$$
$$0.25 \quad \text{denotes} \quad \text{one-quarter unit}$$

and on the number line,

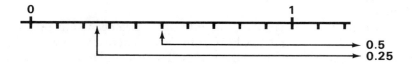

Often, as we mentioned earlier, the zero in the units place is omitted. Thus

$$0.5 \ = .5$$
$$0.25 = .25$$

Most calculators will show the zero in the units place as a way of alerting the user to the fact that a number greater than 0 and less than

1 is being displayed. However, to enter such a decimal into your calculator, you do not have to key in the 0 in the units place.

The comparison of two decimal numbers to determine which is the larger is accomplished by a small modification of the methods given in Section 2.4.

First, compare the whole number parts of the two numbers. The whole number part is just the part to the left of the decimal place. Make this comparison using the principles of Section 2.4.

Example 2.14: Which number is the greater: 23.21 or 19.78?

The whole number part of 23.21 is 23. The whole number part of 19.78 is 19. Because $23 > 19$, it follows that $23.21 > 19.78$.

Second, if the whole number parts are equal, compare the decimal parts as follows. Start at the decimal point and compare the digits to the right. Find the first place to the right of the decimal point at which the digits differ. The number with the larger digit at that place is the greater.

Example 2.15: (The crucial place for comparison is circled.)

$23.②1 < 23.⑨8$ because $2 < 9$
$19.④7 > 19.③9$ because $4 > 3$
$35.2⑧3 > 35.2⑥7$ because $8 > 6$

To use your calculator to determine which of two numbers is the greater, you may use either the rule of subtraction or the rule of division, as described in Section 2.4.

Exercises 2.7

• 1. Read the decimal numbers indicated by these arrows on the number line.

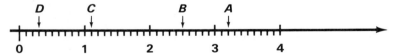

• 2. Indicate with arrows on the number line these numbers. Identify each arrow by the appropriate letter, P, Q. . . .

    P. 1.7   *Q.* 3.8   *R.* 0.3   *S.* 2.1

3. Use a calculator to perform these divisions. Record the answers. Locate the answers with an arrow on the number line.

    •*A.* $5 \div 2$ •*C.* $5 \div 4$ *E.* $4 \div 5$

    B. $18 \div 6$ *D.* $112 \div 25$ •*F.* $7.2 \div 9.6$

In each case, use the rule of division to decide which number is greater.

4. For each of the pairs of numbers in Exercise 3, use your calculator and the subtraction rule to decide which of the pair is greater.

5. For each pair of numbers below, follow Example 2.14 to determine which is the greater.

   •a. 163.753 and 61.3753
   •b. 69.4214 and 69.4124
   c. 84552.67 and 845.5267
   d. 8455.2670 and 8455.267

• 6. Arrange the following numbers in order from the smallest to the largest.

   253.74;   253.47;   2.5347;   .07532468

7. Measure the lengths of the keys in your pocket using a centimetre scale. (Or use beans from your garden instead of keys if you wish!)

8. Use your calculator to determine which number of each pair is the greater.

   a. 8.100 and 8.100000
   •b. 752.16000 and 752.160
   c. 11235.800 and 11235.8

9. Compare each pair of numbers without the use of the calculator.

   a. 9.1 with 9.10
   b. 9.1 with 9.01
   •c. 9.00010 with 9.0010

10. Measure the diameters of these U.S. coins: penny, nickel, dime, quarter, half-dollar, dollar. Use a metric ruler.

11. Compare your measurements from Exercise 10 with measurements others in your class got.

   a. Which of these sets of measurements is most like the measurements others got?

   b. Which of the measurements is most unlike those others got?

   c. Suggest a reason explaining why certain measurements can be expected to be more (less) alike than others.

12. Compare your measurements for Exercise 7 with measurements others in your class got and repeat parts a, b, and c of Exercise 11.

13. Use the calculator to express each of the following as a 7-place decimal. (Recall that the fraction $\frac{a}{b}$ is processed as $a \div b$.)

   •a. $\frac{1}{9}$ d. $\frac{4}{9}$ g. $\frac{7}{9}$

   b. $\frac{2}{9}$ e. $\frac{5}{9}$ h. $\frac{8}{9}$

   •c. $\frac{3}{9}$ •f. $\frac{6}{9}$ •i. $\frac{9}{9}$

14. Divide each of the following by 9 on the calculator and use the results of Exercise 13 to help you write the remainders as common fractions. For example, $25374 \div 9$:

$$25374 \boxed{\div} 9 \boxed{=} 2819.3333 = 2819\tfrac{3}{9}$$

a. 527643	•d. 85539	g. 6247908
•b. 647328	e. 70962	•h. 7358019
c. 173628	•f. 10503	i. 4025772

15. For the number, 25374, the *"sum of its digits"* is

$$2 + 5 + 3 + 7 + 4 = 21.$$

Find the "sum of the digits" for each number in Exercise 14 and divide each sum by 9. Compare each remainder with the corresponding result in Exercise 14. For example, 25374:

$$2 \boxed{+} 5 \boxed{+} 3 \boxed{+} 7 \boxed{+} 4 \boxed{=} 21. \qquad 21 \boxed{\div} 9 \boxed{=} 2.3333333$$

Remainder: $\tfrac{3}{9}$ (Compare with remainder in 25374 ÷ 9).

16. Follow the directions in Exercises 14 and 15 for each of the following. What do you notice about the corresponding remainders?

•a. 528643	d. 85036	g. 6247978
b. 649325	e. 73985	h. 9358613
•c. 173658	•f. 10563	i. 4628772

17. Repeat Exercises 14 through 16, dividing by 3 instead of by 9. Suggest a possible *"sum of digits"* rule: "The remainder obtained on dividing the sum of the digits of a number . . ."

18. Use the calculator to find each to 7-place decimal:

•a. $\frac{1}{13}$	•e. $\frac{5}{13}$	i. $\frac{9}{13}$
•b. $\frac{2}{13}$	•f. $\frac{6}{13}$	j. $\frac{10}{13}$
•c. $\frac{3}{13}$	g. $\frac{7}{13}$	k. $\frac{11}{13}$
•d. $\frac{4}{13}$	h. $\frac{8}{13}$	l. $\frac{12}{13}$

19. Test your "sum of digits" rule from Exercise 17 on the numbers in Exercise 16, using division by 4. (Find the results of $\frac{1}{4}$, . . . , first.)

3

Addition and Subtraction with Numbers in Decimal Notation

3.1 The Old "Big-Ones-from-Little-Ones" Routine

In this chapter we perform two arithmetic operations: addition and subtraction. When we add two numbers together, we usually make a "big one" out of two "little ones." When we subtract one number from another, we usually make little ones from big ones. Let us review the elementary techniques of performing these operations with and without calculators.

3.2 Addition with Whole Numbers

In theory, addition is very simple. Here is a typical problem.

Example 3.1: Mr. Pisces has two fish tanks. The larger tank has eight fish, the smaller tank has six fish. How many fish does Mr. Pisces have?

Figure 3.1
How many fish? Count or add.

Solution: Count all the fish in both tanks! Count the fish in the larger tank first, and continue counting the fish in the smaller tank.
Expressed mathematically,

$$8 + 6 = 14$$

Note that it makes no difference which tank is counted first:

$$6 + 8 = 14 \qquad \text{or} \qquad 8 + 6 = 14$$

Try it on your calculator.

$$8 + 6 = \,?$$

Solution: 8 $\boxed{+}$ 6 $\boxed{=}$ $14.$

Now perform the computation in the *reverse* order.

$$6 + 8 = \,?$$

Solution: 6 $\boxed{+}$ 8 $\boxed{=}$ $14.$

•**Exercise 3.1:**   Complete the following addition table for numbers

$$0, 1, 2, 3, 4, 5, 6, 7, 8, 9$$

Complete the addition table using your calculator (it's good exercise), or complete it using your own mental arithmetic by remembering what the sums are, or by counting out each sum as we did with the fish: $8 + 6 = 14.$

As we practice addition, it will be easier to memorize these basic sums than to use a calculator to compute their values. However, there is nothing wrong with using a calculator. A calculator gives the correct answer if the numbers are entered correctly.

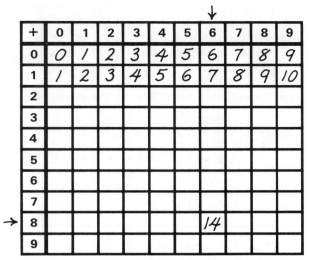

+	0	1	2	3	4	5	6	7	8	9
0	0	1	2	3	4	5	6	7	8	9
1	1	2	3	4	5	6	7	8	9	10
2										
3										
4										
5										
6										
7										
8								14			
9										

Table of Addition

In the square located in a particular row and a particular column, enter the sum of the row and the column number. Thus, in the row labeled 8 under column 6, enter 14: $8 + 6 = 14$.

The addition table is full of patterns.

1. Symmetry: Draw a diagonal line from the top left to the bottom right (see Figure 3.2). Think of the square being folded along this line. Entries that come together have the same value.

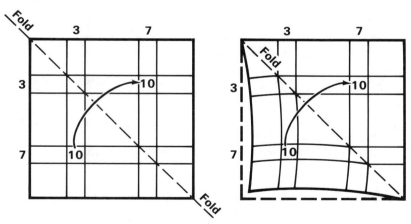

Figure 3.2
Symmetries in the addition table.

2. Pattern from column to column: Look at any column except the last. The entries in the next column are found by adding 1.

3. Pattern from row to row: Look at any row except the last. The entries in the next row are found by adding 1.

4. Patterns along lines parallel to the diagonal from the bottom left up to the top right. (See sketch at right.) The entries along these lines are constant.

Here are some important facts that follow from the table.

$A + B = B + A$ for any two numbers.
$A + 0 = A$ for any number, A.
$0 + A = A$ for any number, A.

Addition becomes more complicated when larger numbers are to be added. Here's a typical problem.

Example 3.2: Attendance figures for the major New Year's Day football game were as shown.

**Rose Bowl
102,815**

**Orange Bowl
95,678**

**Cotton Bowl
89,035**

Altogether, how many people attended these games? (Assume that no one attended more than one game!)

Solution: We could count the people in the **Rose Bowl.** Then continue counting the people in the **Orange Bowl,** and then continue counting those in the **Cotton Bowl.** This is tedious and impractical. If we could count 10 persons in a second, it would take almost eight hours to complete the counting!

We can find the total very quickly on the calculator.

$102{,}815 + 95{,}678 + 89{,}035 = ?$

$(\ 102815\ \boxplus\ 95678\ )\ \boxplus\ 89035\ \boxed{=}\ (\ 198493\ )\ \boxplus\ 89035\ \boxed{=}\ 287528.$

Parentheses are used to show that the first two numbers were added first. Then, that sum is added to the third number. The capability of a calculator to carry out chain operations makes this pause unnecessary.

The order of addition is not important. Verify the following on your calculator.

Numbers like these were added long, long, before electronic calculators were invented. (How many men did Hannibal have crossing the Alps? How many snakes did St. Patrick chase out of Ireland?) How was it done? It is important for you to know how to do this because you may be caught without your calculator some day. Or you may have to add numbers too large to be entered in your calculator.

When numbers exceed the eight-digit capacity of your calculator, your machine displays an overflow sign. This means that the numbers you tried to add overflowed the registers.

Example 3.3: Try to perform

$$87654321 + 23456789.$$

Solution? "Overflow."

By the way, there is an interesting pattern that results from the addition in Example 3.3. Can you find it?

In Section 3.4, we show a way to use your machine to help you add numbers that exceed the capacity of your machine. To do this, we shall have to know the whys and wherefores of addition.

Ways of doing these kinds of calculations (or, more generally, ways of doing any arithmetic operation) are called *algorithms*. Anyone who has used a recipe to prepare, for example, an omelette has used an algorithm. We shall consider algorithms for addition, multiplication, subtraction, and division as well as algorithms for solving other kinds of arithmetic problems.

An omelette recipe, or algorithm, tells in step-by-step procedures, how to cook an omelette. If the algorithm (recipe) is a good one, you will be successful in solving the problem (producing a delicious omelette).

The addition algorithm works as follows: Suppose you memorized (or counted on your fingers) the additions of all pairs of one digit numbers, $0, 1, \ldots, 9$. Indeed, the table you completed in Exercise 3.1 is the table that school children memorize. Equipped with these basic sums, it is easy to see how to add other numbers. Here are some examples.

Example 3.4: $31 + 43 = ?$

Solution:

	Interpretations
$31 + 43 = (30 + 1) + (40 + 3)$	Rearrangement of terms
$= (30 + 40) + (1 + 3)$	$= (3 \text{ tens} + 4 \text{ tens}) + (1 + 3)$
$= 70 + 4$	$= 7 \text{ tens} + 4$
$= 74$	$= 70 + 4 = 74$

The above work may be arranged very neatly in a column where the place or location of the digit keeps the tens and the units separated:

$$31 = 30 + 1$$
$$+43 = 40 + 3$$
$$? \quad 70 + 4 = 74$$

Short form		*Shorter form*
31		31
$+43$		$+43$
4	*First add units:*	74
	$3 + 1 = 4$	
$+70$	*Next add tens:*	
	$3 \text{ tens} + 4 \text{ tens} = 7 \text{ tens}$	
74	*Final addition*	

Example 3.5: $47 + 28 = ?$

Solution:

$$47 + 28 = (40 + 7) + (20 + 8)$$
$$= (40 + 20) + (7 + 8)$$
$$= \quad 60 \quad + \quad 15 \quad = 60 + 10 + 5$$
$$= \quad 70 + 5$$
$$= \quad 75$$

Arranged in vertical form:

$$47 = 40 + 7$$
$$+28 = 20 + 8$$
$$? = 60 + 15 = 60 + 10 + 5 = 75$$

NOTE: Because $7 + 8 = 15 = 10 + 5$, and an additional 10 must be added to the tens column, a "one" is carried. This is sometimes done by indicating a "1" at the top of the tens column.

	First step	Second step	Third step
	47	^{1}47	^{1}47
	28	28	28
	(1)5	5	75

A third example completes our repertoire of tricky problems.

Example 3.6: 76 + 85 = ?

Solution:

$$76 + 85 = (70 + 6) + (80 + 5)$$
$$= (70 + 80) + (6 + 5)$$
$$= \quad 150 \quad + \quad 11$$
$$= 150 + 10 + 1 = 160 + 1 = 161$$

We may arrange the work as follows:

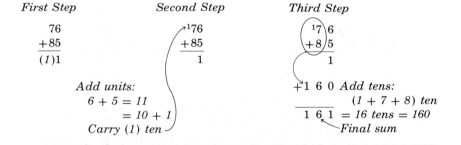

First Step

   76
+85
(1)1

Add units:
6 + 5 = 11
= 10 + 1
Carry (1) ten

Second Step

^{1}76
+85
1

Third Step

^{1}7 6
+8 5
1

+1 6 0 *Add tens:*
(1 + 7 + 8) ten
1 6 1 = 16 tens = 160
Final sum

Exercises 3.2

1. Perform the following additions first "by hand," then on calculator:

•a.   82
  +76

b.   89
 +79

•c.   731
 +258

d.   658
 +561

•e.   865
 +568

f.   35
63
+71

g.   47
69
+81

•h.   803
739
+546

•i.   10035
+99881

•j.   366158
76
8493
+ 90537

• 2. Perform the following addition on the calculator:

$$80605$$
$$+50608$$

Compare your answer with the one you obtained for part e of Exercise 1. Which digits of the answer to Exercise 2 correspond to the "carrying over" in Exercise 1e.

• 3. Perform the following addition on the calculator:

$$80003$$
$$70309$$
$$+50406$$

To which part of Exercise 1 might you compare this answer? Describe the comparison.

• 4. Suppose the shepherd arranged some pots as shown here.

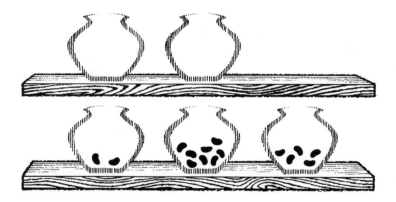

Use beans and pots to explain how to add 285 + 579. Use the pots on the top shelf for "carries."

The pots on the lower shelf represent ones, tens, and hundreds as before. Let the "carries" be shown in the upper pots. Explain the addition 285 + 579 using pots and going through step-by-step addition of ones, tens, hundreds, showing how the carries are merged into the different pots.

5. Consult the financial pages of your newspaper. It should give a list of the most active stocks on the New York Stock Exchange. What was the total number of shares traded for these stocks for the day listed?

6. Solve these problems by guessing an entry in the box so that the indicated sum is correct. Check your guess on a calculator. (In this way, you will be using your calculator as an experimental tool.) You may make as many guesses as you need to find the correct answer.

Sample: $7 + \square = 16.$
　　　　First guess, 5.　　Check: $7 + 5 = 12.$　Tilt!　$(\neq 16).$
　　　　Second guess, 9.　Check: $7 + 9 = 16.$　Eureka!

•a. $2 + \square = 7$　　　　　　d. $\square + 8 = 15$

•b. $3 + 5 + \square = 16$　　　　e. $3 + \square + 9 = 17$

　c. $8 + \square = 11$　　　　　•f. $23 + \square = 41$

7. Find the missing digit. Guess and check!

•a.　 1 3　　　　　•c.　　 8　　　　　　d.　　 7
　　+2 $\square$　　　　　　　$\square$ 2　　　　　　　　 $\square$
　　‾‾‾‾　　　　　　 + 3 $\square$　　　　　　+ $\square$ 9
　　 3 8　　　　　　 ‾‾‾‾‾　　　　　　 ‾‾‾‾‾
　　　　　　　　　　　 5 6　　　　　　 1 1 3

　b.　 1 6
　　+$\square$ 8
　　‾‾‾‾
　　 4 4

• 8. Find the perimeter (distance around) for these figures.

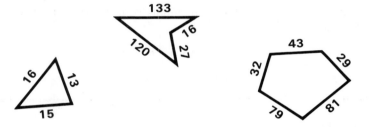

9. Consult the map in Appendix B in the back of the book. Plan three different routes between San Diego and San Francisco. Find the distances for each route. Which is the shortest?

•10. Consult the map in Appendix B in the back of the book. A salesman must leave his plant in San Diego and visit in some order these places: Los Angeles, San Bernardino, Long Beach, Bakersfield. Then he is to return to San Diego. Find the shortest route for him.

11. The Riveras want to erect a fence around three sides of their lot as shown in the adjoining sketch. How much fencing will be needed?

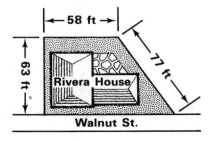

•12. Bonnie and Cindy decided to do some comparison shopping. They each went to a different supermarket and bought the same list of items. Here's what they bought and how much the items cost.

Item	Bonnie's cost in cents	Cindy's cost in cents
quart of milk	68	81
margarine	68	56
dozen eggs	69	78
crackers	78	82
cheese (1 lb.)	217	203
bacon (1 lb.)	148	128
flour (5 lbs.)	83	88
paper towels	39	57
pet food	77	71
coffee (same brand)	338	359

How much did each spend? Who spent less? List the items costing more than $1.00. List the items costing less than 50 cents.

13. Look at the addition table you completed in Exercise 3.1. What is the pattern that expresses the fact

$$A + B = (A + 1) + (B - 1)?$$

[Try some numbers for A and B.]

3.3 Addition with Decimals

Addition takes place in a column designating the same place value. This applies to numbers with decimal points just as well. A familiar instance of this is the addition of a grocery bill. The decimal point shows that the units place (column) is the one immediately to the left of the decimal point.

Calculators using floating-point arithmetic automatically adjust the placement of the decimal point. (This is why it is called *"floating point."*) All we need do is punch the decimal point key at the proper place. Calculators with fixed place arithmetic are not as flexible.

In doing problems by hand, it is best to line up the column of numbers so that all decimal points are in a vertical line. Figure 3.3 shows an example.

```
        ↓
      7 . 2 3 4
   2 3 . 7 5
    0 . 3 3
    0 . 0 0 5
1 0 0 0 . 5 6 1 2
```

Figure 3.3
Some decimal numbers lined up for addition. The place (column) indicated by the arrow is the *ones* place.

Addition is accomplished by lining up the decimal points in a column and proceeding exactly as though the decimal point were not there! It *is* there, though, and you must remember to put it in the answer at the same point. Here are some examples.

Example 3.7: Add 2.5 and 4.1. (*Answer:* 6.6)

Solution by hand:

$$\begin{array}{r} 2.5 \\ +4.1 \\ \hline 6.6 \end{array} \quad \text{Compare with} \quad \begin{array}{r} 25 \\ +41 \\ \hline 66 \end{array}$$

Solution by calculator:

$$2.5 + 4.1 = ?$$

$$2.5 \; \boxed{+} \; 4.1 \; \boxed{=} \; 6.6$$

Example 3.8: $2.25 + 35 cents + $3 = ?

Solution: $2.25 \; \boxed{+} .35 \; \boxed{+} \; 3 \; \boxed{=} \; 5.6$

NOTE: 35 cents = .35 dollars. Key in the decimal point first. Some calculators will record the answer as 5.60. In either case, we interpret the answer as $5.60, or "five dollars and sixty cents."

Solution by hand: List the numbers in a column so that the decimal points are in a vertical line.

2.25	
.35	(35 cents is still .35 dollars!)
+3.00	(It is convenient to write the zeros)
5.60	(Ignore the decimal point in adding. Carry 1 as before. Finally, put decimal point in the same vertical line.)

Example 3.9: 3.25 + 10.1 + 8 = ?

Solution by hand:

$$\begin{array}{r} 3.25 \\ 10.10 \\ +\;\;8.00 \\ \hline 21.35 \end{array}$$

It is convenient to write the extra zeros in 10.10 and 8.00, so that all numbers have the same number of decimal places.

Solution by calculator: $3.25 \; \boxed{+} \; 10.1 \; \boxed{+} \; 8 \; \boxed{=} \; 21.35$

Exercises 3.3

Do these problems both by hand and by calculator.

•1. Dick made four purchases. What is his total bill?

$1.27, $8.99, $12.34,   10 cents.

2. Dorothy made four purchases. What is her total bill?

$1.27, $8.99, $12.34, $10.

3. Add: 15.34 + 36.58

•4. Add: 789.58 + 7.3 + 15.68

5. Add: 3.0
               0.3
               0.03
               0.003
            +0.0003

•6. Make up a problem like Exercise 5 and find the answer.

7. Measure and record the lengths of these triangles in centimetres.

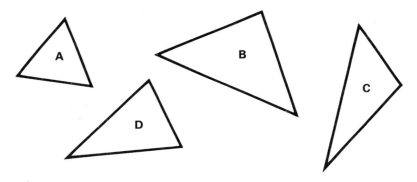

•a. Determine the perimeter of each triangle.
•b. Which triangle has the longest side?
 c. Which triangle has the shortest side?
 d. Order the triangles by increasing size of perimeter.

3.4   What to Do When Addition Overflows the Calculator

Attempting additions like

$$83243678$$
$$+36655213$$

An easier example might give us a clue. Compute:

$$83240000$$
$$+36650000$$

Although these two eight-digit numbers would overflow the calculator, the zeros in each number give no trouble in the addition. The zeros complicate the *use* of the machine, but play no essential part in the computation. The cure is clear: Compute with the first four digits, then adjoin ("tack on") the zeros by hand at the end. Thus,

$$8324 \boxplus 3665 \boxminus 11989.$$

Conclude:

$$83240000$$
$$+\ \ 36650000$$
$$\overline{119890000}$$

This worked out well. Let's try something like this with the original problem. Split the two numbers into two parts, each part consisting of four digits:

$$8324 \quad 3678$$
$$+3665 \quad 5213$$

Add each part with the calculator and record the answers using place-value to tell where to put the answers:

$$3678 \boxplus 5213 \boxminus 8891.$$

Putting our partial sums together gives us:

$$8324 \quad 3678$$
$$+\ \ 3665 \quad 5213$$
$$\overline{11989 \quad 8891}$$

There was no "carry over" from the right half to the left half of the problem. What if the sum of the right hand digits used five digits?

$$8324 \quad 8763$$
$$+3665 \quad 9574$$

As before, split the number into two groups of four digits. As before, add the numbers on the right hand side:

$$8\,763\;\boxed{+}\;9574\;\boxed{=}\;18337.$$

So

$$
\begin{array}{rr}
8324 & 8763 \\
+\,3665 & 9574 \\
\hline
\cdots(1) & 8337
\end{array}
$$

Here, there is a "1" to be carried. Simply add this to the four-digit numbers on the left hand side:

$$8324\;\boxed{+}\;3665\;\boxed{+}\;1\;\boxed{=}\;11990.$$

Finally then,

$$
\begin{array}{rr}
8324 & 8763 \\
+\;\;3665 & 9574 \\
\hline
11990 & 8337
\end{array}
$$

Why did we choose to split the numbers into two parts, each *four* digits long? We didn't have to, but it is convenient to choose a size so that the numbers can be remembered while the numbers are entered in the calculator.

Just as easily as we have added a pair of numbers, we can add a column of large numbers of varying sizes. Here we add numbers that cannot even be entered in an 8-digit calculator.

Example 3.10:

$$
\begin{array}{r}
147653219 \\
209979959 \\
3165534 \\
622030944 \\
325 \\
+\,901580243
\end{array}
$$

Solution:

(1) Split the numbers. (It seems convenient to split the numbers between the fifth and sixth place from the right.)
(2) Add the five-digit numbers of the righthand side.
(3) Record the last five digits of this sum in the last five places of the answer.
(4) Add the carry to the numbers of the left half.
(5) Record this answer as the first (five) digits of the answer.

Here are the details:

(1) The split:

1476	53219
2099	79959
31	65534
6220	30944
	325
+9015	80243

(2) Right half via calculator:

$$5\,3\,2\,1\,9 \boxplus 7\,9\,9\,5\,9 \boxplus 6\,5\,5\,3\,4 \boxplus 3\,0\,9\,4\,4$$
$$\boxplus 3\,2\,5 \boxplus 8\,0\,2\,4\,3 \boxminus 3 \quad 1\,0\,2\,2\,4.$$

(3) Record as last five digits of the answer:

1476	53219
2099	79959
31	65534
6220	30944
	325
+9015	80243
	10224 ←

(4) Carry the "3" and add to the left half:

$$1\,4\,7\,6 \boxplus 2\,0\,9\,9 \boxplus 3\,1 \boxplus 6\,2\,2\,0 \boxplus 9\,0\,1\,5 \boxplus 3 \boxminus 1\,8\,8\,4\,4.$$

(5) Record these as the first five digits of the answer:

1476	53219
2099	79959
31	65534
6220	30944
	325
+9015	80243
18843	10224

Exercises 3.4

1. Use your calculator to compute these large-number sums.

   •a.   52668862
          33068796
       +79653531

   b.   54316992671
          +82921976102

2. Some numbers require more than one splitting. Try these.

   •a.   12659121156412168
           +91325800238737524

   b.   53668736205889132
            66376673663766736
          +87365368720589317

•3. The number of acres in 28 national parks of the United States are listed below.

Park/Location	Acres
Acadia, Maine	29,978.08
Big Bend, Texas	692,304.70
Bryce Canyon, Utah	36,010.38
Carlsbad Caverns, New Mexico	45,846.59
Crater Lake, Oregon	160,290.33
Everglades, Florida	1,258,361.00
Glacier, Montana	999,015.15
Grand Canyon, Arizona	645,295.91
Grand Teton, Wyoming	299,580.45
Great Smoky Mountain, North Carolina–Tennessee	507,159.16
Hawaii	179,950.90
Hot Springs, Arkansas	1,019.13
Isle Royale, Michigan	133,838.51
Kings Canyon, California	453,064.82
Lassen Volcanic, California	103,809.28
Mammoth Cave, Kentucky	50,695.73
Mesa Verde, Colorado	51,017.87
Mount McKinley, Alaska	1,939,319.04
Mount Rainier, Washington	241,571.09
Olympic, Washington	887,986.91
Platt, Oklahoma	911.97
Rocky Mountain, Colorado	254,735.70
Sequoia, California	385,178.32
Shenandoah, Virginia	193,472.98
Wind Cave, South Dakota	27,885.67
Yellowstone, Wyoming–Montana–Idaho	2,213,206.55
Yosemite, California	757,617.36
Zion, Utah	94,241.06

a. Find the total acreage of these national parks.

b. Find the total acreage of the five largest parks.

4. The following table shows the size (in square miles) and the population (1960 census) of each of the Mountain and Pacific states.

State	Area	Population
Alaska	586,400	228,167
Arizona	113,909	1,321,587
California	158,693	15,862,223
Hawaii	6,454	641,794
Idaho	83,557	671,637
Montana	147,138	679,024
Nevada	110,540	291,083
New Mexico	121,666	953,187
Oregon	96,981	1,772,341
Utah	84,916	900,862
Washington	68,192	2,856,963
Wyoming	97,914	331,529

a. Find the total number of square miles in the Mountain and Pacific states of the United States.

b. Find the total population in these states.

3.5 Subtraction

Many processes we carry out are reversible. It is important to know how to reverse them! For example, if we know how to get on the freeway to go to the beach, we must also know how to use the freeway to return from the beach.

Thus it is in addition! We add: $6 + 8 = 14$. We have *"gone from 6 to 14 by adding 8."* How can we go from 14 to 6? That is, how can the box in the display below be filled so that equality holds?

$$6 + \square = 14$$

Just as we can remember how we used the freeway to get to the beach, just so we can remember how we added 8 to 6 to get 14. Thus we remember that 8 goes in the box.

In mathematics, we sometimes haven't already "been there." We must be able to reverse the process of addition without first having made the direct calculation. When we don't know what to do, a little experimentation will help. Don't be afraid to try out numbers on your calculator!

Example 3.11: The down payment on a new car is $525. If Robert Hotrod has already saved $413, how much more must he save to make the down payment?

Solution: Mathematically, we ask to fill in the box so that equality holds. We look for the missing term.

$$413 + \square = 525 \qquad\qquad (*)$$

First you guess: Surely, about $100. Well,

$$413 + 100 = 513; \quad \text{not quite enough; } 513 < 525.$$

Try again:

$$413 + 110 = 523; \quad \text{very close; within 2.}$$

Finally,

$$413 + 112 = 525$$

The operation of determining this missing term is called *subtraction*. We designate the answer to the equation marked (*) as

$$525 - 413$$

The pattern is

$$413 + \boxed{(525 - 413)} = 525$$

From our process of guessing, we know: $525 - 413 = 112$.

Guessing is not a bad way to do the problem at all. The solution to many problems in mathematics and science starts with a good guess. Even with a bad guess! The guesses are refined so that they get better and better. Finally, an answer is obtained. Then, of course, it must be shown that the answer is correct.

Example 3.12: How to perform subtraction on your calculator.

$$525 - 413 = ?$$

Solution: $525 \boxminus 413 \boxminus 112.$

On the calculator, no subtraction problem is more difficult than another. Even decimal points are no problem.

Example 3.13: An item plus tax costs $4.13. A customer gives the clerk $5.25. What is the correct change?

Solution: $5.25 \boxminus 4.13 \boxminus 1.12$

So far, we have avoided such problems as the following:

Subtract 9 from 2.

That is, what is

$$2 - 9?$$

Our theory is not yet prepared for this. But your calculator is! Try it.

Example 3.14: $2 - 9 = ?$

Solution: $2 \boxminus 9 \boxminus -7.$

(For the answer, some machines may show an "overflow" signal instead to indicate that this problem cannot be done on the machine.)

The *minus* $(-)$ sign, or whatever special designation your calculator has, shows that you have subtracted a larger number from a smaller one. Notice that the numerical size, 7, of the answer is what you would obtain in computing $9 - 2$. Thus, a negative number, -7, simply says that you subtracted a larger number from a smaller one. More about that in Chapter 6.

Exercises 3.5

1. Perform the following subtractions on your calculator.

   a. $8 - 6 =$ •d. $5687 - 3292 =$ (*Answer: 2395*)

   •b. $16 - 8 =$ e. $8763 - 25 =$

   c. $231 - 3 =$ •f. $186001 - 9997 =$

2. Do these additions and subtractions in the sequence indicated.

   •a. $8 - 6 + 7 - 3 - 2 =$ (*Answer: 4*)

   •b. $(8 + 7) - (6 + 3 + 2) =$

   c. $5371 - 429 + 33 + 116 - 401 =$

   d. $(5371 + 33 + 116) - (429 + 401) =$

NOTE: Remember that operations within parentheses should be performed first; then use the result to replace the whole expression in parentheses.

3. Each day, Abie's Deli buys ten dozen bagels. Here are the daily net "profits" (in dollars) for bagel sales on 15 days. A plus sign indicates a profit; a minus sign indicates a loss.

$$+12, \quad +9, \quad -3, \quad -14, \quad +7, \quad -8, \quad 0, \quad -18, \quad +17, \quad +10, \quad -2, \quad +6,$$
$$+15, \quad -3, \quad -9$$

•a. All in all, did the 15-day total turn out to be a profit or a loss? By how much?

b. Find the same total, making use of the procedure suggested by Exercise 2.

4. Ms Argent watches the daily prices of her stock in Company G. She records the daily fluctuations in its price for a week. "$+A$" means that the stock went up A dollars; "$-B$" means the stock went down B dollars. What was the net gain or loss for the week?

Monday:	-6	Thursday:	$+5$
Tuesday:	$+3$	Friday:	-3
Wednesday:	$+4$		

5. Two women share an apartment. Each keeps track of how much she spends for the apartment during the month, and they settle the difference at the end of the month. The following is the record of amounts spent by each woman during one month.

Annette		*Doreen*	
Food	$153.18	Food	$ 62.53
Phone bill	32.15	Electric bill	18.47
Water bill	17.42	Newspaper	5.75
Linen laundry	5.23	Light bulb	0.78
Gas bill	28.39	Rent	215.00

a. How much has each woman spent on the apartment during the month?

•b. What is the total amount spent by both women?

•c. If each is to pay half the total amount, how much *should* each pay?

d. In order to achieve this result, which woman should pay money to the other, and how much? Verify that this settlement will leave things even.

•6. Five people share a house. Each person pays some of the grocery and other bills for the household. At the end of the month, they report the following expenditures.

Billie	$79.82	Frankie	$13.00
Mary	$41.17	Tony	$246.99
Jim	$105.02		

They would like to even things up so that each person will have paid one-fifth of the month's expenses. Find one set of payments (each payment being from one person to another) that will accomplish this goal. [NOTE: There are many possible answers. Try to make your answer as simple as possible—that is, involving no more payments than necessary.]

The important thing to remember about subtraction and addition is that the following statements express the same idea.

$$6 + 8 = 14$$

$(14 - 6) = 8$	$6 + (14 - 6) = 14$
$(14 - 8) = 6$	$8 + (14 - 8) = 14$

Which expression is used depends on the point of view.

The relation between subtraction and addition is a very important one to remember. Here is a summary.

> Subtraction is the reverse of addition.
> $A - B = C$ and $A = B + C$ state the same fact.
> $A - B$ is the number such that $B + (A - B) = A$.
> $A - A = 0$ for any number A.

These characteristics of subtraction lead to a hand algorithm for performing subtraction.

Example 3.15:
$$525$$
$$-413$$

Solution: Begin with the "units" digit: $5 - 3 = 2$, because $3 + 2 = 5$.

$$\begin{array}{r} 5\ 2\ \boxed{5} \\ -\ 4\ 1\ \boxed{3} \\ \hline 2 \end{array} \Big\} \ 3 + 2 = 5$$

Next work on the "tens" digit: $2 - 1 = 1$, because $1 + 1 = 2$.

$$\begin{array}{r} 5\ \boxed{2}\ 5 \\ -\ 4\ \boxed{1}\ 3 \\ \hline 1\ 2 \end{array}$$

Finally, work on the "hundreds" digit: $5 - 4 = 1$, because $1 + 4 = 5$.

$$\begin{array}{r} 5\ 2\ 5 \\ -4\ 1\ 3 \\ \hline 1\ 1\ 2 \end{array}$$

The subtraction in Example 3.15 can be carried out without trouble using the shepherd's pots. For example, 525 is represented thus.

To subtract 413 from 525, we can work step-by-step with the ones pot, next the tens, then the hundreds.

Remove 3 pebbles from the ones pot;
        1 pebble from the tens pot;
        4 pebbles from the hundreds pots.

Example 3.16:
$$\begin{array}{r} 231 \\ -3 \\ \hline \end{array}$$

Solution: Begin with the units digits. Trouble! $1 - 3 = ???$ To fix things up, we "borrow" 10 from the tens digit: $231 = 220 + 11$. That is,

$$231 = 200 + 30 + 1$$
$$= 200 + 20 + 10 + 1$$
$$= 200 + 20 + 11$$

$$\begin{array}{rrr} 2 & 2 & 11 \\ - & & 3 \\ \hline 2 & 2 & 8 \end{array} \qquad (8 = 11 - 3)$$

Borrowing is often indicated as follows.

$$\begin{array}{r} {}^{2}2\,\cancel{3}\,{}^{1}1 \\ -3 \\ \hline \end{array}$$

Trying Example 3.16 with the shepherd's pots is not as easy as before. In the ones pots, we cannot subtract (remove) 3 pebbles from 1.

Fortunately, there are some tens. So we can remove 1 pebble from the second pot (1 ten) and trade it for ten ones, dumping these ten in the ones pot.

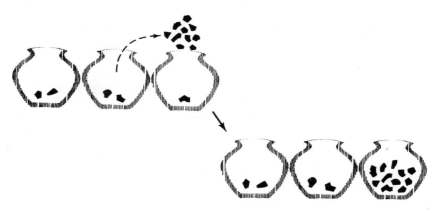

Now, we can complete the subtraction by removing 3 pebbles from the ones pot.

Example 3.17: $28.32 - 7.1 = ?$

Solution: Line up decimal points and proceed as before.

$$\begin{array}{r} 28.32 \\ - 7.10 \\ \hline 21.22 \end{array}$$

Example 3.18: $28.023 - 7.54 = ?$ *(Requires borrowing.)*

Step-by-step solution:

$$
\begin{array}{r}
2\,8\,.\,0\,2\,3 \\
-\ \ 7\,.\,5\,4\,0 \\
\hline
3
\end{array}
\qquad
\begin{array}{r}
{\scriptstyle 7\ \ \ 9}\ \ \ \ \ \ \ \\
2\,\cancel{8}\,.\,\cancel{0}\,{}^{12}\,3 \\
-\ \ 7\,.\,5\,4\,0 \\
\hline
8\,3
\end{array}
\qquad
\begin{array}{r}
{\scriptstyle 7\ \ \ 9}\ \ \ \ \ \ \ \\
2\,\cancel{8}\,.\,\cancel{0}\,{}^{12}\,3 \\
-\ \ 7\,.\,5\,4\,0 \\
\hline
2\,0\,.\,4\,8\,3
\end{array}
$$

↑
In the 2nd column, we cannot subtract 4 from 2

↑
In the 3rd column, we cannot borrow 1 from 0; borrow 1 from 80.

Calculator check: $28.023 \ominus 7.54 \equiv 20.483$.

Exercises 3.6

1. Perform these subtractions. When in doubt, try it on a calculator and see if you can justify or correct your hand calculation.

   a. $7896 - 6002$ •c. $37. - 20.1$ •e. $3.72 - 0.201$

   •b. $302 - 197$ d. $372 - 2.01$ f. $372 - 0.201$

2. The rule "$A - B = C$ is the same fact as $A = B + C$" works even when B is larger than A. With your calculator, verify that

   $$2 - 9 = -7 \qquad \text{and} \qquad 2 = 9 + (-7).$$

   Repeat these operations for the following statements.

   a. $3 - 17 = ?$ b. $1001 - 9876 = ?$

•3. Return to Exercise 12 at the end of Section 3.2 and answer these questions.

   a. How much more did Cindy spend than Bonnie?

   b. On which item was there the biggest difference in expenditures between the two women?

4. Continuing with the information from Exercise 12 for Section 3.2, suppose that Alice buys the same set of items, but buys each item from the store where its cost is least.

   a. What is the total amount that Alice spends?

   b. How much does Alice save in comparison with Bonnie's cost?

5. Return to Exercise 10 for Section 3.2. Of the routes you picked, how much longer was the longest route from San Diego than the shortest route?

Just as we can adapt our procedures for adding numbers that exceed the capacity of the calculator, we can similarly handle subtraction involving such numbers. Again, we first examine simple cases for a clue.

Example 3.19: Subtract: 896500000 − 3400000. That is,

$$896500000$$
$$- \quad 3400000$$

Solution: The adjoined zeros (five of them) play an inessential role in the subtraction algorithm. We may, by hand or by calculator, perform

$$8965 \; \boxminus \; 34 \; \boxminus \; 8931.$$

and then adjoin the five zeros.

$$
\begin{array}{r|l}
8965 & 00000 \\
- \quad 34 & 00000 \\
\hline
8931 & 00000
\end{array}
$$

We immediately extend this technique.

Example 3.20:

$$
\begin{array}{r|l}
8965 & 73862 \\
- \quad 34 & 34651 \\
\end{array}
$$

Solution:

First: $73862 \; \boxminus \; 34651 \; \boxminus \; 39211.$
Second: $8965 \; \boxminus \; 34 \; \boxminus \; 8931.$

Conclusion:
$$
\begin{array}{r|l}
8965 & 73862 \\
- \quad 34 & 34651 \\
\hline
8931 & 39211
\end{array}
$$

We have seen that "borrowing" plays an important role in subtraction. Occasionally, we shall have to "borrow" even using the calculator. Suppose, in the previous example, we had chosen to split the numbers at a different place.

Example 3.21: $896573862 - 3434651 = ?$

$$89657 \quad 3862$$
$$-\quad 343 \quad 4651$$

Solution: We cannot perform $3862 - 4651$ because

$$3862 < 4651.$$

Performing $3862 - 4651$ on our calculator will result in a negative number. So we must "borrow" a one from the next place.

$$89657^{\,6} \quad {}^{1}3862$$
$$-\quad 343 \quad 4651$$

Now, $1\,3862 \ominus 4651 \equiv 9211.$

    Continue as before with $89656 \ominus 343 \equiv 89313.$

Conclusion:

$$89657^{\,6} \quad {}^{1}3862$$
$$-\quad 343 \quad 4651$$
$$\overline{89313 \quad 9211}$$

The reason the above procedure works is, by now, familiar:

$$89657 \quad 3862 = 896570000 + 3862$$

$$= (896560000 + 10000) + 3862$$

$$= 896560000 + 13862.$$

And,

$$3434651 = 3430000 + 4651.$$

So

$$896573862 - 3434651 = (896560000 + 13862) - (3430000 + 4651)$$

$$= (896560000 - 3430000) + (13862 - 4651).$$

In this way, we "borrowed" a "1" from the "7," and the "7" dropped down to "6" as indicated.

Two points may be noted from this example.

1. It may ease your work to split the numbers so that no borrowing is necessary.

2. If that is impossible, simply borrow from the next place to the left.

Example 3.22:

$$\begin{array}{r|l} 1234\overset{4}{\cancel{5}} & {}^{1}61476 \\ -\ 9598 & 73697 \\ \hline \end{array}$$

Solution:

First: "borrow" a "1" from the "5."
Second: *161476* ⊟ *73697* ⊜ *87779*.
Third: *12344* ⊟ *9598* ⊜ *2746*.
Conclusion:

$$\begin{array}{r|l} 1234\overset{4}{\cancel{5}} & {}^{1}61476 \\ -\ 9598 & 73697 \\ \hline 2746 & 87779 \end{array}$$

We have had to make one "borrow" in the above, but we were able to extend the capacity of our calculator. Zeros often appear to be troublesome. Here is an example to show what can be done.

Example 3.23:

$$\begin{array}{r|l} 34500 & 3218 \\ -27898 & 7435 \\ \hline \end{array}$$

Solution: We need to borrow "1" to replace "3218" by "13218." The 1 comes from 500; $500 - 1 = 499$. Thus,

$$\begin{array}{r|l} 3\overset{499}{\cancel{4500}} & {}^{1}3218 \\ -27898 & 7435 \\ \hline 6601 & 5783 \end{array}$$

13218 ⊟ *7435* ⊜ *5783*.

34499 ⊟ *27898* ⊜ *6601*.

So $345003218 - 278987435 = 66015783$.

Exercises 3.7

1. Follow through the explanation of how to subtract large numbers using these examples with small numbers. Verify your results both by full hand calculation and by a single-operation use of the calculator.

•a.   65 74
    −31 23

b.   65 47
    −31 56

c.   65 43
    −31 57

•d.   65 12
    − 7 58

e.   60 02
    −58 76

2. How can you use the calculator to subtract two 9-digit numbers? Experiment with these.

   •a.   987654367   •c.   346375482
        −253142055   −178996344

   b.   912345678
       −765342764

3. Perform these subtractions.

   •a.   8643289756   b.   8640080001
        −5432957889   − 234576430

•4. The population of the United States was 151,325,798 in 1950 and 179,323,175 in 1960. How much did the population increase over those 10 years?

4

Multiplication and Division of Decimal Numbers

4.1 Two Interpretations of Multiplication

Multiplication solves two basic types of problems. Although these problems may at first seem unrelated, they are in fact closely connected. Because they arise under very different circumstances, it is worthwhile to treat them separately.

Problem 1: *Repeated Addition.* Janet's Sport Togs has sweaters on sale for $6. Janet sells 8 sweaters; what are her total receipts?

Solution: $6 \times 8 = 48$.

Problem 2: *Combinations.* Janet's Sport Togs features 8 styles of shirts and 6 styles of jeans. How many different outfits, consisting of jeans and shirt, can she offer to her customer?

Solution: $6 \times 8 = 48$.

We analyze the first problem as follows. First, list the sales:

$$6 + 6 + 6 + 6 + 6 + 6 + 6 + 6 = 48.$$

$$\underbrace{}$$

8 sixes

This is why multiplication is called repeated addition, as in Problem 1 above.

We can analyze the second problem by a diagram. We first number the kinds of jeans: 1, 2, 3, 4, 5, and 6, and number the kinds of shirts: 1, 2, 3, 4, 5, 6, 7, and 8.

Now, make a rectangular array to describe the possibilities. The total number of style combinations equals the total number of x's. This number is 48. Count 'em!

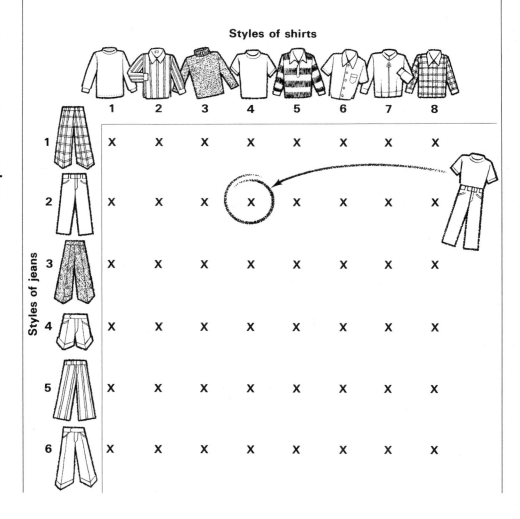

Styles of shirts

Styles of jeans

The connection with the first problem is this. Simply add up the number of combinations in each row. In each row are 6 combinations and there are 8 rows. So the total number is

$$\underbrace{6 + 6 + 6 + 6 + 6 + 6 + 6 + 6}_{\text{8 sixes}} = 48$$

From Chapter 2, we know how to multiply two numbers on a calculator.

Example 4.1: $6 \times 8 = ?$

Solution: $6 \; \boxed{\times} \; 8 \; \boxed{=} \; 48.$

Because of its speed, the calculator appears to regard multiplication as a single combination—the answer is flashed immediately. However, it actually uses one or more special algorithms that involve repeated addition.

The basic multiplication facts may be displayed in a multiplication table. Just as with addition, we need remember only the products of pairs of numbers, $0, 1, 2, \ldots, 9$.

Because $0 \times (\text{anything}) = 0$, we don't have to write these facts in our table. (Check it out on the calculator by multiplying, say, 3×0 or 113×0 or 0×321.)

Exercise: Complete the following table. In each square, enter the product of the number of the row and the column. For example, the entry in row 6 and column 8 is $6 \times 8 = 48$.

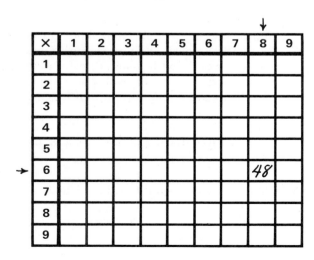

×	1	2	3	4	5	6	7	8	9
1									
2									
3									
4									
5									
6								48	
7									
8									
9									

You may complete this table by using your calculator (it is good exercise), or you may complete it from your own mental arithmetic. As you continue to practice multiplication, you will find it easier to memorize these basic products than to use a calculator. However, there is nothing wrong in using a calculator—especially if you cannot remember whether $7 \times 8 = 56$ or 65.

The multiplication table is full of patterns.

1. *Symmetry.* Draw a diagonal line from the top left corner to the bottom right corner. Think of the square being folded along this line. (See Figure 3.2 in Chapter 3.) Entries which come together have the same value. This is because

$$A \times B = B \times A$$

for any two numbers, A and B.

2. Patterns from column to column: Look at any column. Compare the entries in this column with the entries in the next column. The difference of two such numbers in any row is constant. Find a similar pattern between rows.

Here are some important facts about multiplication.

$A \times 0 = 0 \times A = 0$ for any number, A.
$A \times 1 = 1 \times A = A$ for any number, A.
$A \times B = B \times A$ for any two numbers, A and B.

The truth of the law

$$A \times B = B \times A$$

for multiplication can be seen from the point of view of combinations (Problem 2). Indeed, in the example cited about shirts and jeans, the total number of combinations is 48. It makes no difference whether we think of matching shirts (8) with jeans (6) or of matching jeans (6) with shirts (8). That is, as 8×6 or as 6×8.

This law is a bit more surprising when multiplication is viewed as repeated addition:

$$8 \times 6 = 6 + 6 + 6 + 6 + 6 + 6 + 6 + 6,$$
$$6 \times 8 = 8 + 8 + 8 + 8 + 8 + 8.$$

Why, indeed, should 8 sixes equal 6 eights? But it does!

• 1. Complete the table on page 81 in full.

2. Verify these products on your calculator.

a. $3 \times 16 = 48$

•b. $16 \times 3 = 48$

c. $16 \times 14 = 224$

•d. $8 \times (16 \times 14) = 1792$

•e. $(8 \times 16) \times 14 = 1792$

f. $1111999 \times 1 = 1111999$

•g. $87654321 \times 0 = 0$

h. $(337 \times 1) \times 45 = 15165$

i. $1 \times 2 \times 3 \times 4 \times 5 \times 6 \times 7 \times 8 \times 9 \times 10 = 3628800$

• 3. What happens if you try 54321×6789 on your calculator?

• 4. Find the largest number, L, you can enter so that your calculator will compute $L \times L$.

5. Auto expenses for driving for a charity may be deducted from your income tax. Mrs. Tosio drove 1,534 miles last year for her charity. She may deduct 7 cents per mile. How much may she deduct for her driving expenses?

6. Fallriver has about 1,235 families. The elementary schools of Fallriver accommodate 2,400 pupils. If each of these families has two children in the next few years, do you think Fallriver will need a new elementary school?

• 7. Along the California coast, there are about 250 commercial sport fishing boats. If each boat catches at least 175 fish a day, at least how many fish are caught each day off the California coast? In a week? In a year (365 days)?

8. The area of a rectangle is given by the product of its width and its length. (Here is an application of multiplication.)

a. Find the areas of the rectangles in the following figures.

b. Find the perimeters of these rectangles. (The perimeter is the total distance around the edges of the figure—in other words, the sum of the lengths of the four sides of the rectangle.)

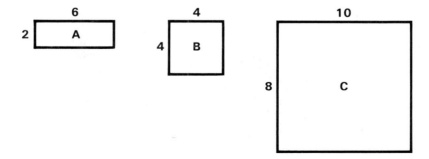

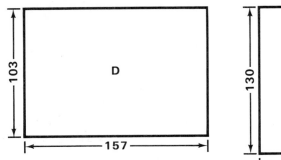

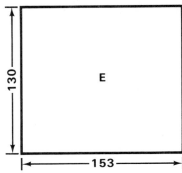

• 9. The seating chart of an auditorium looks like this.

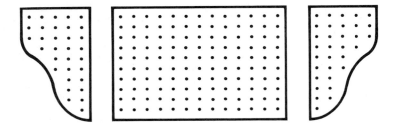

    a. Without doing a full computation, decide whether there is an odd or an even number of seats.

    b. How many seats are there in the central section?

    c. How many seats are there in the left side section?

    d. How many seats are there in total?

    e. Is this number even or odd? (Check with your guess in part a.)

10. The volume of a box is the product of its width, length, and height, as shown in the sketch below.

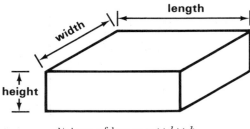

Volume of box $= w \times l \times h$

a. Measure the dimensions in inches of the following boxes and compare their volumes in cubic inches:
   a shoe box?
   a box containing a roll of film;
   a small cereal box?
   a large economy-size box of cereal;
   a large carton big enough to hold a case of something.

Make rough estimates for a box large enough to contain your car.

b. Measure the same boxes in centimetres and compute their volumes in cubic centimetres (cm^3).

•11. There are 12 inches in a foot, 3 feet in a yard, and 1,760 yards in a mile.

a. Find the number of feet in a mile.

b. Find the number of inches in a mile.

•12. There are 10 millimetres in a centimetre. There are 100 centimetres in a metre. There are 1,000 metres in a kilometre.

a. Find the number of centimetres in a kilometre.

b. Find the number of millimetres in a kilometre.

13. There are 60 seconds in a minute. There are 60 minutes in an hour. There are 24 hours in a day, and 365 days in a year.

a. How many minutes are there in a day?

b. How many minutes are there in a year?

c. How many seconds are there in an hour?

d. How many seconds are there in a year?

e. Find how many hours you are old. (You may estimate your age in a whole number of years and ignore leap years.)

f. How many more hours do you think you will live?

14. There are 2.54 centimetres in an inch. There are 36 inches in a yard. How many centimetres are there in a yard?

•15. There are about 0.45359 kilograms in a pound. How many kilograms would a 135 pound person weigh?

16. If a man, 170 centimetres tall, weighs 135 kilograms, do you think he is underweight or overweight?

•17. A certain city ordinance requires a public meeting place to have at least 50 cubic feet of space for each person in the room. At least how large should an auditorium be to hold an audience of 1,250?

18. In 1975, there were about 133 million motor vehicles registered in the United States. If all these vehicles were lined up bumper-to-bumper, about how long would the line of vehicles be? Would it reach from coast to coast?

4.1 Two Interpretations of Multiplication

4.2 Calculator Computation of Product of Decimal Numbers

On a calculator, addition is as easy for decimal numbers as it is for whole numbers. So it is with multiplication.

Example 4.2: $2.5 \times 4 = ?$

Solution: $2.5 \; \boxed{\times} \; 4 \; \boxed{=} \; 10.$
Compare with: $25 \; \boxed{\times} \; 4 \; \boxed{=} \; 100.$

Example 4.3: $7.832 \times 16.43 = ?$

Solution: $7.832 \; \boxed{\times} \; 16.43 \; \boxed{=} \; 128.67976$
Compare with: $7832 \; \boxed{\times} \; 1643 \; \boxed{=} \; 12867976.$

As you can see, the sticky point in multiplying decimals is the location of the decimal point in the product. Floating-point calculators make this quite easy because they are programmed to do it correctly. We shall give an explanation in Section 4.9, but for now we would like to suggest that you try these exercises to see if you can find a pattern in the scheme for locating the decimal point in the answer. To have a chance to see the pattern, be sure to do and record the answers to both columns of problems as they appear in your calculator.

Incidentally, these problems afford a good opportunity to use your constant (K) switch if your calculator has one. Take a minute to read your manual to see how to use it.

Exercises 4.2

1. Compare the problems on the left with those on the right. In these problems, one factor is kept the same integer.

•a.	$4 \times 2 =$	$4 \times 2 =$
•b.	$4 \times 21 =$	$4 \times 2.1 =$
•c.	$4 \times 32 =$	$4 \times 3.2 =$
•d.	$4 \times 99 =$	$4 \times 9.9 =$
•e.	$4 \times 213 =$	$4 \times 2.13 =$
•f.	$4 \times 1238 =$	$4 \times 12.38 =$
•g.	$4 \times 9999 =$	$4 \times 99.99 =$
•h.	$4 \times 2137 =$	$4 \times 2.137 =$
i.	$4 \times 8012 =$	$4 \times 8.012 =$

j. $4 \times 99999 =$ $4 \times 99.999 =$

k. $4 \times 21111 =$ $4 \times 2.1111 =$

l. $4 \times 03462 =$ $4 \times 0.3462 =$

m. $4 \times 436794 =$ $4 \times 43.6794 =$

n. $4 \times 999999 =$ $4 \times 99.9999 =$

2. In this exercise, the constant multiplier is 43 or 4.3. How do the answers on the left compare with those on the right?

•a. $43 \times 2 =$ $4.3 \times 2 =$

•b. $43 \times 21 =$ $4.3 \times 2.1 =$

•c. $43 \times 32 =$ $4.3 \times 3.2 =$

•d. $43 \times 99 =$ $4.3 \times 9.9 =$

•e. $43 \times 213 =$ $4.3 \times 2.13 =$

•f. $43 \times 1238 =$ $4.3 \times 12.38 =$

•g. $43 \times 9999 =$ $4.3 \times 99.99 =$

h. $43 \times 2137 =$ $4.3 \times 2.137 =$

i. $43 \times 8012 =$ $4.3 \times 8.012 =$

j. $43 \times 99999 =$ $4.3 \times 99.99 =$

k. $43 \times 21111 =$ $4.3 \times 2.1111 =$

l. $43 \times 03462 =$ $4.3 \times 0.3462 =$

m. $43 \times 436794 =$ $4.3 \times 43.6794 =$

n. $43 \times 999999 =$ $4.3 \times 99.9999 =$

3. In these problems, the constant multiplier is 456 or 4.56. Compare the answers on the left with those on the right.

•a. $456 \times 2 =$ $4.56 \times 2 =$

b. $456 \times 21 =$ $4.56 \times 2.1 =$

•c. $456 \times 32 =$ $4.56 \times 3.2 =$

d. $456 \times 99 =$ $4.56 \times 9.9 =$

•e. $456 \times 213 =$ $4.56 \times 2.13 =$

f. $456 \times 1238 =$ $4.56 \times 12.38 =$

•g. $456 \times 9999 =$ $4.56 \times 99.99 =$

h. $456 \times 2137 =$ $4.56 \times 2.137 =$

•i. $456 \times 8012 =$ $4.56 \times 8.012 =$

j. $456 \times 99999 =$ $4.56 \times 99.999 =$

•k. $456 \times 21111 =$ $4.56 \times 2.1111 =$

l. $456 \times 03462 =$ $4.56 \times 0.3462 =$

•m. $456 \times 436794 =$ $4.56 \times 43.6794 =$

n. $456 \times 999999 =$ $4.56 \times 99.9999 =$

4.2 Calculator Computation of Product of Decimal Numbers

4. How many places would the product have if—

•a. one factor has 2 places the other has 3? (That is, a 2-place decimal times a 3-place decimal.)

b. one factor has 2 places and the other has 1?

•c. one factor has 3 places the other has none? (That is, the other is an integer.)

d. one factor has 0 places and the other has 0?

e. one factor has 5,004 places and the other has 2,349 places? (Use your calculator.)

4.3   The Distributive Law

One of the most useful rules in mathematics is the *distributive law*. To describe the rule in rough terms, we say that it connects addition and multiplication. In this section, we shall use it to explain the algorithm for multiplying numbers with several digits. But first, let us look at an example of this rule. Please keep in mind that, when parentheses are used in mathematical statement, it means that the terms enclosed by the parentheses are to be combined first.

Example 4.4:   Is $(2 + 4) \times 8$ the same as $(2 \times 8) + (4 \times 8)$?

First: $(2 + 4) \times 8 = ?$

Solution: $(2 \boxplus 4) \boxtimes 8 \rightarrow 6 \boxtimes 8 \boxminus 48.$

Second: $(2 \times 8) + (4 \times 8) = ?$

Solution: $(2 \boxtimes 8) + (4 \boxtimes 8) \rightarrow 16 \boxplus 32 \boxminus 48.$

Thus we have verified that

$$(2 + 4) \times 8 = (2 \times 8) + (4 \times 8).$$

Now look again at the form and particularly at the positions of the numbers 2, 4, and 8. A diagram will help show what is going on. You may find it helpful to *"let your fingers do the walking along the arrows"* to see how the positions of the numbers change.

$$((2) + (4)) \times (8) = (2 \times 8) + (4 \times 8).$$

It is also important to see how the operation symbols, $+$ and $\times$, change positions.

$$(2 \; \boxed{+} \; 4) \; \boxed{\times} \; 8 = (2 \times 8) + (4 \times 8).$$

You have probably used the distributive law many times without realizing it. The next example shows what may seem like a familiar situation.

Example 4.5: Four couples (8 persons) plan an evening out, with a show and snack afterward. The show will cost $4 each. The snack is estimated to cost $2 each. How much will the total bill be?

Solution: The cost per person is thus $4 + $2. There are eight in the party:

$$(4 \; \boxed{+} \; 2) \times 8 = 6 \; \boxed{\times} \; 8 \; \boxed{=} \; 48; \text{ or } \$48.$$

That was all well and good. However, that is not the way the money is paid out. First, comes the show. The bill there is $4 per person, and there are 8 persons.

$$\text{Total cost for the show is: } 4 \; \boxed{\times} \; 8 \; \boxed{=} \; 32.$$

Next comes the snack. The bill there is $2 per person, and there are 8 persons.

$$\text{Total snack bill: } 2 \; \boxed{\times} \; 8 \; \boxed{=} \; 16.$$

The total expense for the evening, as it was paid out, is

$$(4 \; \boxed{\times} \; 8) + (2 \; \boxed{\times} \; 8) = 32 \; \boxed{+} \; 16 \; \boxed{=} \; 48.$$

Thus the distributive law tells us what we already knew—that an evening out costs as much action by action as it does on the planning board!

Next, we present an important geometrical interpretation of the distributive law. This physical model will help you remember the law and give a geometric proof for it as well.

Example 4.6: A toy store displays red and white balls in a rectangular case as shown. "R" stands for red and "W" for white. How many balls are in the case?

		2			4		
	R	R	W	W	W	W	
	R	R	W	W	W	W	
	R	R	W	W	W	W	
	R	R	W	W	W	W	
8	R	R	W	W	W	W	
	R	R	W	W	W	W	
	R	R	W	W	W	W	
	R	R	W	W	W	W	

Solution:

Total # balls

= # columns × # rows = (# red columns + # white columns) × 8

=   6   ×   8   =   (2   +   4) × 8 = 48.

On the other hand,

Total # balls = # red balls + # white balls

= (# red columns × 8)   +   (# white columns × 8)

=   (2 × 8)   +   (4 × 8)

=   16   +   32 = 48.

Hence,

(2 + 4) × 8 = (2 × 8) + (4 × 8).

We may replace the array of boxes with a region and think in terms of area. Recall that the area of a rectangle is the product of its width and length.

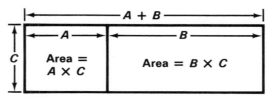

Figure 4.1
A geometric proof of the distributive law.
The total area = the sum of two smaller
areas: $(A + B) \times C = (A \times C) + (B \times C)$.

The distributive law is an important aid to your calculator. Let us do another example, this time changing the numbers for variety!

Example 4.7: Calculate $(30 \times 109) + (6 \times 109)$.

Solution: Again recall that the parentheses indicate that the products, 30×109 and 6×109, are to be computed separately, and then added. Do this on your calculator and record your answers:

$$30 \times 109 = 3270 \quad \text{(record)};$$

$$6 \times 109 = 654.$$

We indicate this as follows:

$$(30 \boxed{\times} 109) + (6 \boxed{\times} 109) = 3270 \boxed{+} 654 \boxed{=} 3924.$$

If your machine has a memory, use it instead of writing down the first product, 3,270.
Now calculate $(30 + 6) \times 109$:

$$(30 \boxed{+} 6) \boxed{\times} 109 \boxed{=} 3924.$$

Notice that the second sequence of machine operations is shorter than the first, and no memory was needed.

The distributive law applied here,

$$(30 \times 109) + (6 \times 109) = (30 + 6) \times 109,$$

says that the two routines in Example 4.7 yield equivalent answers. The distributive law can be extended to more than two addends.

Example 4.8: $(2 + 3 + 4) \times 5 = (2 \times 5) + (3 \times 5) + (4 \times 5)$.

Solution: $9 \times 5 = 10 + 15 + 20$
$45 = 45$

Notice that the "5" is distributed amongst the 2, 3, and 4.

Check via the calculator:

$$(2 \boxed{+} 3 \boxed{+} 4) \times 5 \rightarrow 9 \boxed{\times} 5 \boxed{=} 45.$$

$$(2 \boxed{\times} 5) + (3 \boxed{\times} 5) + (4 \boxed{\times} 5) = 10 \boxed{+} 15 \boxed{+} 20 \boxed{=} 45.$$

The extended distributive law can be used to explain multiplication as repeated addition.

Example 4.9: 4×5 as repeated addition.

Solution: $4 = 1 + 1 + 1 + 1$, so

$$4 \times 5 = (1 + 1 + 1 + 1) \times 5.$$

By the distributive law,

$$(1 + 1 + 1 + 1) \times 5 = (1 \times 5) + (1 \times 5) + (1 \times 5) + (1 \times 5)$$
$$= 5 \quad + \quad 5 \quad + \quad 5 \quad + \quad 5.$$

Therefore,

$$4 \times 5 = (1 + 1 + 1 + 1) \times 5$$
$$= \underbrace{5 + 5 + 5 + 5}_{4 \text{ times}}$$

Exercises 4.3

1. Use your calculator to verify the distributive law by first computing the lefthand side and then the right. The total is indicated in each problem.
   - •a. $(2 \times 3) + (2 \times 4) = 2 \times 7$ $(= 14)$
   - b. $(37 \times 30) + (37 \times 1) = 37 \times 31$ $(= 1147)$
   - •c. $(1024 \times 53) + (1024 \times 37) = 1024 \times 90$ $(= 92160)$
   - d. $1137 \times 102 = (1137 \times 100) + (1137 \times 2)$ $(= 115974)$
   - •e. $(8 \times 5) + (8 \times 6) + (8 \times 8) = 8 \times (5 + 6 + 8) = 8 \times 19$ $(= 152)$

2. Billy's Boutique sold scarves at \$3 each during last week. At this price, she sold one boxful (72 scarves). This week, the same scarves are on sale for \$2 each and she sold two boxfuls, each containing 72 scarves. How much were gross sales in these two weeks?

   Answer: $(72 \times 3) + (72 \times 2) + (72 \times 2) = 72 \times (3 + 2 + 2)$.

•3. Grandburg's Service Club has four lodges. Each lodge sold Christmas trees at \$3.50 a tree. Here is a table showing how many trees each lodge sold. Complete the table to show how much money each lodge took in. Find the total amount received by the whole Service Club. Do this in two ways and show the connection with the distributive law.

Lodge	Number of Trees Sold	Receipts
Red Lodge	120	_____
Blue Lodge	150	_____
Gold Lodge	80	_____
Silver Lodge	230	_____
Totals for Club	_____	_____

The distributive law makes it possible to give an algorithm for multiplying any two decimal numbers using place-value notation. It is important to understand how this works for two reasons. You shouldn't have to rely on the calculator to compute everything. And you may need to multiply numbers whose product is too big for your calculator.

Our place-value system of recording numbers makes it easy to compute products by 10, 100, 1000, and so on. The basic relation is exemplified in the following table.

□ × 10	□ × 100	□ × 1000
2 × 10 = 20	2 × 100 = 200	2 × 1000 = 2000
23 × 10 = 230	23 × 100 = 2300	23 × 1000 = 23000
234 × 10 = 2340	234 × 100 = 23400	234 × 1000 = 234000

Notice that the answer is written by adjoining to the first factor as many zeros as these are in the multiplier. To see why this is so, let us just consider the simple case of 23 × 10. It will be instructive to find out how our shepherd friend might handle this kind of problem.

To represent 23, the shepherd fills his pots like so:

2 beans in the tens pot;

3 beans in the ones pot.

Let's double the number of beans in each pot; what would they tally?

4 beans in the tens pot;

6 beans in the ones pot.

Forty-six—twice as much as the shepherd had before! Likewise, three times the number of beans account for 3 times as many sheep.

Now, a bit of magic, please. The shepherd finds himself with *ten times* as many beans in each pot.

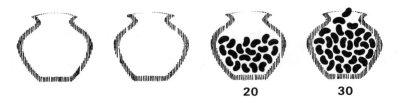

20 beans in the tens pot;
30 beans in the ones pot.

The pots are overflowing!

So, starting with the beans in the tens pot, the shepherd counts out 10 beans and trades these for 1 bean in the next (hundreds) pot.

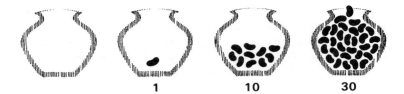

This is better, but still not according to his aim of having less than 10 beans in each pot. So he trades in the other 10 beans in the tens pot for another bean in the hundreds (10-tens) pot.

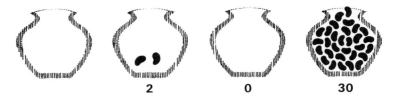

The rest of the story is clear. For every ten beans in the ones pot, he trades for 1 bean in the tens pot. Because there are 3 tens (or thirty) beans, he dumps out all 30 beans from the ones pot and replaces them with 3 beans in the tens pot.

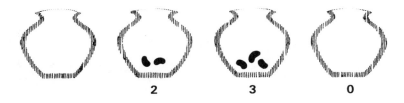

Now, as we sit back and contemplate, it strikes us that, beginning with the 2 beans in the tens pot and 3 beans in the ones pot, all we have to do is shove each pot (contents and all) one place to the left. Then we place an empty pot for the ones, and say, "Mission accomplished!"

The efficiency of our modern day notation is such that the mushrooming-pots and trading-of-beans story we told above can be described in a few sweet lines. The stages, the reasons, and the final relationship displayed below become almost self-evident!

Here is the story:

$$23 \times 10 = (20 + 3) \times 10$$
$$= (20 \times 10) + (3 \times 10) \qquad [\textit{by the distributive law}]$$
$$= \quad 200 \quad + \quad 30$$
$$= 230.$$

Example 4.10(a): $234 \times 100 = ?$

Solution:

$$234 \times 100 = (200 + 30 + 4) \times 100$$
$$= (200 \times 100) + (30 \times 100) + (4 \times 100)$$
$$\qquad\qquad\qquad [\textit{by the distributive law}]$$
$$= \quad 20000 \quad + \quad 3000 \quad + \quad 400$$
$$= 23400.$$

It is troublesome to always count zeros and be sure that all the digits have been entered in the correct places. Doing this by hand is made easier by using a format that takes full advantage of the place-value system of notation.

Example 4.10(b): $234 \times 100 = ?$

Solution:

$$
\begin{array}{r}
234 \\
\times\,100 \\
\hline
400 \\
3000 \\
+\,20000 \\
\hline
\end{array}
$$

$(= 4 \times 100)$
$(= 30 \times 100)$
$(= 200 \times 100)$

Usually we make even more use of place-value notation to obtain a shorter form:

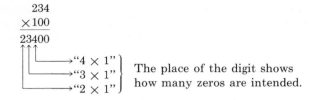

$$234$$
$$\times 100$$
$$\overline{23400}$$

→ "4 × 1"
→ "3 × 1"   } The place of the digit shows
→ "2 × 1"   how many zeros are intended.

Example 4.10(c): Compute 234 × 100 on your calculator.

Solution: $234\ \boxed{\times}\ 100\ \boxed{=}\ 23400$.

In summary, the product of any whole number and 10, or 100, or 1000, may be written simply by adjoining to the whole number the corresponding number of zeros.

This rule still holds for numbers too large for your calculator. What happens when you try to compute the product

$$123456 \times 1000000$$

on your calculator? Overflow! By hand, the answer is 123456000000; too big for eight-digit calculators!

Using the distributive law, it is easy to explain how to multiply any two whole numbers.

Example 4.11: $23 \times 8 = ?$

Solution:

$$23 \times 8 = (20 + 3) \times 8 = (20 \times 8) + (3 \times 8) \quad [\textit{distributive law}]$$
$$= (2 \times 8 \times 10) + (3 \times 8)$$
$$= (16 \times 10) + 24$$
$$= 160 + 24$$
$$= 184$$

A tabular form makes it possible to employ the place-value notation to keep track of the 10's, 100's, and so on.

$$2\,\overcircle{3}$$ First perform
$$\times\ \overcircle{8}\ \longrightarrow 3 \times 8$$
$$\overline{2\ 4}$$

$$2\,3$$ Next perform
$$\times\ 8\ \longrightarrow 20 \times 8$$
$$\overline{2\ 4}$$
$$+1\ 6\ 0$$

Finally, add: $1\ 8\ 4$

As a shortcut, the "2" in the 24 is sometimes "carried" mentally, or with a notation as shown below.

$$^2 2\ 3$$
$$\times\ \ 8$$

(1 8) 4   (18) $= 16 + 2$ No extra 0's are shown because the place indicates the proper number.

By repeating this scheme, we can multiply larger numbers.

Example 4.12: $523 \times 8 = ?$

Solution:

$$523 \times 8 = (500 + 20 + 3) \times 8 = (500 \times 8) + (20 \times 8) + (3 \times 8)$$
$$= (5 \times 8 \times 100) + (2 \times 8 \times 10) + (3 \times 8)$$
$$= (40 \times 100) + (16 \times 10) + 24$$
$$= 4000 + 160 + 24$$
$$= 4184$$

In vertical form:

$$523$$
$$\times\ \ 8$$
$$24 = 8 \times\quad 3$$
$$160 = 8 \times\ \ 20$$
$$4000 = 8 \times 500$$
$$4184 \quad\text{add}$$

Short form (dropping zeros):

$$523$$
$$\times\ \ 8$$
$$24$$
$$16$$
$$40$$
$$4184$$

Again, notice that the use of place-value notation makes the additional zeros unnecessary.

Still shorter form (using carrying):

$$^{15\,2}2\ 3$$
$$\times\qquad 8$$
$$4\ 1\ 8\ 4$$

$3 \times 8 = 24$; carry "2" for 20
$20 \times 8 = 160$; + 20 which was carried; carry "1" for 100
$500 \times 8 = 4000$; + 100 which was carried

Finally, we can multiply large numbers by repeating these steps.

Example 4.13: $523 \times 48 = ?$

Solution:

$$523 \times 48 = (523 \times 40) + (523 \times 8)$$
$$= \underline{(523 \times 4)} \times 10 + \underline{(523 \times 8)}$$

Done in previous example

Done like 523×8 as before

Long form:

```
      52 3
    × 4 8
      2 4  =   8  × 3
     16 0  =   8  × 20
    400 0  =   8  × 500        add to obtain 523 × 8
     12 0  = 4  0  × 3
     80 0  = 4  0  × 20
   2000 0  = 4  0  × 500       add to obtain 523 × 40
   2510 4           add all
```

These zeroes come from these!

Short form:

```
     523
   ×  48
    4184  = 523 × 8     (as before)
  + 2092  = 523 × 4     (and moved one place to the left to
   25104                indicate multiplication by 10)
```

The calculator makes life easier:

$523 \times 48 = ?$

523 ⊠ 48 ⊟ 25104

$48 \times 523 = ?$

48 ⊠ 523 ⊟ 25104

Exercises 4.4

1. In these problems, use your calculator to check the answers or to form the products or sums indicated.

• a.
```
    12
  ×  4
    48
```

c.
```
    16
  ×  9
        = 9 × 6
        = 9 × 10
   144
```

• b.
```
      4
   × 12
      8 = 2 × 4
     40 = 10 × 4
 add: 48
```

d.   9
×16
$= 6 \times 9$
$= 10 \times 9$
144

f.   13
×38
$= 8 \times 13$
$= 30 \times 13$
494

•e.   38
×13
$= 3 \times 38$
$= 10 \times 38$
494

2. An ice cream cone vendor holds his cones in a rack with holes in a rectangular array. There are 5 rows, each containing 12 holes.

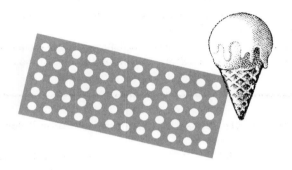

•a. How many cones does the rack hold? ($12 \times 5 = ?$)

b. If he sells 7 trays full during his shift, how many cones has he sold? (Use your calculator.)

Remark: In solving part a, you were using the combination interpretation of multiplication, while in b you are using naturally the repeated addition interpretation.

3. A section is a square mile of land. A Midwest subdivider expects to lay it out in spacious square lots measuring 12 lots to the mile. How many lots will there be to a section?

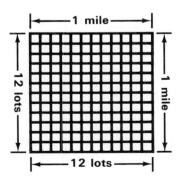

•4. A certain car holds 23 gallons of gas and (on the average) gets 14 miles per gallon. How far can the car travel on a tank of gas?

5. The Los Angeles Strings, a tennis team, expect 3,500 paying customers for each home match.

a. How many paying customers will they have in 33 matches if their expectations are met?

b. If 2,150 customers buy $4 tickets and 1,350 buy $6 tickets for each match, what would the gate receipts be for each match?

c. What would be the gate receipts for the entire season of 33 matches?

6. At some expense, the owner of the tennis club in Exercise 5 can convert the arena so that there will be 1,350 seats at $4 and 2,150 at $6.

a. How much more will the gate receipts be for each match if there are 3,500 paying customers on this arrangement?

b. How much more are the gate receipts for a season of 33 matches, each filled to capacity of 3,500?

c. If the arena conversion costs $100,000, and if the owner keeps 0.6 of the gate, would be get back his investment after 3 seasons?

4.5 What to Do When Multiplication Overflows

Try to compute 43897 × 3675 on an 8-digit calculator. Many calculators will react like this:

$$43897 \; \boxed{\times} \; 3675 \; \boxed{=} \circ 1.6132147$$

A symbol like "∘" denotes overflow. The situation is very much as it was in adding two numbers that caused an overflow. (Reread Section 3.4.) As before, our clue about what to do comes from a simpler problem.
Consider

$$43800 \times 3675.$$

We know that, if we were to compute

$$438 \times 3675,$$

then we could obtain 43800 × 3675 by simply adjoining two zeros. (Recall Example 4.10.)
Thus, we may calculate on our machines:

$$438 \; \boxed{\times} \; 3675 \; \boxed{=} \; 1609650.$$

and conclude, without further computation, that

$$43800 \times 3675 = 160965000.$$

Now return to the original problem. Because

$$43897 = 43800 + 97,$$

then

$$43897 \times 3675 = (43800 + 97) \times 3675.$$

Applying the distributive law,

$$43897 \times 3675 = (43800 + 97) \times 3675$$

$$= (43800 \times 3675) + (97 \times 3675).$$

Now we can do both parts of the right side of the equality on our calculators.

We've done	$43800 \times 3675 =$	160965000
And, by calculator,	$97 \times 3675 =$	356475
Now add:	$43897 \times 3675 =$	161321475

to obtain the full and correct answer.

This addition may be made by hand (quickest), or by machine using the method of Section 3.4.

Look back at our first attempt at multiplying these two numbers on the calculator. It is worth noting that the displayed "answer" for this product gives us a lot of information, despite the fact that the product exceeds 8 digits and we obtain an "overflow" signal. The displayed result was

$$\circ \; 1.6132147$$

and we have just computed the correct answer as 161321475. The displayed digits are correct—as far as they go—but the decimal point is in the wrong place. In fact, the "overflow" signal indicates that (1) a "false" decimal point has been inserted, and (2) some digits are missing from the answer.

To obtain the correct answer, we must move the "false" decimal point *eight places to the right.* This gives us

$$16132147\square.$$

where $\square$ represents an unknown digit. However, we can easily find this digit; it is the ones digit of $7 \times 5 \, (= 35)$, so the missing digit must be a "5." Therefore, we can deduce the correct answer (161321475.) from the overflow form of the display and a little mental arithmetic.

Example 4.14: $1265813 \times 7836 = ?$

Solution: First try it on an 8-digit calculator. The final display is

$$\circ \; 99.189106$$

in the overflow form. Moving the decimal point eight places to the right, we obtain

$$99189106\square\square.$$

The two missing digits must be the last two digits of $13 \times 36 = 468$ (which we can easily obtain on the calculator), so the answer must be

$$9918910668.$$

Now let's confirm the answer to Example 4.14 using the distributive principle. This time, we'll display the calculation in the usual vertical format. Please note the similarity to the format of Example 4.12. We begin by splitting the larger factor.

$$
\begin{array}{r}
126\ 5813\,(= 1260000 + 5813) \\
\times \qquad 7836 \\
\hline
4555\ 0668\,(= 5813 \times 7836) \\
+\,987336\ 0000\,(= 126 \times 7836 \times 10000) \\
\hline
991891\ 0668 = (1260000 + 5813) \times 7836
\end{array}
$$

Example 4.15: $43662477 \times 52343399 = ?$

Solution: These two factors are so large that we don't get much help from the calculator. Try it! The final display is the overflow

$$\circ \; 22854424.$$

The false decimal point is at the far right. When we move it eight places farther to the right, we have eight unknown digits in the answer. It is not as easy to determine these missing digits as it was in the previous two examples.

Our solution is patterned after example 4.13. We split both factors and proceed, letting the calculator do the hard work!

$$
\begin{array}{r r l}
4366 & 2477 & \\
\times\,5234 & 3399 & \\
\hline
841 & 9323 & (= 2477 \times 3399) \\
1484\quad 0034 & 0000 & (= 4366 \times 3399 \times 10000) \\
1296\quad 4618 & 0000 & (= 2477 \times 5234 \times 10000) \\
+\,2285\quad 1644\quad 0000 & 0000 & (= 4366 \times 5234 \times 10000 \times 10000) \\
\hline
2285\quad 4424\quad 5493 & 9323 & = (4366000 + 2477) \times (52340000 + 3399)
\end{array}
$$

The following is a schematic diagram to show where these products came from and how they were computed.

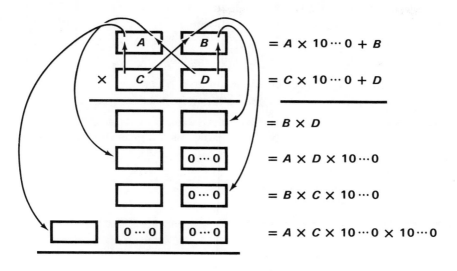

Add the partial products for the answer.

In this example, we have calculated the missing 8 digits to be "549323." Notice that if we had tried to uncover these missing digits by multiplying 2477×3399 ($= 841\ 9323$), we would not have been too successful.

If instead, we tried the product of the last 5 digits of each number,

$$62477 \times 43399,$$

we would have gotten

$$\circ\ 27.114393$$

Using 77×99 to get the last two digits for this answer, we would have gotten

$$27\ 1143\ 9323.$$

Comparing the last 8 digits for this with the one we got above,

$$5493\ 9323,$$

we can see we would still be off by a considerable amount. Using the present scheme, however, we did find that our calculator was still handy for the intermediate task of getting the partial product.

Exercises 4.5

1. Use a calculator to help you compute these products.

   •a. 21436877 × 9   c. 7896347 × 536

   b. 65042 × 78975   •d. 1139163 × 8670

2. Using some of the data from Exercise 13 in Section 4.1—

   •a. compute the number of seconds in a year;

   b. find how many seconds old you are by your nearest birthday.

•3. Light travels approximately 186,284 miles per second. How far will light travel in a year of 365 days? (The answer is a unit of distance called a *"light year"* by astronomers.)

4. The nearest star outside our solar system is *Proxima Centauri*. It is 4.3 light years away. How many miles is that?

4.6 What Is Division?

The relation between division and multiplication is very much like that between subtraction and addition. To begin, division is the reverse operation of multiplication. We multiply

$$6 \times 8 = 48.$$

We have gone from 6 to 48 by multiplying by 8. How can we go back from 48 to 6? How can we fill in the box below so that equality holds?

$$6 \times \square = 48.$$

It is easy, for we remember that we multiplied 6 by 8 to get 48. Thus, we know that 8 goes in the box.

However, in mathematics, we sometimes have not already *"been there,"* and so we must be able to reverse the process of multiplication without first having made the direct calculation.

When we don't know just what to do, a little experimentation will help! Don't be afraid to try out numbers on your calculator!

Example 4.16: After making the down payment on a car, Robert Hotrod has $936 balance to pay. This is to be paid off in 18 installments. How much is each installment?

Solution: Expressed mathematically, we wish to fill in the box below so that equality holds. We look for the missing factor:

$$18 \times \square = 936. \hspace{3cm} (*)$$

You may guess: Surely, about $50. Well,

$$18 \times 50 = 900.$$

Not quite enough. (Check this and other products on your calculator.) Try again. Try 60:

$$18 \times 60 = 1080.$$

Much too much! Try again—a number between 50 and 60. (Why?) Say, try 52.

$$18 \times 52 = 936; \hspace{1cm} \text{BINGO!}$$

(Notice that if $\square$ is to be a whole number, the last digit would be such that its product with 8 should end in the digit 6.)

We indicate the answer to the original question (*) above as

$$936 \div 18.$$

In symbols,

$$18 \times \boxed{(936 \div 18)} = 936.$$

From our process of guessing, we know that

$$936 \div 18 = 52.$$

Guessing is often a good way to start a problem. We can learn from poor guesses. For example, the trial of 60 may have been a poor guess because it was so far off the mark. But it told us to make the next estimate closer to 50.

Division can be performed on your calculator with the $\div$ key.

$$936 \div 18 = ?$$

$$9\,36 \div 18 = 52.$$

On a calculator, no division problem is much harder than another—even decimals. Here is an example involving decimals and division.

Example 4.17: Robert Hotrod takes his new car for a gas-mileage run. He fills the tank and drives 331.7 miles, stopping at a service station where he again fills the tank. It takes 12.4 gallons of gas to fill the tank. Thus, he drove 331.7 miles on 12.4 gallons of gas. How many miles did he go on each gallon (on the average)?

Solution: We need to find the missing factor in

$$
\begin{bmatrix} \text{amount of} \\ \text{gas used} \end{bmatrix} \times \begin{bmatrix} \text{miles driven} \\ \text{on each gallon} \end{bmatrix} = \begin{bmatrix} \text{total miles} \\ \text{driven} \end{bmatrix}
$$

$$
12.4 \quad \times \quad \square \quad = \quad 331.7
$$

The answer is written symbolically as $331.7 \div 12.4$.
On your calculator,

$$
331.7 \; \boxplus \; 12.4 \; \boxminus \; 26.75
$$

The answer may be interpreted thus.

Robert Hotrod got 26.75 miles per gallon.

The relationship between multiplication and division may be summarized as follows.

1. Division is the reverse of multiplication.
2. If $B \neq 0$, then $A \div B$ is the number such that $B \times (A \div B) = A$.
3. If $B \neq 0$, the equations $A \div B = C$ and $B \times C = A$ express the same fact.
4. If $B \neq 0$, then $B \div B = 1$ for all numbers, B.
5. $A \div 1 = A$ for all numbers, A.
6. If $B \neq 0$, then $(A \times B) \div B = A = (A \div B) \times B$.

WARNING: No number can be divided by zero (0).

Try $1 \div 0$ on your calculator. Your calculator will complain by showing some symbol (perhaps the overflow symbol or a blinking) to indicate that an improper operation has been attempted. An interesting pattern develops if you try to divide 1 by successively smaller and smaller numbers. You may have some fun with Exercise 4 at the end of this section.

The third statement in the summary list points out that there is a sure-fire test for division.

Example 4.18:   Does $1470 \div 35 = 40$?

Solution: Property 3 of the summary states that

If $B \neq 0$, the equations $A \div B = C$ and $B \times C = A$ express the same fact.

Let's try it out. Please pay close attention to what numbers are being substituted for (used in place of) the letters.

"$A \div B = C$ and $B \times C = A$ state the same fact."

"$1470 \div 35 = 40$ and $35 \times 40 = 1470$ state the same fact."

This means that $1470 \div 35 = 40$ is true only if $35 \times 40 = 1470$ is true. We can check to see if $35 \times 40 = 1470$ is true. Do it any way you please—on your calculator, if you wish. Find that $35 \times 40 = 1400$, not 1470. Thus, $1470 \div 35$ does not equal 40. In symbols,

$$1470 \div 35 \neq 40.$$

It will often be convenient to make such a check on divisions. Property 3 will also be useful in later discussion about fractions and about other properties of numbers.

As we have said, no division problem is more difficult than any other on the calculator (provided we can enter them in the machines.) But having the ability to make these divisions so easily puts a responsibility on you. It is important to know just when to divide—and when not to!

Property 6 will be especially useful to us. It is true because division is the reverse of multiplication.

$$A \xrightarrow{\times B} A \times B \xrightarrow{\div B} (A \times B) \div B = A$$

because the division just undoes the previous multiplication.

Similarly,

$$A \xrightarrow{\div B} A \div B \xrightarrow{\times B} (A \div B) \times B = A$$

because the multiplication just undoes the previous division.

Exercises 4.6

1. This exercise is designed to show that $A \div B = C$ and $B \times C = A$ and $A \div C = B$ mean the same thing. Do these calculations on your calculator. For each part, identify the numbers for A, B, and C.

$B \times C$	$A \div B$	$A \div C$
•a. $3 \times 4 =$	•e. $12 \div 3 =$	•i. $12 \div 4 =$
b. $13 \times 16 =$	f. $208 \div 13 =$	j. $208 \div 16 =$
•c. $173 \times 57 =$	•g. $9861 \div 173 =$	•k. $9861 \div 57 =$
d. $4.3 \times 5.7 =$	h. $24.51 \div 4.3 =$	l. $24.51 \div 5.7 =$

2. Use your calculator. First perform the division indicated at the left; then use the answer to perform the multiplication indicated at the right.

   a. $12 \div 3 =$ (answer)   (answer) $\times 3 =$

   •b. $69 \div 23 =$ (answer)   (answer) $\times 23 =$

   c. $6001 \div 17 =$ (answer)   (answer) $\times 17 =$

3. Check by making an appropriate multiplication to see whether the given answers are correct.

   a. $25 \div 6 = 4$

   •b. $81 \div 11 = 8$

   c. $56704 \div 1234 = 456$

   •d. $29.61 \div 6.3 = 4.7$

   e. $254.8 \div 3.25 = 7.84$

4. Compute these divisions on your calculator. Record the answers. What is the pattern appearing here?

   a. $2 \div 1 =$

   b. $2 \div 0.5 =$

   c. $2 \div 0.1 =$

   d. $2 \div 0.05 =$

   •e. $2 \div 0.01 =$

   f. $2 \div 0.005 =$

   g. $2 \div 0.001 =$

5. In each part, use your calculator to find a number to put in the box so that equality holds in the first equation; then complete the second equation.

   a. $8 \times \square = 56$

   b. $18 \times \square = 144$

   •c. $1013 \times \square = 144859$

   $\square \times 7 = 56$

   $\square \times 8 = 144$

   $\square \times 143 = 144859$

6. Make up five problems of the type indicated in Exercise 6, and verify using the calculator.

7. You should be able to supply the answer to each of the following "by inspection"—that is, without having to do any multiplications or divisions (even in your head). Check each answer with your calculator.

   a. $(2 \times 3) \div 3 = \square$

   b. $(3 \times 2) \div 2 = \square$

   c. $(5 \times 3) \div 3 = \square$

   •d. $(3 \times 5) \div 5 = \square$

   •e. $(2149 \times 528) \div 528 = \square$

   f. $(1357 \times 2468) \div 2468 = \square$

   •g. $(796 \times 2806) \div 2806 = \square$

   h. $(752 \times 1637) \div 1637 = \square$

8. Supply the answer to each of the following "by inspection" or by mental arithmetic. Then check your answer with the calculator.

   •a. $(6 \div 2) \times 2 = \square$

   b. $(6 \div 3) \times 3 = \square$

   •c. $(42 \div 6) \times 6 = \square$

   d. $(42 \div 7) \times 7 = \square$

   •e. $(111 \div 37) \times 37 = \square$

   f. $(91 \div 13) \times 13 = \square$

   •g. $(5332114 \div 1234) \times 1234 = \square$

   h. $(12545856 \div 1472) \times 1472 = \square$

9. The Society of Positrons (SOPS) decided to raise $41,250 for a new atomic-particle machine by selling charged particles for $0.30 (or 30 cents) each.

   •a. How many charged particles will they have to sell?

   b. If SOPS has 500 members, how many particles must each sell to do a fair share?

•10. A car travels 300 miles at an average speed of 60 miles per hour (mph). Another car travels 300 miles at 50 mph. How much longer does the trip take the second car?

11. Gasper Gasgulp made a test run in his new Spitfire Sixteen and found that he used 23.2 gallons of gas to go 150.8 miles. How many miles per gallon did he average on this run?

12. Refer to the information in problem 4 in Exercises 3.4.

    •a. Find the population per square mile (population ÷ number of square miles) for each of the Western states of the United States. (This gives the *"population density"* for each state.)

    b. Which of the Western states is most dense? (Calculate to two places beyond the decimal point.)

    •c. Which of the Western states is the least dense?

    d. Find the difference in density between the most and the least dense of the Western states.

13. Property (6) of the summary in this section is that

$$(A \times B) \div B = A.$$

    a. Guess the correct answer to the following:

$$(A \times B) \div A = ?$$

    b. Give reasons to justify your guess in part a.

    c. Check your guess by performing the operations

$$(23 \times 47) \div 23 = ?$$

14. The statement of property (6) of the summary includes the requirement that $B \neq 0$.

    •a. Explain why it is all right to have $A = 0$ in the statement.

    •b. Try $A = 0$ and $B = 1234$ on your calculator for

$$(A \times B) \div B.$$

    c. Using the same values of A and B, compute

$$(A \times B) \div A.$$

    Does the result lead to any modification of your guess in part a of Exercise 13?

4.7 Fractions

A fraction is a quotient of whole numbers. Some fractions that are used frequently are $\frac{1}{2}, \frac{1}{4}, \frac{3}{4},$ and $\frac{1}{3}$. Fractions such as $\frac{7}{13}, \frac{6}{11},$ and $\frac{355}{113}$ are not used as much.

In each case,

$$\frac{A}{B} \text{ is the number } (A \div B).$$

So, with a calculator, it is easy to convert any fraction into a decimal.

Example 4.19: Find the decimal equivalents of these fractions.

$$\frac{1}{2} = \qquad \frac{1}{4} = \qquad \frac{2}{4} = \qquad \frac{3}{4} = \qquad \frac{1}{3} =$$

$$\frac{2}{3} = \qquad \frac{4}{3} = \qquad \frac{7}{13} = \qquad \frac{6}{11} = \qquad \frac{355}{113} =$$

Recall property 3 from Section 4.6 (page 106). The equations $A \div B = C$ and $B \times C = A$ express the same fact. In terms of fractions, this means that

$$\frac{A}{B} = C \text{ and } B \times C = A \text{ express the same fact.}$$

The fraction idea changes the look of several others of those properties also. For example,

Property 2   If $B \neq 0$, then $\dfrac{A}{B}$ is the number such that $B \times \dfrac{A}{B} = A$.

Property 4   If $B \neq 0$, then $\dfrac{B}{B} = 1$.

In particular, Property 6 means that

$$\frac{A \times B}{B} = A \qquad \text{and} \qquad A = \frac{A}{B} \times B.$$

When we try to verify this last property on our calculator, we often experience the effect of a "round-off" error.

Example 4.20: Try to verify

$$\frac{1}{2} \times 2 = 1, \qquad \frac{1}{3} \times 3 = 1, \qquad \frac{3}{7} \times 7 = 3.$$

Solutions: $(1 \boxplus 2) \boxtimes 2 \boxminus 1.$ (check!)

$(1 \boxplus 3) \boxtimes 3 \boxminus 0.9999999$ (slight error)

$(3 \boxplus 7) \boxtimes 7 \boxminus 2.9999998$ (slight error)

If the above errors occur in your calculator, they are due to the fact that your calculator has to limit the number of digits in the division, $1 \div 3$ (or $3 \div 7$.) The final digits are lost, and so when the result is multiplied by 3 (or 7 as the case may be) the final result is slightly less than the correct answer. This is one of the reasons why it is important to understand how arithmetic really works and how your calculator really works. These matters are discussed fully in Chapter 8.

Fractions occur most frequently in questions like the following.

Example 4.21: How many minutes in one-half of a hockey period?

Solution: $\frac{1}{2}$ of $20 = \frac{1}{2} \times 20 = (1 \times 20) \div 2 = 10$.

Example 4.22: A recipe for 3 dozen cookies calls for $4\frac{1}{2}$ (four and a half) cups of sugar. How much sugar should be used in cutting the recipe to make one dozen cookies?

Solution: $4\frac{1}{2} = 4 + \frac{1}{2} = 4.5$

$\frac{1}{3}$ of $4.5 = \frac{1}{3} \times 4.5 = (1 \times 4.5) \div 3 = ?$

$4.5 \boxed{\div} 3 \boxed{=} 1.5 = 1\frac{1}{2}$.

Thus $1\frac{1}{2}$ cups of sugar is the correct amount.

Exercises 4.7

•1. Use your calculator to compute this table of fractions and decimal equivalents.

$\frac{1}{2} =$	$\frac{2}{2} =$	$\frac{3}{2} =$	$\frac{4}{2} =$	$\frac{5}{2} =$	$\frac{6}{2} =$		
$\frac{1}{4} =$	$\frac{2}{4} =$	$\frac{3}{4} =$	$\frac{4}{4} =$	$\frac{5}{4} =$	$\frac{6}{4} =$	$\frac{7}{4} =$	$\frac{8}{4} =$
$\frac{1}{8} =$	$\frac{2}{8} =$	$\frac{3}{8} =$	$\frac{4}{8} =$	$\frac{5}{8} =$			
$\frac{6}{8} =$	$\frac{7}{8} =$	$\frac{8}{8} =$	$\frac{9}{8} =$	$\frac{10}{8} =$			

•2. What patterns appear in the computations in Exercise 1? Without calculator, determine the following.

$\frac{7}{2} =$	$\frac{9}{4} =$	$\frac{11}{8} =$
$\frac{8}{2} =$	$\frac{10}{4} =$	$\frac{12}{8} =$

3. Find the round-off error in your calculator.

   a. $\frac{2}{3} \times 3 = (2 \boxed{\div} 3) \boxed{\times} 3 \boxed{=} ?$

   b. $\frac{4}{9} \times 9 = (4 \boxed{\div} 9) \boxed{\times} 9 \boxed{=} ?$

   c. $\frac{6}{7} \times 7 = (6 \boxed{\div} 7) \boxed{\times} 7 \boxed{=} ?$

4.8   Order, Multiplication, and Division

If A's Tire Shop sells Rollo Radials at $47.50 a tire and B's Tyre Shoppe sells the same tire for $42.50, then we know that 4 tires bought from A will cost more than 4 tires bought from B.

This obvious, but immensely important, fact generalizes to the principle:

If $A > B$, and if C is any positive number, then $A \times C > B \times C$.

This fact means that, if C is a positive number, the product $A \times C$ increases as A increases.

Example 4.23: Fix $C = 2.3$. Use your calculator to compute $A \times C$ for these different values of A.

$A \times C$	$A \times C$
$0.5 \times 2.3 = 1.15$	$8.11 \times 2.3 =$
$1 \times 2.3 = 2.3$	$8.11002 \times 2.3 =$
$1.5 \times 2.3 =$	$8.1100208 \times 2.3 =$
$4 \times 2.3 =$	$8.1100209 \times 2.3 =$
$8 \times 2.3 =$	$8.110021 \times 2.3 =$
$8.1 \times 2.3 =$	

Is it true that as A increases, so does $A \times C$? Check carefully the last three answers. Your calculator may round off to 8 places and so the answers may appear equal! Actually, the product does increase with increasing A. The problems involved with round-off errors are discussed in Chapter 8.

A consequence of this rule is that

If a positive number, C, is multiplied by a number. M, greater than one, then $M \times C$ is greater than C.

In symbols,

If $M > 1$ and $C > 0$, then $(M \times C) > (1 \times C)$.

On the other hand,

If $M < 1$ and $C > 0$, then $(M \times C) < (1 \times C)$.

Similar rules hold for division. However, because we are less familiar with these comparisons with respect to division, we must be especially careful in treating these problems. We separate these problems into two types.

1. *Same Divisor.* If a case of Chugalug Beer (consisting of four six-packs) costs $7.44, and a case of Gulpo Beer costs $6.12, which beer costs less per six-pack?

We know that, because $6.12 < 7.44$, it follows that

$$(6.12 \div 4) < (7.44 \div 4);$$

and indeed we find that

$$1.53 < 1.86.$$

Notice that the order (that is, the direction of the order symbol $<$) is the same after division as it was before.

In general,

If $A < B$, and if C is a positive number,

then $A \div C < B \div C$.

A consequence of this rule is that, as A increases, so does $A \div C$.

Example 4.24: Fix $C = 2.3$. Use your calculator to compute $A \div C$ for different numbers A, as shown.

$A \div C$	$A \div C$
$0.5 \div 2.3 =$	$4.6 \div 2.3 =$
$1 \div 2.3 =$	$10.5 \div 2.3 =$
$2 \div 2.3 =$	$10.52 \div 2.3 =$
$2.3 \div 2.3 =$	$10.520001 \div 2.3 =$
$2.7 \div 2.3 =$	$10.520002 \div 2.3 =$
$4 \div 2.3 =$	$10.520003 \div 2.3 =$

Again, check your answers to the last three examples. Your 8-place calculator may give the same answer as it is forced to round off its answer to 8 places. But it is true that the quotients increase with increasing A.

An important observation from these examples, and one that we have already used in comparing numbers by division, is that

If $A < C$, and if $C > 0$, then $A \div C < 1$; and

if $A > C$, and if $C > 0$, then $A \div C > 1$.

2. *Different Divisors.* Doughnuts cost 72¢ a dozen. Four men buy a dozen doughnuts and split the cost. Six women buy a dozen doughnuts and split the cost. Who pays more, a man or a woman?

Again, we know the answer. Because $4 < 6$, it follows that

$$72 \div 4 > 72 \div 6;$$

and indeed,

$$18 > 12.$$

Notice that, in this example, the order (the direction of the order symbol) is *reversed from* $<$ to $>$.

In general,

If C is a positive number, and if $A < B$,

then $C \div A > C \div B$.

Example 4.25: Fix $C = 36$. Use your calculator, if necessary, to compute $C \div A$ for the different numbers, A, as shown.

$C \div A$	$C \div A$	$C \div A$
$36 \div 1 =$	$36 \div 4 =$	$36 \div 12 =$
$36 \div 2 =$	$36 \div 6 =$	$36 \div 18 =$
$36 \div 3 =$	$36 \div 9 =$	$36 \div 36 =$

In this example, we purposely restricted the divisors so that the quotient would be a whole number. This will highlight the pattern you should observe.

As the divisor *increases* from 1 to 36, the quotient *decreases*. When $A = 6$ (that is, for $36 \div 6$), we have $36 \div 6 = 6$. At this point, the divisor and quotient are equal.

Here, $36 \div 6 = 6$ or, equivalently, $6 \times 6 = 36$. We may ask: "Is there always such a number, R, so that $C \div R = R$?" The answer is, "Yes, but it is not always easy to find."

Let us try $C = 12$:

$C \div A$	$C \div A$
$12 \div 1 =$	$12 \div 4 =$
$12 \div 2 =$	$12 \div 6 =$
$12 \div 3 =$	$12 \div 12 =$

At which point does the divisor equal the quotient? It does not appear in this table. But we can get some information on where it lies.

Because $12 \div 3 = 4$, the quotient is greater than the divisor here. Whereas, in $12 \div 4 = 3$, the quotient is less than the divisor. In symbols,

$12 \div 3 = 4 \quad \text{and} \quad 4 > 3$

while

$12 \div 4 = 3 \quad \text{and} \quad 3 < 4.$

Therefore, we can guess that for some number between 3 and 4, the "break-even" point occurs. Beyond the "break-even" point, the divisor is greater than the quotient. Let us try these divisors:

$$12 \div 3.2 = \qquad 12 \div 3.6 =$$

$$12 \div 3.4 = \qquad 12 \div 3.8 =$$

We can argue as before that the "break-even" point lies between 3.4 and 3.6. Try again.

$$12 \div 3.46 =$$

$$12 \div 3.47 =$$

By continuing in this way, you can eventually show that the "break-even" point is, to within the 8-place accuracy of your calculator,

$$3.4641016.$$

Stated mathematically,

$$12 \div 3.4641016 = 3.4641016; \text{ or}$$

$$12 = 3.4641016 \times 3.4641016.$$

Actually, because of the round-off error, your 8-place calculator may show:

$$3.4641016 \times 3.4641016 = 11.999999 \doteq 12.$$

(The symbol $\doteq$ means "is approximately equal to.")

Another application of this effect on the comparison of the divisor and quotient is to determine whether a whole number, like 29, has a whole number divisor with a whole number quotient.

Your reaction may be, "Of course it has. $29 \div 1 = 29$!" Yes, but that and the related $29 \div 29 = 1$ are the only ways a whole number can divide 29 and have a whole number quotient. Division by 1 and division by the whole number itself of course always hold for any whole number. To make the question interesting, we exclude these types from discussion. We have just seen above that 36 has a large number of integer divisors with whole number quotients. But, as we shall see, 29 has none.

Now theoretically, it is a simple matter to decide. All one has to do is to check whether the number in question, say 29, when divided by 2, 3, $\cdots$, 28 have whole number quotients. We have only 27 numbers to try. But suppose we were interested in whether 90909089 had this property. (It does!) It would take a lot of checking; in fact, 90909087 different divisions would have to be tried. Any fact that could reduce this amount of checking would be very valuable. The application we promised earlier provides just such a fact and cuts the work by a substantial factor.

Incidentally, we shall have occasion to refer to whole numbers that have the property illustrated by 29 and 90909089 above. So it will be handy to refer to such numbers by a short and descriptive name. Accordingly, we introduce the following definition.

> Whole numbers greater than 1 having no whole number divisors with whole number quotients (except 1 and the number itself) are called *primes*.

Prime numbers are discussed more fully in Chapter 11.

Example 4.26: We want to show that 29 is a prime. What happens when we divide 29 by whole numbers (other than 0)?

$$29 \boxplus 2 \boxminus 14.5 \qquad \text{not a whole number.} \quad (2 < 14.5)$$
$$29 \boxplus 3 \boxminus 9.6666666 \qquad \text{not a whole number.} \quad (3 < 9.6666666)$$
$$29 \boxplus 4 \boxminus 7.25 \qquad \text{not a whole number.} \quad (4 < 7.25)$$
$$29 \boxplus 5 \boxminus 5.8 \qquad \text{not a whole number.} \quad (5 < 5.8)$$
$$29 \boxplus 6 \boxminus 4.8333333 \qquad \text{not a whole number.} \quad (6 > 4.8333333!)$$

Now we can stop. Notice that we have passed the "break-even" point without finding a whole number divisor whose quotient is a whole number. We can conclude at this point, that no whole number except 1 and 29 divides 29 and has a whole number quotient.

Why were we able to stop at division by 6? Well, 6 is greater than the "break-even" point, and if a divisor B is greater than the "break-even" point, then

$$(29 \div B) < B.$$

If $29 \div B$ were a whole number, say Q, then

$$29 \div Q = B.$$

However, because $Q < B$, we would have tested and found Q before. So 29 is a prime.

Exercises 4.8

1. Insert the proper sign ($<$ or $>$) in the box to indicate the order of the numbers. You should be able to do this without calculation!

- •a. $(17 \times 28) \ \square \ (19 \times 28)$
- b. $(6 \times 70) \ \square \ (6 \times 59)$
- c. $(3.78 \times 7.9) \ \square \ (3.78 \times 8.1)$
- •d. $(8937 \times 3.14) \ \square \ (9738 \times 3.14)$
- •e. $(65 \times 73) \ \square \ (73 \times 85)$
- •f. $(12 \div 4) \ \square \ (21 \div 4)$
- g. $(9.3 \div 20) \ \square \ (9.29 \div 20)$
- •h. $(36.8 \div 14.3) \ \square \ (38.6 \div 14.3)$

i. $(9873 \div 762) \square (10015 \div 762)$ m. $(110 \div 37) \square (110 \div 73)$

•j. $(1345 \div 0.78) \square (999 \div 0.78)$ n. $(43.1 \div 15.2) \square (43.1 \div 15.7)$

k. $(18 \div 6) \square (18 \div 3)$ •o. $(3.78 \div 2) \square 1$

•l. $(18 \div 2) \square (18 \div 9)$ p. $(3.78 \div 4) \square 1$

•q. $(1234 \div 47) \square (1240 \div 36)$ [HINT: Compare each side with $1240 \div 47$.]

2. Use your calculator to make a table for $17 \div A$ for all whole number values of A from $A = 2$ to $A = 16$. Make a guess at the "break-even" point for divisor and quotient.

•3. By computing with your calculator for some values of A, find a number A such that $28.7296 \div A = A$. (There is an exact 2-place decimal that will meet this requirement.)

4. Does 101 have a whole number divisor N between 2 and 100 such that $101 \div N$ is a whole number? How large a whole number N need be tried?

4.9 Division by 10

In studying multiplication (Section 4.4), we saw that the special features of the decimal place-value system of notation can be exploited to record products by 10, 100, 1000, and so on. Thus,

$$23 \times 10 \ \ = 230 \qquad 2.3 \times 10 \ \ = 23$$

$$23 \times 100 \ = 2300 \qquad 2.3 \times 100 \ = 230$$

$$23 \times 1000 = 23000 \qquad 2.3 \times 1000 = 2300$$

Division by 10, 100, 1000, and so on is just as easy. This is because division is so intimately related to multiplication. Thus, from the products above, we have their counterparts in division.

$$230 \div 10 \ \ \ = 23 \qquad 23 \div 10 \ \ \ \ = 2.3$$

$$2300 \div 100 \ \ = 23 \qquad 230 \div 100 \ \ = 2.3$$

$$23000 \div 1000 = 23 \qquad 2300 \div 1000 = 2.3$$

Here are some more examples. Confirm them with your calculator. In the table below, concentrate on the position of the decimal point. What happens when we multiply by 10? What happens when we divide by 10?

$$7.25 \times 10 = 72.5 \qquad 72.5 \div 10 = 7.25$$

$$72.5 \times 10 = 725 \qquad 725 \div 10 = 72.5$$

$$0.725 \times 10 = 7.25 \qquad 7.25 \div 10 = 0.725$$

It is convenient to express these calculations as "moving the decimal point."

Multiplication by 10 "moves" the decimal point *one* place to the right.

Division by 10 moves the decimal point *one* place to the left.

Here is the same table, with arrows showing the "shifting":

$$7.25 \times 10 = 72.5 \qquad\qquad 72.5 \div 10 = 7.25$$

$$72.5 \times 10 = 725. \qquad\qquad 725. \div 10 = 72.5$$

$$0.725 \times 10 = 7.25 \qquad\qquad 07.25 \div 10 = 0.725$$

We can use these laws to explain the rule (about locating the decimal point in a product) that we discovered by experimentation in the exercises for Section 4.2.

REMARK: In practice, we often see the notation "0.725" written with reference to the number ".725". This originated from engineering practice when plans had to be blue-printed. Some grades of blue-print may not be fine enough to show distinctly decimal points when they should occur, and some may leave extra specks on the print that may be mistaken for decimal points. The consistent use of a "0" preceding any decimal number less than 1 calls attention to the decimal point that is meant to precede all digits. Presumably this characteristic is kept in calculators for the same purpose.

Example 4.27: Where does the decimal point go in 2.6×3.2?

Solution:

Write $2.6 = (2.6 \times 10) \div 10 = 26 \div 10$
$3.2 = (3.2 \times 10) \div 10 = 32 \div 10$
Hence, $2.6 \times 3.2 = (26 \div 10) \times (32 \div 10)$
$ = (26 \times 32) \div (10 \times 10)$ More about this
$ = 832 \div 100 $ in Chapter 8.
$ = 8.32$

Verify this calculation on your calculator:

$$2.6 \;\boxed{\times}\; 3.2 \;\boxed{=}\; 8.32$$

SUMMARY: To multiply two decimal numbers—

1. Multiply each number by 10 enough times to get an integer.
2. Perform the multiplication of the integers.
3. Divide (shift decimal places to the left) by as many factors of 10 as you multiplied by in step 1.

It may seem more familiar to do Example 4.27 in the more conventional vertical form. Follow the arrows!

$$
\begin{array}{r}
2.6 \\
\times\, 3.2 \\
\hline
5\,2 \\
7\,8 \\
\hline
8.3\,2
\end{array}
\quad \rightarrow \quad
\begin{array}{r}
26 \\
\times\, 32 \downarrow \\
\hline
52 \\
78 \\
\hline
832.
\end{array}
\quad
\begin{array}{l}
(= 2.6 \times 10) \\
(= 3.2 \times 10) \\
\\
\\
(= 2.6 \times 3.2 \times 10 \times 10)
\end{array}
$$

The decimal point in 832 is moved two places to the left because we multiplied by 10 two times.

***

Example 4.28: 2.62×7.8

Solution:

$$2.62 = (2.62 \times 10 \times 10) \div (10 \times 10) = 262 \div 100$$
$$7.8\ = (7.8 \times 10) \div 10 \qquad\qquad = 78 \div 10$$

Hence

$$2.62 \times 7.8 = (262 \div 100) \times (78 \div 10)$$
$$= (262 \times 78) \div (100 \times 10)$$
$$= (262 \times 78) \div 1000$$
$$= 20.436$$

***

Note that, because we *divided* by 1000, we moved the decimal point three places (3 zeros in 1000) to the *left*.

Here's the problem in vertical form:

$$
\begin{array}{r}
2\,.\,6\,2 \\
\times\ 7\,.8 \\
\hline
2\ \ 09\ 6 \\
18\ \ 34 \\
\hline
20\,.43\ 6
\end{array}
\quad \rightarrow \quad
\begin{array}{r}
262 \\
\times\ 78 \downarrow \\
\hline
2096 \\
1834 \\
\hline
20436.
\end{array}
\quad
\begin{array}{l}
(= 2.62 \times 10 \times 10) \\
(= 7.8 \times 10) \\
\\
\\
(= 2.62 \times 7.8 \times 10 \times 10 \times 10)
\end{array}
$$

The decimal point in 20436 is moved three places to the left because we multiplied by 10 three times.

Although our procedure for hand calculation of two decimals will always place the decimal point at the proper place, sometimes it will seem that the calculator does not shift the decimal by the same rule. For example, try the following:

$$7.125 \times 1.24 = ?$$

7.125 ⊠ 1.24 ⊟ 8.835

By our rule, we expect the decimal point in the answer to be *five* places from the right; here, the calculator has just *three* places! If we find the product by hand calculation, we will find

$$7.125 \times 1.24 = 8.83500$$

The calculator simply drops the unnecessary zeros at the end.

Exercises 4.9

1. The purpose of these exercises is to gain mastery in locating the decimal point in a product. Perform these calculations by hand. However, you may use your calculator to compute products as they occur in various stages of the problems. (To get the most value from these exercises, perform the multiplication on the calculator using integers.) After you finish, check your computation by doing the problem on a calculator, keying in the decimals as they occur, if possible.

- •a. 4.1×3.2
- •f. $6.53 \times .57$
- b. 4.11×3.2
- •g. 0.2×0.3
- •c. 4.003×6.75
- h. 0.02×0.003
- •d. 41.3×6.75
- i. $.7893 \times .6324$
- e. 4.2×20
- •j. $0.24 \times 7.5 \times 4.65$
- •k. 0.47893×0.56324 (Be sure you have the correct number of decimal places in your answer.)

4.10 Division as Repeated Subtraction

Just as multiplication can be viewed as repeated addition, so can its reverse operation, division, be viewed as repeated subtraction. Recall that

$$4 \times 3 = 12 \qquad \text{can be written} \qquad 0 + \underbrace{3 + 3 + 3 + 3}_{\text{four 3's}} = 12 \qquad (*)$$

Likewise

$$12 \div 3 = 4 \qquad \text{can be written} \qquad 12 - \underbrace{3 - 3 - 3 - 3}_{\text{four 3's}} = 0$$

REMARKS: In the equation marked (*), it is clear that

$$0 + 3 + 3 + 3 + 3$$

could have been written without the 0. There are two important reasons for inserting the zero here. (1) In clearing the machine, the "0." appears automatically and any addition is started at this point. (2) The 0 here finds a parallel to the 0 remainder in the equation

$$12 - 3 - 3 - 3 - 3 = 0.$$

Just as in the special case ($12 \div 3$) above, to find $A \div B$ in general, we may start subtracting B repeatedly from A until we reach zero, then count how many times we need to subtract.

Example 4.29: $39 \div 13 = ?$

Solution:

$$
\begin{array}{r}
39 \\
-13 \\
\hline
26 \\
-13 \\
\hline
13 \\
-13 \\
\hline
0 \\
\end{array}
$$

26 ← subtracted once

13 ← subtracted twice

0 ← subtracted three times

Therefore, $39 \div 13 = 3$.
That is,

$$39 - 13 - 13 - 13 = 0$$

or,

$$39 = 13 + 13 + 13$$

from which,

$$39 = 13 \times 3; \quad \text{so } 39 \div 13 = 3.$$

Example 4.30: $208 \div 16 = ?$

Solution: Use your calculator; the repeated subtraction as this method of solution will require a bit of time! To save space, the subtraction is shown in horizontal form.

$208 - 16 = 192$	$128 - 16 = 112$	$64 - 16 = 48$
$192 - 16 = 176$	$112 - 16 = 96$	$48 - 16 = 32$
$176 - 16 = 160$	$96 - 16 = 80$	$32 - 16 = 16$
$160 - 16 = 144$	$80 - 16 = 80$	$16 - 16 = 0$
$144 - 13 = 128$		

Thus,

$$208 - \underbrace{16 - 16 - \cdots - 16}_{\text{13 times}} = 0$$

or

$$208 - (16 \times 13) = 0.$$

So

$$208 \div 16 = 13.$$

Although it is illuminating to see how division can be viewed from this standpoint, the above procedure is a very tedious operation to do by hand—or even with the aid of your calculator when the number of subtractions is large. However, it is one that electronic circuits can be programmed to do electronically very rapidly. Therefore, this process is emphasized to provide an insight into machine functioning, because refinements of this method are, in fact, used to perform the division calculations inside your calculator.

Exercises 4.10

1. Perform these divisions by the method of repeated subtraction. Check the answers with your calculator.

a. $14 \div 7$

•b. $27 \div 9$

c. $18 \div 3$

•d. $378 \div 126$

e. $7590 \div 1265$

•f. $1436889 \div 478963$

NOTE: If your machine has a constant (K) key, you will probably want to learn to use it for your work on part f.

4.11   Division and the Distributive Law

We can now consider the evening social of the four couples (8 persons) of Example 4.5 from a slightly different angle.

Example 4.31: Four couples went to a show and a snack afterward. The total cost for the show was $32 and the total snack bill was $16. What was the expense per person for that evening?

Solution: The total bill for the evening was $32 + 16$. This is to be shared equally by 8 persons.
   Cost per person: $(32 + 16) \div 8$,

$$(32 \; \boxplus \; 16) \div 8 = 48 \; \boxdiv \; 8 \; \boxeq \; 6 \qquad \text{or } \$6.$$

We could also settle the bill separately.

Cost per person for show: $32 \; \boxdiv \; 8 \; \boxeq \; 4$
Cost per person for snack: $16 \; \boxdiv \; 8 \; \boxeq \; 2$
Total cost: $4 \; \boxplus \; 2 \; \boxeq \; 6$

Thus $(32 + 16) \div 8 = (32 \div 8) + (16 \div 8).$

In general, there is a distributive law for division.

If $C \neq 0$, then $(A + B) \div C = (A \div C) + (B \div C)$.

Sometimes a complicated division problem may be made into two easier ones (ones with smaller numbers) by writing the number to be divided as the sum or difference of two smaller numbers. To make things easy, choose one of the numbers so that it may be easily divided. Multiples of 10, 100, and so on, of the divisor readily suggest themselves.

Example 4.32: $98 \div 7 = ?$

Solution: Write $98 = 70 + 28$.

$$98 \div 7 = (70 + 28) \div 7 = (70 \div 7) + (28 \div 7)$$
$$= \quad 10 \quad + \quad 4$$
$$= \quad 14.$$

Example 4.33: $153 \div 17 = ?$

Solution: Write $153 = 170 - 17$.

$$153 \div 17 = (170 - 17) \div 17$$
$$= (170 \div 17) - (17 \div 17)$$
$$= \quad 10 \quad - \quad 1$$
$$= \quad 9$$

Notice that this distributive law applies to division on the right side of the sum. It does not apply, for example, to $6 \div (3 + 2)$. In general, it does not apply to $A \div (B + C)$.

Example 4.34: Check that $308 \div (4 + 7) \neq (308 \div 4) + (308 \div 7)$.

Solution:

$$308 \div (4 \boxplus 7) = \mathit{308} \boxdiv \mathit{11} \boxeq \mathit{28.}$$

On the other hand,

$$(\mathit{308} \boxdiv 4) + (\mathit{308} \boxdiv 7) = 77 \boxplus 44 \boxeq \mathit{121}$$

Therefore, $308 \div (4 + 7) \neq (308 \div 4) + (308 \div 7)$.

Exercises 4.11

1. Use the distributive law to simplify the following divisions.

   a. $102 \div 3$ [Try $102 = 99 + 3$]

   •b. $95 \div 5$ [Try $95 = 100 - 5$]

   c. $42 \div 7$ [Try $42 = 35 + 7$]

   •d. $147 \div 7$

   •e. $117 \div 13$

   f. $121 \div 11$

2. Let d represent any of the integers 0, 1, 2, 3, 4, 5, 6, 7, 8, or 9. Consider the number dd. (For example, if $d = 7$, $dd = 77$. And so on.) Show that

$$dd \div 11 = d.$$

   (You can show this by calculating each separate case and showing that the statement is true in each one. It is more enlightening, however, to find a general reason why the statement must be true. HINT: Use your knowledge of the place-value system to express the value of dd in terms of d.)

3. Give an explanation of the distributive law for division in terms of areas of rectangles, in a manner similar to the argument used in Figure 4.1.

•4. Compare $120 \div (3 + 5)$ with $(120 \div 3) + (120 \div 5)$.

•5. Show that $6 \div (3 + 2) \neq (6 \div 3) + (6 \div 2)$.

6. Compare $A \div (B + C)$ with $(A \div B) + (A \div C)$ when $A = 0$.

4.12   Division with Remainder

Anyone who has tried to divide 8 candy bars equally between three children knows that division doesn't always come out even.

Figure 4.2
How shall 8 candy bars be divided among 3 children? Your calculator says

$$8 \div 3 = 2.6666666$$

Does that help?

Let us see what happens on a calculator.

$$8 \div 3 = 2.6666666$$

A calculator gives its answer in decimals. That's fine, but it is hard to hand out 2.6666666 bars of candy to each child!

In this section, we shall regard division as repeated subtraction to solve problems. As it turns out, these results are often useful in themselves. They can be used to explain more fully the meaning of decimal numbers. Moreover, division with remainder has far-reaching consequences and applications in mathematics.

We shall continue to try to divide 8 candy bars among three children. We begin by handing out 3 bars, one to each child.

Figure 4.3
Each child gets 1 candy bar. Five bars remain to be divided.

Mathematically, we subtract: $8 - 3 = 5$.

Because 5 is larger than 3, we may subtract again. So we give one more bar to each child. At this stage, $5 - 3 = 2$.

Figure 4.4
Each child gets a second bar. Now 2 bars remain to be divided.

But now, no further subtraction of 3 can be made (without using negative numbers). We have,

$$8 - 3 - 3 = 2 \quad \text{or} \quad 8 = \underbrace{3 + 3}_{\text{two 3's}} + 2. \quad \text{That is, } 8 = 6 + 2$$

or

$$8 = (3 \times 2) + 2.$$

Now divide both sides by 3:

$$8 \div 3 = [(3 \times 2) + 2] \div 3.$$

Use the distributive law on the right-hand side of the equation:

$$[(3 \times 2) + 2] \div 3 = [(3 \times 2) \div 3] + (2 \div 3)$$

$$= \quad 2 \quad + (2 \div 3).$$

So $\qquad\qquad\qquad\qquad 8 \div 3 = 2 + (2 \div 3).$

As a result of these steps, the division of $8 \div 3$ has been reduced to the division of the smaller number, $2 \div 3$.

Check out the results on your calculator:

$$(2 \div 3) + 2 = ?$$

$$(2 \;\div\; 3 \;+\; 2 \;=\; 2.6666666$$

This is exactly what we got before when we divided 8 by 3 on the calculator:

$$8 \;\div\; 3 \;=\; 2.6666666$$

Let us summarize these stages of passing out the candy bars mathematically:

1 bar to each child: $\quad 8 = (3 \times 1) + 5 = \mathbf{3} + 5$

2 bars to each child: $\quad 8 = (3 \times 2) + 2 = \mathbf{6} + 2.$

We split 8 into two smaller numbers, one of which is a multiple of 3 (in bold type). So far, what we have done is almost parallel to what we did in Section 4.11 when we split a division problem into two easier ones by writing the number to be divided as the sum of two smaller numbers. The difference here is that one of the smaller numbers cannot be divided by 3 a whole number of times.

Even so, we have reduced our problem to dividing two candy bars among three children. That is really easier—in fact, just as easy as dividing one candy bar among three children. Just split each candy bar into three equal pieces. Then there is one piece of each candy bar per child. So, in this case, each child gets two pieces (or two-thirds) of a candy bar.

Each of the remaining candy bars has been split into 3 pieces.

Figure 4.5
Eight candy bars have been divided equally among 3 children.

Example 4.35: $13 \div 5 = ?$

Solution: We begin with repeated subtraction.

$$
\begin{array}{rl}
13 & \\
- \ 5 & \text{one subtraction} \\
\hline
8 & \\
- \ 5 & \text{two subtractions} \\
\hline
3 & \text{STOP}
\end{array}
$$

Thus
$$13 - 5 - 5 = 3 \quad \text{or} \quad 13 = \underbrace{(5 + 5)}_{\text{two 5's}} + 3$$

$$= (5 \times 2) + 3$$

So
$$13 \div 5 = [(5 \times 2) + 3] \div 5$$
$$= [(5 \times 2) \div 5] + (3 \div 5)$$
$$= \quad 2 \quad + (3 \div 5)$$

We complete this calculation on a calculator. It is important to note that $3 \div 5$ is enclosed in parentheses; so we begin with $3 \div 5$ first, and then add 2.

$$(3 \div 5) + 2 = ?$$
$$(\ 3 \div 5\ ) + 2 = 2.6$$

A calculator computation of the original problem, $13 \div 5$, also yields 2.6:

$$1\ 3\ \boxdot\ 5\ \boxdot\ 2.6$$

The general algorithm for division with remainder is as follows. Given a number N and a divisor D, we may subtract D repeatedly until we reach a number R that is less than D. Then we stop. If Q is the number of times we have subtracted D, then we have

$$N - \underbrace{D - D - \cdots - D}_{Q \text{ times}} = R$$

In other words,

$N = (D \times Q) + R,$ where R is some whole number less than D.

R is called the *remainder;*
Q is called the *quotient.*

Note that it is possible for Q to be zero.

Example 4.36: $2 \div 3 = ?$

Solution: Here, $N = 2$ and $D = 3$.

$$2 = (3 \times 0) + 2$$

$R = 2$
$Q = 0$

It is also possible that $R = 0$. When this happens, we know that D divides N evenly.

Example 4.37: $39 \div 13 = ?$

Solution: Here, $N = 39$; $D = 13$.

$$39 - 13 = 26$$
$$26 - 13 = 13$$
$$13 - 13 = 0.$$

So,

$$39 - 13 - 13 - 13 = 0 \qquad \text{or} \qquad 39 = (13 \times 3) + 0.$$

Thus, $Q = 3$ and $R = 0$.

$36 \div 13.$ **Solution:** $36 = (13 \times 2) + 10$
$$Q = 2; \; R = 10$$

$8 \div 13$ **Solution:** $8 = (13 \times 0) + 8$
$$Q = 0; \; R = 8$$

$67 \div 13$ **Solution:** $67 = (13 \times 5) + 2$
$$Q = 5; \; R = 2$$

$130 \div 13$ **Solution:** $130 = (13 \times 10) + 0$
$$Q = 10; \; R = 0$$

In the above examples, you may find Q by repeated subtraction. It may be possible also to anticipate the result by experimenting with integer multiples of the divisor (in these cases, 13) until you find the largest multiple of 13 less than (or equal to) the number. Here is a table of some multiples of 13.

$13 \times 1 = 13$ $13 \times 4 = 52$ $13 \times 7 = 91$

$13 \times 2 = 26$ $13 \times 5 = 65$ $13 \times 8 = 104$

$13 \times 3 = 39$ $13 \times 6 = 78$ $13 \times 9 = 117$

Thus, in treating $67 \div 13$ (Example 4.38), we see that 13×5 is smaller than 67, and that 67 is smaller than 13×6. So we have

$$67 = (13 \times 5) + R.$$

Because R turns out to be 2,

$$67 = (13 \times 5) + 2.$$

The major application of division with remainder for school children is the awkward and burdensome algorithm to perform long division. We do not have to rely on this algorithm because our calculators are faster, more accurate, and (for most problems) give more significant data than we would have patience to obtain by hand. Nevertheless, it seems worthwhile to give a short description of how that algorithm works. In particular, in the next section, we shall see how we can exploit this algorithm in order to perform division of numbers too large for our calculators.

Example 4.39: Divide 139 by 7. That is, find the quotient Q and the remainder R so that

$$139 = (7 \times Q) + R, \qquad \text{where } R \text{ is less than 7.}$$

Solution: The problem is made easier by the decimal notation:

$$139 = 130 + 9$$

So

$$(130 + 9) \div 7 = (130 \div 7) + (9 \div 7).$$

The problem of dividing 130 by 7 is made easier by realizing that we need really only divide 13 by 7, and then multiply by 10. Thus

$$13 = (7 \times 1) + 6.$$

So

$$130 = 13 \times 10 = [(7 \times 1) + 6] \times 10$$

$$= (7 \times 10) + 60$$

and

$$139 = 130 + 9 = (7 \times 10) + 69$$

We have completed the first step of the algorithm. We would be done if 69 were less than 7. It is not; so we now work on 69.

$$69 = (7 \times 9) + 6$$

Now we put these two steps together:

$$139 = (7 \times 10) + 69$$

$$= (7 \times 10) + (7 \times 9) + 6$$

Using the distributive law,

$$139 = [7 \times (10 + 9)] + 6$$

$$= (7 \times 19) + 6$$

Remainder

Quotient

In the elementary school, the above steps are usually laid out in the following scheme:

First step:
$$7\overline{)139} \qquad 13 = (7 \times ①) + ⑥$$
$$\underline{-\ 7}$$
$$⑥$$

The "1" in the partial quotient placed over the "13" (in the tens' place) indicates the factor of 10 that is understood. In fact.

the "1" over the "13" represents 1×10

the "7" under the "13" represents 7×10

the "6" for remainder represents 6×10

and of course,

the "13" represents 13×10,

where 139 is split into $130 + 9$.

We continue the algorithm by adding 9 to the remainder. We say we *"bring down the 9"* from "139": $60 + 9 = 69$. Now we continue to divide by 7:

Second step:
$$7\overline{)69} \qquad 69 = (7 \times 9) + 6$$
$$-63$$
$$6$$

The reasoning is more evident in the *"platform method"* that is often taught in school now.

$$
\begin{array}{r}
7\overline{)139} \\
-\ 70 \quad | \quad 10 \qquad 130 = (7 \times 10) + 60 \\
\hline
69 \\
-\ 63 \quad | \quad +\ 9 \qquad 69 = (7 \times 9) + 6 \\
\hline
6 \quad | \quad 19
\end{array}
$$

For division, the calculator gives the answer in decimals rather than by the quotient Q and the remainder R. Sometimes the remainder in division is as important as the quotient.

Example 4.40: Chas. Brown Peanuts, Inc. has a surplus of \$2,170,065 to be distributed equally in whole numbers of dollars for each of 162,841 shares of stock held in the company. The excess after distribution goes back to the company for growth. How much is kept for growth?

Instead of solving the above problem right away, we shall follow through one involving smaller numbers. It will be seen that to find the whole number quotient Q and the remainder R using the calculator requires pencil and paper or a little trick.

Example 4.41: Find the quotient and remainder when 139 is divided by 7.

Solution: Begin by calculating:

$$139 \div 7 = 19.857142$$

From this computation, we know that the quotient is 19. We have

$$19 < 19.857142 < 20$$

Multiplying each of these numbers by 7,

$$19 \times 7 < 19.857142 \times 7 < 20 \times 7$$
$$\downarrow \qquad\qquad \downarrow \qquad\qquad \downarrow$$
or, $\qquad$ 133 $\quad<\qquad$ 139 $\qquad<\quad$ 140

Hence, to find the remainder, we want to fill the box for

$$139 = (19 \times 7) + \boxed{\text{Remainder}}$$
and so $\qquad 139 - (19 \times 7) = \boxed{\text{Remainder}}$

This shows that we only have to compute 19×7 and subtract it from 139. We can do this by clearing our machine and starting over:

$$139 - (19 \times 7) \rightarrow 139 - 133 = 6.$$

Observe that, in calculating $139 - (19 \times 7)$, we first compute $19 \times 7 (= 133)$ and record it. The, 139 is entered and the result, 133, is re-entered (or recalled from memory if the machine has one). To sidetrack this extra work, we can find the difference the other way around:

$$(19 \times 7) - 139 = -6.$$

The sign in the answer is *"negative"* because 139 is greater than 19×7. The difference is 6. We discuss this more fully in Chapter 6.

Now this is all well and good. However, it is inelegant. We had to clear 19.857142, remember the integer part, 19 and re-enter it, and then enter 7. If that can be avoided, it should be. One reason is that each time you must enter a number in the machine, there is a big chance for error in reading the number when keying it in. Another is that you must either remember the number you want to re-enter or write it down on paper. Remembering a number with more than five digits is not easy—even for a few seconds; and once distracted from your work, you must start over again.

Here is what to do to eliminate the necessity for clearing your calculator. The trick is to add 10000000 to the number in the display, and then subtract it. As you will see, only the whole number part of the number stays in the display when addition is performed, and the original integer part is kept when 10000000 is subtracted. There are seven zeros after the 1—an easy number to count out as you key in 10000000.

Here is how it goes. We pick up the action and the display at

$$(\boxed{19.857142} \; \boxed{+} \; \boxed{10000000}) \; \boxed{-} \; \boxed{10000000} \; \boxed{=} \; \boxed{19.}$$

Now we are at exactly the same position that we were before we had to clear the machine and enter 19. At this stage, we continue as before.

Example 4.42: What is the quotient and remainder when 13746 is divided by 37?

Solution:

$$\boxed{13746} \; \boxed{\div} \; \boxed{37} \; \boxed{=} \; \boxed{371.513 51}$$

Continue: $(\boxed{371.51351} \; \boxed{+} \; \boxed{10000000}) \; \boxed{-} \; \boxed{10000000} \; \boxed{=} \; \boxed{371.}$

Continue: $(\boxed{371} \; \boxed{\times} \; \boxed{37}) \; \boxed{-} \; \boxed{13746} \; \boxed{=} \; \boxed{-19.}$

Ignore the $(-)$ sign. Hence,

$$13749 = (371 \times 37) + 19.$$

In the example, we paused between steps to show the different stages of the calculation. We indicate below the actual sequence of the operations. The whole number part, Q, of the quotient is read right after "-10000000" (see vertical arrow).

$$\boxed{13746} \; \boxed{\div} \; \boxed{37} \; \boxed{+} \; \boxed{10000000} \; \boxed{-} \; \boxed{10000000} \overset{\downarrow}{} \boxed{\times}$$
$$\boxed{37} \; \boxed{-} \; \boxed{13746} \; \boxed{=} \; \boxed{-19.}$$

Here is an application: converting minutes to hours.

Example 4.43: Suppose it takes 51 minutes to drive from Pasadena to San Bernardino and 34 minutes to drive from San Bernardino to Palm Springs. How long does it take to drive from Pasadena to Palm Springs via San Bernardino?

Solution: In minutes: $\boxed{51} \; \boxed{+} \; \boxed{34} \; \boxed{=} \; \boxed{85.}$
How many hours and how many minutes?

Because there are 60 minutes in an hour, we should divide 85 by 60. The calculator says:

$$\boxed{85} \; \boxed{\div} \; \boxed{60} \; \boxed{=} \; \boxed{1.4166666}$$

The decimal form of the answer is not adequate to tell us how many hours and how many minutes. We need to find a quoient and a remainder.

$$85 = (60 \times 1) + 25.$$

Thus, 85 minutes = 1 hour and 25 minutes.

Exercises 4.12

1. How would you divide 18 doughnuts among 4 hungry hikers?

•2. A family of 7 has a dog, Colonel, who loves hot dogs. For lunch one day, 23 hot dogs were prepared. The family decides to divide the hot dogs evenly to all persons, but rather than splitting any hot dogs, they will give the leftover ones to the dog. How many hot dogs does Colonel get?

3. Your house is to be painted. It will take 38 hours. If the painters work 8 hours a day, on how many days will the painters work on your house?

4. Carry out the following long divisions and indicate the remainder for each. (You may use a calculator on the individual steps.)
    •a. $6\overline{)483}$ b. $16\overline{)483}$ •c. $274\overline{)483}$

5. For the following pairs of integers (A, B) express A in the form

$$A = (B \times Q) + R, \quad \text{where } 0 \le R < B.$$

   (That is, find the quotient Q and the remainder R for $A \div B$.)
    •a. $A = 23$ $B = 4$ $Q = ?$ $R = ?$
    b. $A = 78$ $B = 11$ $Q =$ $R =$
    •c. $A = 94$ $B = 17$ $Q =$ $R =$
    d. $A = 5641$ $B = 13$ $Q =$ $R =$

6. Find the solution for Example 4.40.

7. Find the quotient and remainder when each of the following is divided by 8131.
    •a. 91351785   •c. 91351790
    b. 91351787   d. 91351793
   Explain the trouble in comparing these answers on an 8-digit calculator.

•8. By examining the numbers in Exercise 7, suggest five other numbers that would give the same "quotient" when divided by 8131.

9. Find another pair of integers (A, B) like those in Exercise 7 so that, on your calculator, $A \div B$ and $(A + 1) \div B$ appear to have the same integral quotient.

4.13   Dividing Numbers Too Large for Your Calculator

From the process of finding the quotient and remainder when dividing two whole numbers, we will derive a way to divide numbers too large for your calculator. We must first follow the long (by hand) algorithm for division.

Example 4.44: $123456192 \div 732 = ?$

Solution: The long-hand scheme is

$$732\overline{)123456192}.$$

We divide as much as we can, showing the split by a dashed line:

$$7\ 3\ 2\ \overline{)\ 1\ 2\ 3\ 4\ 5\ 6\ 1\ 9\,|\,2}$$

Now we perform $12345619 \div 732$ on our calculator:

$$\mathit{123456 19} \boxplus \mathit{732} \boxed{=} \mathit{16865.599}$$

and, by the methods of the last section,

$$12345619 = (16865 \times 732) + 439,$$

which we indicate in the usual fashion:

$$
\begin{array}{r}
1\ 6\ 8\ 6\ 5 \\
7\ 3\ 2\ \overline{)\ 1\ 2\ 3\ 4\ 5\ 6\ 1\ 9\,|\,2} \\
\underline{1\ 2\ 3\ 4\ 5\ 1\ 8\ 0\,|} \\
4\ 3\ 9
\end{array}
$$

Next, we "bring down" the "2":

$$
\begin{array}{r}
1\ 6\ 8\ 6\ 5 \\
7\ 3\ 2\ \overline{)\ 1\ 2\ 3\ 4\ 5\ 6\ 1\ 9\,|\,2} \\
\underline{1\ 2\ 3\ 4\ 5\ 1\ 8\ 0\,|} \\
4\ 3\ 9\ 2
\end{array}
$$

Now $\mathit{4392} \boxplus \mathit{732} \boxed{=} \mathit{6}$

That is, $4392 = 732 \times 6,$

And thus

$$
\begin{array}{r}
1\ 6\ 8\ 6\ 5\ 6 \\
7\ 3\ 2\ \overline{)\ 1\ 2\ 3\ 4\ 5\ 6\ 1\ 9\ 2} \\
\underline{1\ 2\ 3\ 4\ 5\ 1\ 8\ 0} \\
4\ 3\ 9\ 2 \\
4\ 3\ 9\ 2
\end{array}
$$

Computations become even more complicated when the divisor has more than 8 digits as well. The same principles apply, but it is rarely necessary to carry out all the details. Good estimates can be made with the aid of scientific notation, as we shall see in Chapter 7. Here is an example of how the method might work.

Example 4.45: $987654321 \div 123456789 = ?$

Solution: Write the first number as 98765432.1 × 10 and the second number as 12345678.9 × 10.

Thus, ignoring the (.1) and the (.9), the quotient is approximately

$$(98765432 \times 10) \div (12345678 \times 10)$$

$$\boxed{=}\ \mathit{98\ 765\ 432}\ \boxed{\div}\ \mathit{1\ 2\ 3\ 4\ 5\ 6\ 78}\ \boxed{=}\ \mathit{8.0000006}$$

Thus, the original quotient is about 8. In fact,

$$123456789 \times 8 = 987654312$$

so that

$$987654321 = (123456789 \times 8) + 9. \quad [21 - 12 = 9]$$

The bold type in the two numbers indicates the digits that differ from the original number 987654321.

Exercises 4.13

1. Compute the quotient and remainder in these divisions. Use your calculator as much as you can.
   - •a. 712456355 ÷ 7
   - b. 9908037625 ÷ 155
   - •c. 123456789 ÷ 1234

2. Estimate these quotients.
   - •a. 1278947631 ÷ 111455556
   - b. 1278947632 ÷ 111455556
   - c. 1278947632 ÷ 11145555678

Geometric Applications

5.1 Areas of Rectangles and Squares

We measured lengths on several earlier occasions. To do this, we picked a unit of length (an inch, a foot, or a centimetre) and based our measurements on this unit. If we mark off exactly 5 unit lengths end-to-end, we say the length is 5 units long.

Length is a useful measure, but it is only on one dimension. Sometimes we need to measure a two-dimensional region. If we want to buy enough paint to paint a room (but not too much), we need to know the area of the room. Knowing the length or even the distance around the room is not enough.

Just as we chose a *"unit length"* to measure lengths, we need to choose a *"unit area"* to measure area. Suppose we want to find the area of a rectangle 3 centimetres by 4 centimetres.

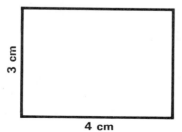

Figure 5.1
What is the area of this rectangle?

There is a simple rule:

The area is the product of the lengths of the two sides.

Why does this work? In Figure 5.1, mark off the length into four parts, each part 1 centimetre long. We say we have subdivided the side into four parts of equal length.

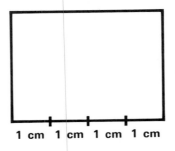

Figure 5.2
One side subdivided into
four equal parts.

These points of division determine four strips, each 1 centimetre wide.

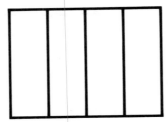

Figure 5.3
Each part of the subdivision
determines a strip.

Similarly, subdivide the width. We choose to do so in three equal parts because the rectangle is 3 centimetres wide.

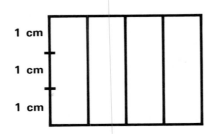

Figure 5.4
Subdivision of the other side
into three equal parts.

These points of division determine three horizontal strips. These horizontal strips cross the vertical strips to form rows of squares, each 1 centimetre on a side.

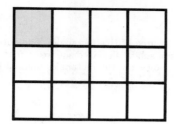

Figure 5.5
Subdivision into $3 \times 4 = 12$ squares,
each 1 centimetre on a side.

Now the rule for calculating areas is clear. There are

3 rows, each row having 4 squares.

So there are

3×4 squares.

Each square unit measures 1 centimetre on each side, and hence is 1 square centimetre (denoted *"1 cm²"* for short). So the 3-by-4 rectangle measures

(3×4) or 12 square centimetres.

Please refer to Chapter 4. Which form of multiplication is this?

Example 5.1: What is the area of a square 3 inches on each side?

Solution: A square is a rectangle in which the length and width are equal. The length and width of a "3-inch square" are each 3 inches. The rule for computing areas of rectangles gives $3 \times 3 = 9$ square inches.

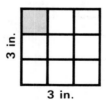

3 in.

3 in.

In the examples so far, the measures of all the sides are whole numbers. What if the lengths of the sides involve fractions?

Example 5.2: The width of a rectangle is 3 feet and its length is $4\frac{1}{2}$ feet. Find its area.

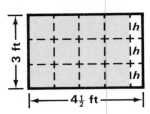

Figure 5.6
What is the area of this rectangle? The dotted lines show a subdivision into squares 1 foot on a side. Left over are the rectangles marked *h*.

Solution: The rule gives

$$3 \times 4\frac{1}{2} = 3 \times \left(4 + \frac{1}{2}\right) \quad \text{and} \quad 3 \boxed{\times} 4.5 \boxed{=} 13.5$$

The distributive law helps to explain why the rule works.

$$3 \times \left(4 + \frac{1}{2}\right) = (3 \times 4) + \left(3 \times \frac{1}{2}\right) \tag{*}$$

The area of the shaded region in Figure 5.6 is (3×4) square feet. This is the term in the first set of parentheses on the right side of the equation marked (*).

The term in the second member on the right side of equation (*) is

$$3 \times \frac{1}{2} = \frac{3}{2}.$$

Each rectangle marked *h* in Figure 5.6 is one foot on one side and one-half foot on the other side. Two of these can be "pushed together" to make up one square foot (see Figure 5.7). So it is reasonable to think of each as $\frac{1}{2}$ of a square foot.

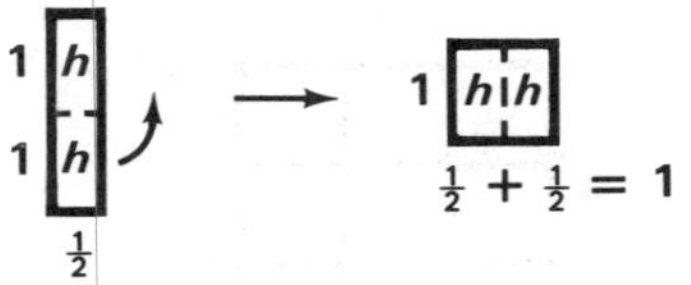

Figure 5.7
Each square foot contains 2 *h*-rectangles.
$2 \times h = 1$ means $h = \frac{1}{2}$.

Two regions of the same size add to 1 unit. Each region is $\frac{1}{2}$ unit in area. Because there are 3 of these, we have $\frac{3}{2}$ or $1\frac{1}{2}$ square feet in addition to the first 12 square feet. In summary, altogether the area is

$$3 \times 4\frac{1}{2} = 12 + 1\frac{1}{2} = 13\frac{1}{2} \text{ (square feet)}.$$

To get another viewpoint, we change the direction of the argument slightly.

Example 5.3: Find the area of a rectangle whose dimensions are $3\frac{1}{2}$ metres by $4\frac{1}{2}$ metres.

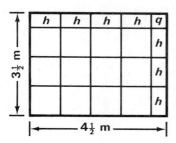

Figure 5.8
What is the area of this rectangle? The dotted lines show a subdivision into squares 1 metre on a side. Left over are the rectangles marked h and the small squared marked q.

Solution:

$$3\frac{1}{2} \times 4\frac{1}{2} = ? \qquad 3.5 \,\boxed{\times}\, 4.5 \,\boxed{=}\, 15.75$$

We have computed the area of the small rectangle marked h. It is $\frac{1}{2}$ square metre (or 0.5 square metre). What is the area of the small square q?

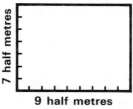

Figure 5.9
Each square metre contains 4 q-squares.
$4 \times q = 1$ means $q = \frac{1}{4}$.

Instead of marking off in 1 metre intervals, mark off the length and width in half-metre intervals. What does this change do to each square metre? Each square is subdivided into four smaller squares. Four of these make up one square metre. So one q is $\frac{1}{4}$ square metre.

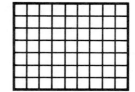

9 half metres

Figure 5.10
Subdividing a $3\frac{1}{2}$ by $4\frac{1}{2}$ rectangle into q-squares. How many q-squares are there?

The width, $3\frac{1}{2}$ metres, is marked off into seven half-metre intervals. That is,

$$3\frac{1}{2} \div \frac{1}{2} = 7.$$

The length, $4\frac{1}{2}$ metres, is marked off into nine half-metre intervals. That is,

$$4\frac{1}{2} \div \frac{1}{2} = 9.$$

Figure 5.10 shows that there are $7 \times 9 = 63$ quarter (q) squares. Thus the area is

$$63 \times \frac{1}{4} = \frac{63}{4} = \frac{60 + 3}{4} = \frac{60}{4} + \frac{3}{4}$$

$$= 15 \text{ (square metres)} + 3 \text{ (quarters of a}$$
$$\text{square metre)}$$

$$= 15 + \frac{3}{4} \text{ (square metres)}$$

$$= 15\frac{3}{4} \text{ (square metres)}.$$

In general, we define (once and for all) the area of a rectangle by the formula

$$\text{area} = \text{length} \times \text{width}.$$

This holds whether or not the measures of the sides are integers. However, it is usual to use the same *unit* for the measure of the sides. When this is done, the area is in square units.

Of course, the product can be found directly with a calculator.

Example 5.4: What is the area of a rectangle $4\frac{1}{2}$ metres long and $3\frac{1}{2}$ metres wide?

Solution: $4\frac{1}{2} \times 3\frac{1}{2} = ?$

$$4.5 \;\boxed{\times}\; 3.5 \;\boxed{=}\; 15.75$$

The area is 15.75 square metres. Do you recognize that $0.75 = \frac{3}{4}$? Try it:

$$3 \;\boxed{\div}\; 4 \;\boxed{=}\; 0.75$$

Example 5.5: Find the area of a rectangle 14 inches wide, 2 feet long.

Solution: If we were to multiply 14 × 2 to get the area, the "unit areas" would not be squares. Instead, they would be rectangles, 1 inch wide by 1 foot long.

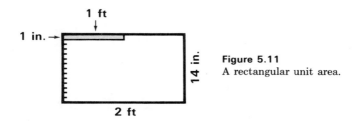

Figure 5.11
A rectangular unit area.

Although it is possible to use such units, the ones customarily agreed upon are squares. We should express both the length and width either in inches or in feet.

To use inches: Convert 2 feet to inches. Because 1 foot is 12 inches, 2 feet is 24 inches. So

area (in square inches) = $\boxed{1\,4}\;\boxed{\times}\;\boxed{2\,4}\;\boxed{=}\;\boxed{3\,3\,6.}$

To use feet: Convert 14 inches to feet.

$$14 \text{ inches} = \frac{14}{12} \text{ feet} = \frac{7}{6} \text{ ft. } (= 1\frac{1}{6} \text{ ft.})$$

So area (in square feet) =

$\boxed{(}\;\boxed{7}\;\boxed{\div}\;\boxed{6}\;\boxed{)}\;\boxed{\times}\;\boxed{2}\;\boxed{=}\;\boxed{2.3333332}$

To compare these two answers, observe that a square foot measures 1 foot, or 12 inches, on each side.

12 in.

Therefore, 1 square foot is 12 × 12 = 144 square inches. Thus, the number of square feet in 336 square inches may be computed:

$\boxed{3\,3\,6}\;\boxed{\div}\;\boxed{1\,4\,4}\;\boxed{=}\;\boxed{2.3333333}$

You may notice a difference in the seventh decimal place due to round-off error in your calculator.

Exercises 5.1

•1. For each figure below, make necessary measurements to the nearest tenth centimetre and calculate the area and the perimeter of each.

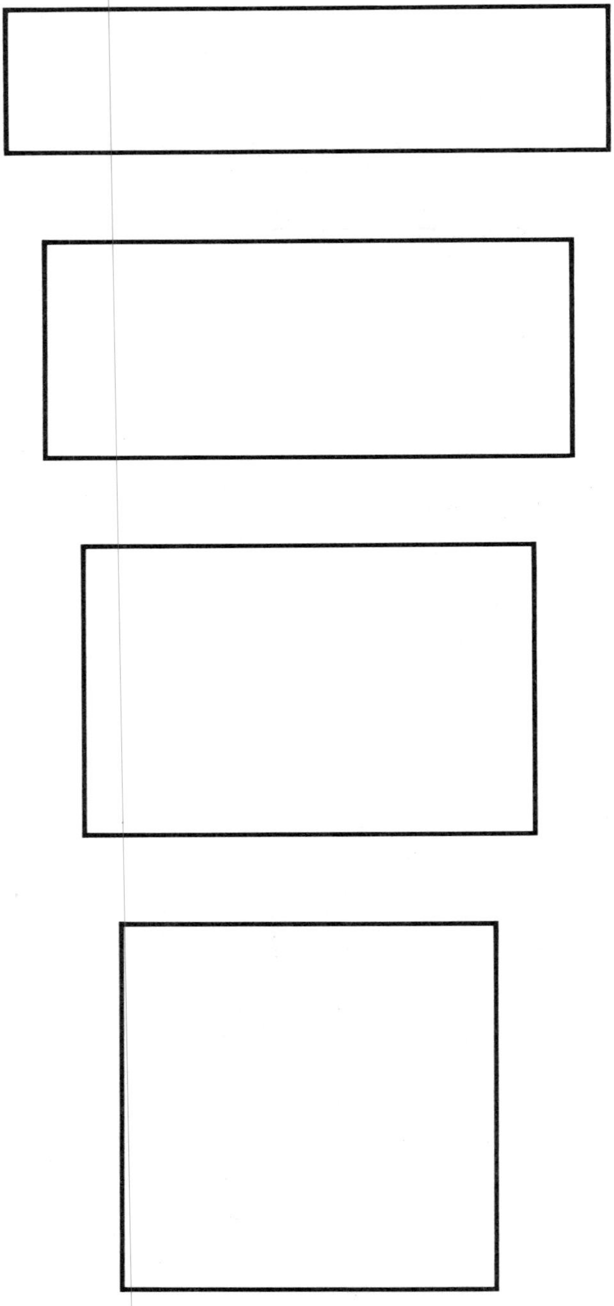

2. Verify that each of the rectangles has the same perimeter. Which has the largest area?

3. Draw rectangles with the following dimensions:

Length	Width
a. 8.5 cm	3 cm
•b. 11.72 cm	3.73 cm
•c. $5\frac{1}{8}$ in.	$4\frac{7}{8}$ in.

4. By computing the areas of the rectangles forming these regions, find the total area. (Dimensions are in metres.)

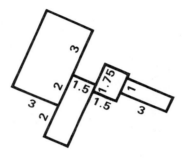

•5. What is the area of the walk around this rectangular pool? (Dimensions are in feet.)

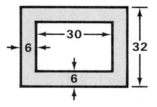

5.2   Areas of Triangles and Other Polygons

A triangle consists of three points and the line segments joining them. A triangular region is one whose boundary is a triangle, and thus it has three sides. What is the area of a triangle?

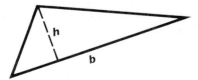

Figure 5.12
What is the area of a triangle?
$$\text{Area} = \frac{\text{base} \times \text{height}}{2}.$$

To find the area, choose any side; call it the "base" of the triangle. Measure the length of the base. Next measure the distance from the other point to this side; call it the "height" of the triangle. The area is given by the formula

$$\text{Area} = \frac{(\text{length of base}) \times (\text{length of height})}{2}.$$

CAUTION: In measuring the height, be careful to draw the shortest line from the point to the base. This line will always be perpendicular (at right angles) to the base.

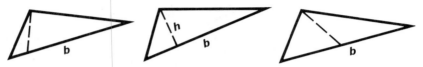

Figure 5.13
The height of a triangle is found by drawing a line from a point to the side opposite (base) so that the line is perpendicular to the base. The center figure shows a correctly drawn height.

Notice that the altitude splits the triangle into two pieces. In Figure 5.14, the dotted line at the top is parallel to the base and the vertical dotted lines are parallel to the altitude.

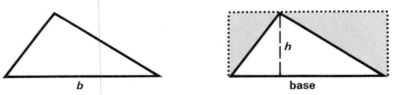

Figure 5.14
"Completing" the triangle.

The dotted lines, together with the base, form a rectangle. The area of the rectangle is given by

$$\text{Area of rectangle} = \text{length} \times \text{width}$$

$$= \text{base} \times \text{altitude}.$$

The word *altitude* is used two ways: it refers to the segment from a point to the opposite base; it also refers to the *length* of this segment.

Because the shaded portions are exactly the same size as the unshaded portions, the area of the rectangle is twice the area of the triangle. Or,

the area of the triangle is half the area of the rectangle.

So the formula is

$$A = \frac{1}{2} \times h \times b = \frac{h \times b}{2}.$$

Example 5.6: What is the area of a triangle whose altitude is 3 inches and whose base is 4 inches?

Solution:

Area $= \dfrac{3 \times 4}{2} = $ 3 ⊠ 4 ⊞ 2 ⊟ 6.

Area $= 6$ square inches.

Many figures, including a rectangle, may be thought of as constructed from triangles, and so their areas may be computed. One such figure is a *parallelogram,* as shown in Figure 5.15.

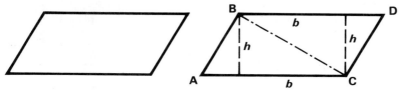

Figure 5.15
A parallelogram is a four-sided figure whose opposite sides are parallel. A diagonal line splits the parallelogram into two triangles, ABC and BCD, with same base and height. Area of parallelogram = base × height.

The area of the parallelogram is thus twice the area of one of the triangles:

$$\text{Area} = 2 \times \frac{\text{base} \times \text{height}}{2} = \text{base} \times \text{height}.$$

Example 5.7: Find the area of a parallelogram whose base measures 4 ft. 9 in. and whose height is 2 ft. 3 in.

Solution (*in inches*):
base $\quad = 4$ ft. $+ 9$ in. $= (4 \times 12) + 9$ inches
$\quad\quad\; = ($ 4 ⊠ 12 $)$ ⊞ 9 ⊟ 57 (inches).
height $= 2$ ft. $+ 3$ in. $= (2 \times 12) + 3$ in.
$\quad\quad\; = ($ 2 ⊠ 12 $)$ ⊞ 3 ⊟ 27 (inches).
Area $\quad = 57 \times 27$ square inches
$\quad\quad\; = $ 57 ⊠ 27 ⊟ 1539 (square inches).

Solution (*in feet*):

base $= 4$ ft. $+ 9$ in. $= 4 + \frac{9}{12}$ feet

$= 4 + (9 \boxdot 12) \rightarrow 4 \boxplus 0.75 \boxminus 4.75$ (feet).

height $= 2$ ft. $+ 3$ in. $= 2 + \frac{3}{12}$ feet

$= 2 + (3 \boxdot 12) \rightarrow 2 \boxplus 0.25 \boxminus 2.25$ (feet).

Area $= 4.75 \times 2.25$ square feet

$= 4.75 \boxtimes 2.25 \boxminus 10.6875$ (square feet).

Conversion check:

$1539 \boxdot 144 \boxminus 10.6875$ (square feet).

$\downarrow$ $\downarrow$

sq. in.   sq. in. per

sq. ft.

Example 5.8: Mr. Hy Finanz's house is on a lot fronting Walnut Street as shown in the accompanying sketch. The 63-foot side is perpendicular to the street, and the 58-foot boundary in the back is parallel to Walnut. If his land is worth a dollar per square foot, what is the value of his lot?

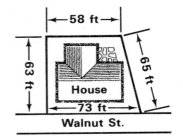

The lot is in the shape of a trapezoid. Let us examine the area of trapezoids in general before answering this question. We shall return to the solution of this problem.

A *trapezoid* is a quadrilateral (four-sided figure) with two sides being parallel (see Figure 5.16). The parallel sides are called its *bases*.

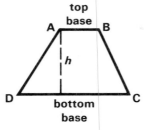

Figure 5.16
A trapezoid.

$$\text{Area} = \frac{(\text{top base} + \text{bottom base}) \times \text{height}}{2}.$$

To find its area, again draw a diagonal line to make two triangles. Suppose, as in Figure 5.17, the length of the top base is 3 inches, the bottom base 8 inches, and the altitude, 5 inches.

First, draw diagonal $\overline{DB}$. The trapezoid is split into two triangles.

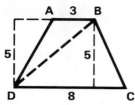 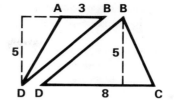

Figure 5.17
Trapezoid split into two triangles. Triangle ABD has base 3 in. and altitude 5 in.; so its area is $\frac{1}{2} \times 3 \times 5$. Triangle BDC has base 8 in. and altitude 5 in.; so its area is $\frac{1}{2} \times 8 \times 5$.

Altogether, the area is

$$\left(\frac{1}{2} \times 3 \times 5\right) + \left(\frac{1}{2} \times 8 \times 5\right) = \left(\frac{1}{2} \times 5\right) \times 3 + \left(\frac{1}{2} \times 5\right) \times 8$$

$$= \left(\frac{1}{2} \times 5\right) \times (3 + 8)$$

$$= \frac{1}{2} \times 5 \times 11 = \frac{55}{2} = 27.5 \text{ (sq. in.)}.$$

In general, if the top base is b units and the bottom base is B units, then for the one triangle we have

$$\frac{1}{2} \times b \times h = \frac{1}{2} \times h \times b,$$

and for the other triangle,

$$\frac{1}{2} \times B \times h = \frac{1}{2} \times h \times B.$$

Together, we have

$$\left(\frac{1}{2} \times h \times b\right) + \left(\frac{1}{2} \times h \times B\right) = \left(\frac{1}{2} \times h\right) \times (b + B).$$

Hence, the formula:

$$A = \frac{(b + B) \times h}{2}.$$

We can now solve Example 5.8.

Example 5.8 (continued): Find the value of Mr. Finanz's lot at a dollar per square foot (see page 148).

Solution: From the given measurements, we have to identify the bases and the altitude of the trapezoid. Here, the bases are 58 feet and 73 feet; the altitude is 63 feet.

$$(58 \boxplus 73) \boxtimes 63 \boxdiv 2 \boxminus 4126.5$$

The area of the lot is 4126.5 square feet. So, at a dollar per square foot, the value is $4,126.50.

The areas of other figures whose sides are straight lines may be found in this manner. Such figures are called *polygons*. Computations for such figures are suggested in the exercises.

Exercises 5.2

1. For each figure, make any necessary measurements in centimetres to the nearest tenth, then calculate the area.

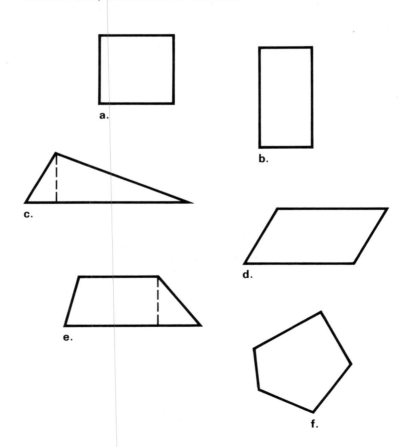

a.

b.

c.

d.

e.

f.

2. Use any necessary measurements in the drawings to calculate the area of each region. (The symbols " and ' are used to represent inches and feet, respectively.)

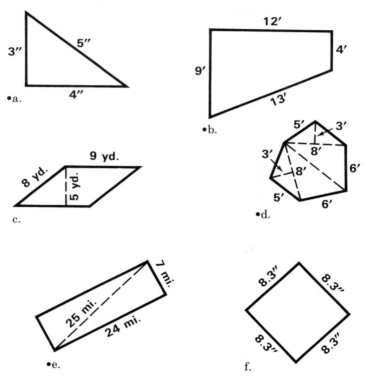

•a.

•b.

c.

•d.

•e.

f.

3. Draw each figure with the given dimensions. Then find the area of each.
   a. rectangle: length = 11.72 cm; width = 3.73 cm.
   •b. triangle: base = $3\frac{1}{4}$ in.; altitude = $5\frac{3}{8}$ in.
   •c. trapezoid: bases = 7.13 cm, 4.58 cm.
                altitude = 3.27 cm.
   d. parallelogram: sides = 3.5 in.; 6.5 in.
                height = 2 cm.

4. There are two possible parallelograms for part d of Exercise 3. Sketch both parallelograms, and find the area of the second one if you had not included both in your answer before.

5. Mr. Urb Anlife got a rectangular city lot that has a 40 foot frontage and measures 40' by 120'.

   •a. If he plans to build a rectangular house as wide as his lot, how long must it be to occupy 1,500 square feet of land?

   •b. If the city code requires that the number of square feet for a house divided by the number of square feet in the lot cannot be greater than 0.65, would the house Mr. Urb Anlife is planning violate this code?

   c. What is the largest area of land he would be allowed to build on under the code in part b?

•6. If the length of each base of a trapezoid is b units, and the altitude is h units, give a simple formula for its area.

7. The area of a triangle is given by the formula

$$A = \frac{1}{2} \times b \times h,$$

where b stands for the length of the base and h for the altitude.

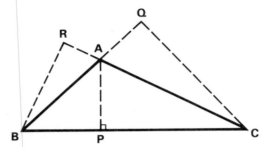

In the triangle ABC shown, if side $\overline{BC}$ is taken to be the base, then the corresponding altitude is $\overline{AP}$; if $\overline{AB}$ is to be the base, then the corresponding altitude is $\overline{CQ}$; and if $\overline{AC}$ is the base, $\overline{BR}$ is the altitude. Measure the lengths in centimetres and calculate the area of ABC using each side as the base in turn.

5.3 Area of Circles and Other Nonpolygonal Figures

It is fundamentally much more difficult to find the areas of regions whose boundaries are not all straight lines. A circle is an important example of this problem. It is remarkable that there is a simple formula for the area of a circle. But before discussing circles, let us look at an arbitrary figure and discuss the general principle used to find its area.

Figure 5.18
What is the area of this glob?

We must approach this problem by approximation. In Figure 5.19, we have set a grid with unit squares on top of the figure.

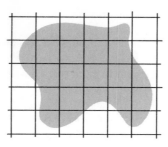

Figure 5.19

Now we count the number of square units in the grid that are entirely within the boundary. (See heavy outline inside the glob in Figure 5.20a.)

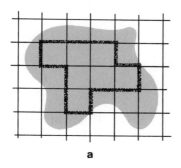

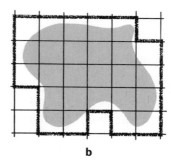

a

b

Figure 5.20
Area of glob in Figure 5.18. The area is at least _____ square units. The area is at most _____ square units.

The area of the region is *at least* this size.

Now we count the number of square units that just barely enclose the region. (See heavy outline in Figure 5.20b.) The area of the region is *at most* this size.

So the area is

between 7 and 26 square units.

These estimates are rough. We can do better with a finer grid. Suppose each of the squares in the grid is subdivided into four squares. Now place this over the glob as before.

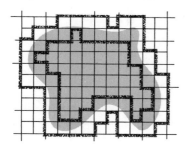

Figure 5.21
Each large square has been subdivided into four equal squares.

For these new units, as you can count,

the area is at least 47 square units and at most 95 square units.

Because there are four new squares in each old square, in terms of the old units,

47 new units = $\boxed{4\,7}\,\boxed{\div}\,\boxed{4}\,\boxed{=}\,\boxed{1\,1.75}$ square units;

95 new units = $\boxed{95}\,\boxed{\div}\,\boxed{4}\,\boxed{=}\,\boxed{23.75}$ square units.

So in terms of the original units (using the finer screen or mesh), the area is

between $11\frac{3}{4}$ and $23\frac{3}{4}$ square units.

Contrast these measures with the original estimates:

between 7 and 26 square units.

By using finer grids, better estimates are found. We might even try subdividing each side of the unit square into 10 parts. Then, each unit would be subdivided into

(10 × 10) or 100 new squares.

By refining these techniques, we can estimate the area of a circle. The area of a circle of radius 1 is, by definition, a number called π. (This is a Greek letter called *pi*—pronounced "pie.") We can get approximations for π from this geometric point of view. The radius of a circle is the distance from its center to any point on the circle.

Example 5.9: Here is a circle 1 inch in radius. What is its area in square inches?

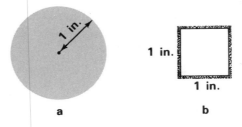

a   b

Figure 5.22

Solution: Figure 5.22b shows a sketch of a unit square. Essentially, we wish to compare the size of Figure 5.22a with Figure 5.22b. "How many times the area of the square is the area of the circle?"

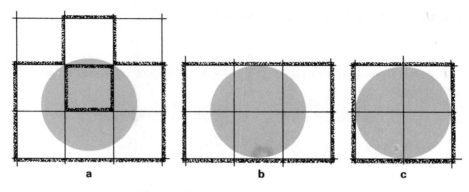

Figure 5.23
First estimates for the area of a circle.

From the grid in Figure 5.23a, we see that the area of the circle is

at least 1 square inch and at most 7 square inches.

This is surely not a very good estimate! By moving the grid a bit (Figure 5.23b,) we get a better estimate. The area is

at most 6 square inches.

By another repositioning of the grid, the area is

at most 4 square inches.

Now, we have the area

between 1 square inch and 4 square inches.

Let us try some subdivisions. Each side of the grid in Figure 5.24 is half as long as the unit square in Figure 5.22. So there are (2 × 2) or 4 unit squares in each old unit.

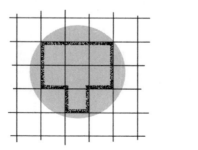

Figure 5.24
Second estimates for area of circle.

The area of the circle is

at least 7 new units and at most 21 new units.

Because there are 4 new units in each old unit, in terms of the old units the area is

at least $(7 \div 4)$ old units and at most $(21 \div 4)$ old units.

That is,

at least $\dfrac{7}{4} = 1\dfrac{3}{4}$ sq. in. and at most $\dfrac{21}{4} = 5\dfrac{1}{4}$ sq. in.

Before, we found the area to be at least 1 sq. in.; now, we know it is at least $1\frac{3}{4}$ sq. in. So this is a bit more information than we had. Before, we also found the area to be at most 4 sq. in.; here, we have the area at most $5\frac{1}{4}$ sq. in. If it is at most 4 sq. in., it is automatically no more than (at most) $5\frac{1}{4}$ sq. in. So

$$1\dfrac{3}{4} \text{ sq. in.} < \text{area of circle} < 4 \text{ sq. in.}$$

Let us try to improve the situation still more. Subdivide each side of the unit square into 10 parts (so there are now (10×10) or 100 square units in 1 square inch).

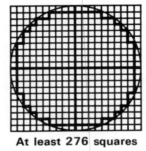

At least 276 squares   **At most 332 squares**

Figure 5.25
Third estimates for area of circle.

Thus

$$276 \text{ new units} < \text{area} < 332 \text{ new units,}$$

or, in square inches,

$$\dfrac{276}{100} < \text{area} < \dfrac{332}{100}$$

That is,

$$2.76 \text{ sq. in.} < \text{area} < 3.32 \text{ sq. in.}$$

How much have we improved in the estimates? To find out, we find the difference between the low and high values for each and arrange the information in the following table.

Unit	High	Low	Difference
1 sq. in. (no subdivisions)	4	1	3 sq. in.
$\frac{1}{4}$ sq. in.	4	$1\frac{3}{4} = 1.75$	2.25 sq. in.
$\frac{1}{100}$ sq. in.	3.32	2.76	0.56 sq. in.

So there has been decided improvement! But for this, we have to pay a price. The counting of squares is getting to be unwieldy. How can we make things easier?

Fortunately, the circle is a symmetrical figure, and we can take advantage of this fact. As shown in Figure 5.26, we only need to count the squares in a quarter of the circle and multiply this number by four.

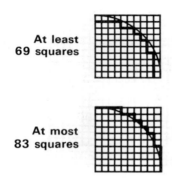

At least 69 squares

At most 83 squares

Figure 5.26
The area of the circle is four times the area of a quarter circle.

Moreover, we can separate the squares in convenient blocks and use the calculator to compute each rectangular block as indicated in Figure 5.27 for the low estimate.

$$
\begin{array}{rcl}
4 \times 9 &=& 36 \\
2 \times 8 &=& 16 \\
1 \times 7 &=& 7 \\
1 \times 6 &=& 6 \\
1 \times 4 &=& \underline{4} \\
&& 69
\end{array}
$$

Figure 5.27
Counting squares.

Example 5.10: Find the area of a circle $\frac{1}{2}$ inch in radius.

Solution: Using 100 subdivisions in a square inch and with the same techniques as above, we estimate that the area is

Figure 5.28

between 60 squares and 88 squares,

or

between 0.60 sq. in. and 0.88 sq. in.

Notice that 60 to 88 squares are almost the same numbers of squares as there were in each of the quarter circles in Figure 5.26.

Example 5.11: Find the area of a circle with a 2″ radius.

Solution: Take quarter circles as shown in Figure 5.29.

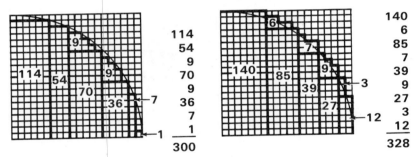

Figure 5.29

The area is between 1200 squares and 1312 squares, or

12 sq. in. $<$ area $<$ 13.12 sq. in.

Notice that the estimate for a quarter of this circle,

between 300 squares and 328 squares

or

3 sq. in. $<$ area $<$ 3.28 sq. in.

is almost the same as the area of the circle of 1 inch radius.

Examples 5.10 and 5.11 seem to indicate that by doubling the radius of the circle, the area is increased four-fold. It can be proved that this is indeed the case:

the area of a circle is a multiple of the square of its radius.

In fact, it can be shown that

the area of a circle is equal to π times the square of its radius.

In symbols,

$$A = \pi \times r^2$$

or $$A = \pi \times r \times r.$$

As this is the case, then when $r = 1$,

$$A = \pi \times 1^2$$

$$= \pi,$$

which agrees with our definition for the area of a circle 1 unit in radius. Thus we have an experimental way to estimate π. From Example 5.9, π is between 2.76 and 3.32. By using other methods, to 8 digits,

$$\pi \doteq 3.1415927$$

For calculators, an excellent approximation is

$$\pi \doteq \frac{355}{113} = 355 \div 113$$

$$355 \; \boxed{\div} \; 1 \, 1 \, 3 \; \boxed{=} \; 3.1415929$$

correct to 6 decimal places.

Example 5.12a: Find the area of a circle with a 6 inch *radius*.

Solution: $r = 6$; $\pi \doteq 355 \div 113$

$$6 \; \boxed{\times} \; 6 \; \boxed{\times} \; 355 \; \boxed{\div} \; 1 \, 1 \, 3 \; \boxed{=} \; 1 \, 1 \, 3.09734$$

Therefore, the area of the circle is about 113.097 sq. in.

Example 5.12b: Find the area of a circle with a 7 foot *diameter*.

Solution: Because the diameter is 7, the radius, $r = 7 \div 2 = \frac{7}{2}$. What we want is

$$A = \pi r^2 = \pi \times \frac{7}{2} \times \frac{7}{2}$$

$$7 \; \boxed{\div} \; 2 \; \boxed{\times} \; 7 \; \boxed{\div} \; 2 \; \boxed{\times} \; 355 \; \boxed{\div} \; 113 \; \boxed{=} \; 38.484513$$

So the area rounded to two decimals is 38.48 sq. ft.

Example 5.13: A cow is tied to a stake at the outside corner of a barn. If the barn measures 20 ft by 30 ft and the tether is 25 ft (Figure 5.30), over what area can she graze?

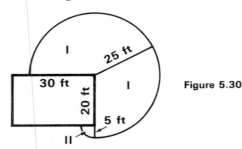

Figure 5.30

Solution: From Figure 5.30, we see that we can split the grazing area into two regions:

    I: three-quarters of a circle with radius 25 ft;
    II: one-quarter of a circle with radius 5 ft ($= 25 - 20$).

To calculate I:

$$3 \; \boxed{\div} \; 4 \; \boxed{\times} \; 25 \; \boxed{\times} \; 25 \; \boxed{\times} \; 355 \; \boxed{\div} \; 113 \; \boxed{=} \; 1472.6216$$

To calculate II:

$$1 \; \boxed{\div} \; 4 \; \boxed{\times} \; 5 \; \boxed{\times} \; 5 \; \boxed{\times} \; 355 \; \boxed{\div} \; 113 \; \boxed{=} \; 19.634955$$

Now add:

$$1472.6216 \; \boxed{+} \; 19.634955 \; \boxed{=} \; 1492.2565$$

So the cow can graze over about 1492 square feet of land.

Finding the perimeters of regions whose boundaries are not straight lines is even more difficult than finding their areas! The distance around a circle is called its *circumference,* and it is given by the formula

$$C = \pi \times d \quad \text{or} \quad C = 2 \times \pi \times r$$

1. Measure the radii of these circles and compute their areas in square centimetres.

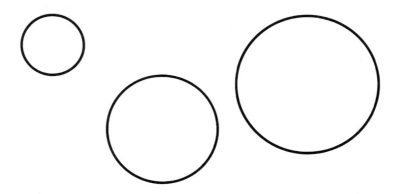

2. Using the grid technique, find the area of this ellipse in square centimetres.

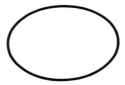

•3. A figure consists of a semicircle having two smaller semicircles cut from it.

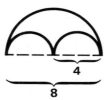

a. Find the area if the radius if the large semicircle is 8 cm and the radius of each smaller semicircle is 4 cm.

b. Find the area if the large semicircle has radius 11 cm and each of the smaller has radius $\frac{11}{2}$ cm.

c. From the results in parts a and b, if the radius of the large semicircle is twice the radius of each smaller one, is the area cut away greater than the area that is left?

d. Find the area of the "cashew nut" that is formed by flipping one of the smaller semicircles upside down. (Try to do this the easy way.)

4. Suppose Mrs. O'Leary's cow in Example 5.13 were tied to the middle of a side of the barn instead of at the corner. The tether is still 25 feet long and the barn is still the same size.

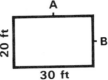

•a. Find the area over which she can graze if she were tied at A, where A is the midpoint of a long side.

b. Find the area over which she can graze if she were tied at B, where B is the midpoint of a short side.

•5. Suppose you know that the area of a circle is 120.7016 cm² (sq. cm) and that its circumference is 38.936. Find the radius of this circle without using π in the computations. (HINT: From the relations

$$A = \pi \times r^2 \quad \text{and} \quad C = 2 \times \pi \times r,$$

get a comparison between the measures of the area and the circumference.)

6. Find the area of the moon-shaped region shaded in the figure at the right. It is bounded by a semicircle having the diagonal of the square as diameter and by a quarter-circle

with a side of the square for radius. Such figures are called lunes. HINT: From Chapter 12, if the side of a square measures 4 cm, then the diagonal, d, would be such that

$$d^2 = 4^2 + 4^2 = 32,$$

so
$$d = 4\sqrt{2}.$$

5.4 Surfaces—in the Plane and in Space

When we talked about area before as a measure for flat space, we were about half right. Geometric regions in a plane are measured by areas. But we also mentioned an application like painting a house. Here, even though each area concerned lies in a plane, many planes may be involved. In other words, not all the measurements are in a single plane throughout.

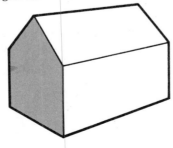

Figure 5.31
What is the area of this house?

In our "house" in Figure 5.31, each slanted roof is in a different plane; each wall is in a different plane; the floor is in yet another plane. Our house is in 3-D! Even so, in this case, each panel is flat, and the areas can be calculated if we have the right measurements. Let's start with the simple task of painting a room first.

Example 5.14: A room, in the form of a rectangular box 14 feet by 20 feet by 9 feet, is to be painted (see Figure 5.32). Estimate the area to be painted.

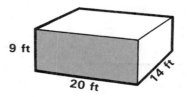

Figure 5.32

To make things simple, we shall not worry about windows and doors at this time.

Before working on this example, let us talk about it a bit. First, we must decide whether we are only painting the four walls. Is the ceiling to be included? What about the floor?

At this point, let us decide to paint the ceiling also, but not bother with the floor.

Next, we need to visualize the areas drawn on paper. Some people have a knack of visualizing from such sketches. Others need more details. Let us redraw the room (at the left, Figure 5.33), labeling the different corners of the room. Next, slit the vertical seams AE, BF, CG, DH, and drop the walls to the floor.

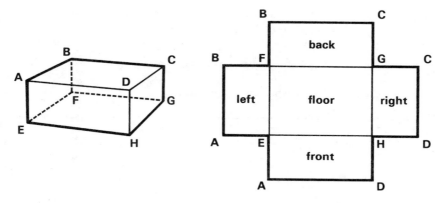

Figure 5.33
Room with the walls knocked down.

What happened to the ceiling? We were so busy with the walls, we forgot all about it! Notice in the sketch at the left, that the ceiling can be hinged

to the left wall (at AB);
or to the right wall (at DC);
or to the front wall (at AD);
or to the back wall (at BC).

It really makes no difference. Let's hinge it to the right wall at CD and flop it on its back.

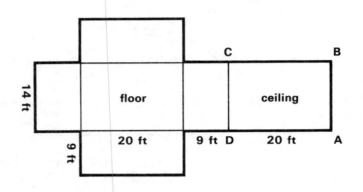

Figure 5.34

Example 5.14 (continued): Now we are ready to pick up on the original problem. Some of the panels come in pairs, so we can do both at the same time in these cases. This accounts for the extra factor of 2 in the calculations:

the front and back: $(9 \times 20) \times 2 = 360$.
the left and right: $(9 \times 14) \times 2 = 252$.
the ceiling: $14 \times 20 = 280$.

total: $360 + 252 + 280 = 892$.

So 892 square feet of space is to be painted.

Example 5.15: A gallon of paint covers about 400 square feet in area. If enamel paint costs $11 per gallon and prime coating costs $6.60 per gallon, how much would it cost to get enough paint and prime coat to give the room in Example 5.14 one coat each? Suppose paint can be bought only in gallon size in this store.

Solution: Because the area is 892 sq. ft., two gallons of each would not be enough:

$$2 \ \boxed{\times} \ 400 \ \boxed{=} \ 800.$$

and
$$800 < 892.$$

So we must have 3 gallons each.
 At $11 per gallon,

$$3 \ \boxed{\times} \ 11 \ \boxed{=} \ 33. \qquad \text{(for the enamel)}$$

and at $6.60/gal.,

$$3 \ \boxed{\times} \ 6.60 \ \boxed{=} \ 19.80 \qquad \text{(for the primer)}$$

The total is

$$\$33 + \$19.80 = \$52.80.$$

It may be found that, in practice, 1 gallon of paint does not cover exactly 400 square feet. But these calculations give a good estimate of how much paint will be used.

Now that we have painted a room, we're ready for bigger things—like painting the whole house!

Example 5.16: The entire outside of a house is to be painted the same color. Disregarding doors and windows for this purpose, what is the area to be covered if the dimensions are as in Figure 5.35?

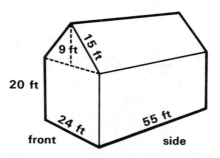

Figure 5.35

Solution: The easy things first:

The two sides: $(20 \boxtimes 55) \boxtimes 2 \boxminus 2200.$

The lower half of the front and back: $(20 \boxtimes 24) \boxtimes 2 \boxminus 960.$

The gables are triangles with altitude 9 and base 24: $1 \boxplus 2 \boxtimes 24 \boxtimes 9 \boxtimes 2 \boxminus 216.$

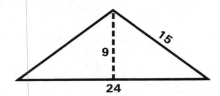

Each gable: $\frac{1}{2} \times (24 \times 9) \times 2 = 216$

The roof consists of two rectangles, each 55 feet long and 15 feet wide: $(55 \boxtimes 15) \boxtimes 2 \boxminus 1650.$

Therefore, the total area is

$$2200 \boxplus 960 \boxplus 216 \boxplus 1650 \boxminus 5026.$$

At 400 sq. ft./gal.,

$$5026 \boxdiv 400 \boxminus 12.565$$

or about 13 gallons of paint for each coat.

Occasionally, we have to worry about curved surfaces such as the sides of a can.

Example 5.17: A large can of fruit juice is in the form of a cylinder with radius 2 in. and height $6\frac{11}{16}$ in. How much tin goes into making one such can?

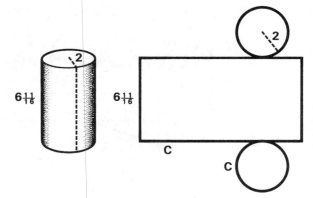

Figure 5.36

Solution: In our imagination, we open the top and bottom of the can and also slit it along the dashed line in Figure 5.36. Then, we flatten out the can by unrolling it.

Thus, the side wall of the can is unrolled into a rectangle whose width is equal to the height of the can. The length is exactly the distance around the circle—that is, the *circumference*.

There is a simple relationship between the circumference and the diameter:

$$\text{circumference} = \text{diameter} \times \pi.$$

That is, the distance around the circle is a bit more than three times the distance across. In this case,

$$\text{circumference} = \text{diameter} \times \pi = 4 \times \pi.$$

On your calculator, using $\pi \doteq 355 \div 113$, the area of the side is

$$(\boxed{1\,1} \boxed{\div} \boxed{1\,6} \boxed{+} \boxed{6}) \boxed{\times} \boxed{4} \boxed{\times} \boxed{3\,5\,5} \boxed{\div} \boxed{1\,1\,3} \boxed{=} \boxed{8\,4.0\,3\,7\,6\,1}$$

The areas of the top and bottom are

$$2 \times r \times r \times \pi = \boxed{2} \boxed{\times} \boxed{2} \boxed{\times} \boxed{2} \boxed{\times} \boxed{3\,5\,5} \boxed{\div} \boxed{1\,1\,3} \boxed{=} \boxed{2\,5.1\,3\,2\,7\,4\,3}$$

Adding these together,

$$\text{total area} = 109.17035$$

Therefore, the total area is about 109.17 square inches.

Try this experiment. Cut out a circle of, say, 4 in. radius. Cut out the quarter-circle shown shaded in Figure 5.37. Tape the edges AC and BC together. You should get a cone similar to that shown on the right.

Figure 5.37
Making a model of a cone. What is its area?

Example 5.18: A cone is made by cutting away $\frac{1}{3}$ of a circular region from a sheet of tin. If the slant height measures $6''$, how much tin goes into making this funnel?

Solution: Because the slant height corresponds to the radius in Figure 5.37, we want to calculate the area of $\frac{2}{3}$ of a circle with radius 6. The area of the entire circle is given by the formula,

$$A = \pi \times r^2.$$

We want $\frac{2}{3}$ of this area.

Area of cone = area of circle $\times\ (2 \div 3)$

$$= 355\ \boxed{\div}\ 1\ 1\ 3\ \boxed{\times}\ 6\ \boxed{\times}\ 6\ \boxed{\times}\ 2\ \boxed{\div}\ 3\ \boxed{=}\ 75.398226$$

So the area is about 75 sq. in.

Although the last two examples are "curved" in 3-space, the regions are both capable of being flattened out. A shape that *cannot* be flattened out physically is a sphere. Parts of a sphere, such as a hemisphere (half-sphere) and a spherical wedge also cannot be flattened out. (A spherical wedge is like a slice of orange.)

Formulas for the surface area of a sphere have been determined:

$$A = 4 \times \pi \times r^2 = \pi \times d^2.$$

Example 5.19: The radius of a balloon is 45 feet. How many square yards of material go into making this balloon?

Solution: Because the radius is 45 ft., in yards it is

$$45\ \boxed{\div}\ 3\ \boxed{=}\ 15.$$

So $r = 15$ and

$$\text{area} = 4 \times \pi \times 15^2$$

Hence

$$\text{area} = 4\ \boxed{\times}\ 355\ \boxed{\div}\ 1\ 1\ 3\ \boxed{\times}\ 15\ \boxed{\times}\ 15\ \boxed{=}\ 2827.4334$$

So the balloon used about 2827.43 square yards of material.

Exercises 5.4

•1. A box measures 2.3 cm by 3.4 cm by 4.5 cm. Find the surface area of this closed box.

2. Find the area of the box in Exercise 1 as an open box—
   •a. sitting on the 3.4 cm by 4.5 cm face;
   b. sitting on the 2.3 cm by 4.5 cm face;
   •c. sitting on the 2.3 cm by 3.4 cm face.

3. Sketch the flattened-out box in Figure 5.33 with the ceiling hinged to the back wall at BC. Label all the corners with the appropriate letters and the measurements of each side.

4. In Example 5.15, suppose that, when you buy 5 gallons or more of the enamel, instead of $11 per gallon, the price is reduced to $10.25/gal. Explain why it would be more economical to just give the room two coats of enamel rather than one coat each of enamel and primer.

•5. Find the surface area of the house in Example 5.16 if the slanted roof is not to be included.

6. Suppose part of the tin can in Example 5.17 may be replaced by less expensive cardboard. Due to consideration of strength, the entire can cannot be replaced by cardboard, but either the two round ends can be replaced or the side wall. Which would be the more economical replacement?

7. A piece of filter paper for coffee is made in the shape of a cone from $\frac{5}{12}$ of a circular region pasted together similar to Example 5.18.

    a. Find the area of the filter if the slant height is 14 cm.

    b. Suppose the paper can filter the coffee at the rate of 90 square centimetres per minute. How long would it take to filter coffee filled to the brim?

•8. A balloon is inflated from a radius of 6 feet to 7 feet. How much has the surface area been increased?

9. A 40-watt tungsten light bulb (everyday type) can throw out 206 candlepower of light per square centimetre. If the bulb is 3 cm in radius, how many candlepower can it emit? Assume the light bulb to be spherical in shape.

5.5 Measuring in Three Dimensions

The objects we handle have thickness as well as length and width. Geometric objects having length, width, and thickness are called geometric *solids*. A "little brown jug" is a solid. How much a little brown jug holds is a question of the *volume* of a solid. At this time, we want to derive some simple formulas for volumes.

As before, we start with the simpler shapes and work our way to more complicated shapes. One simple shape is a shoe box (see Figure 5.38). We refer to it as a *rectangular solid* or a rectangular box.

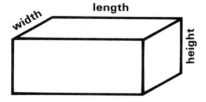

Figure 5.38
A rectangular solid. What is its volume?

Most rooms that we live in are rectangular boxes. In these cases, all the walls, the ceiling, and the floor are in the shape of rectangles. Each of these is called a *face* of the rectangular solid. Opposite faces have exactly the same measurements. Along the edges, we measure the length, width, and the height of the box.

To find the *volume* of such figures, follow a procedure like the one for area. To fix ideas, think of a rectangular block of cheese—say, 5 inches long, 4 inches wide, 3 inches high (see Figure 5.39).

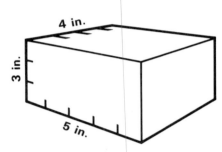

Figure 5.39
A rectangular solid measuring 5 in.
× 4 in. × 3 in. What is its volume?

Because the block measures 5 inches across, we can mark off the length at 1 inch intervals and think about slicing it in the fashion of Figure 5.40.

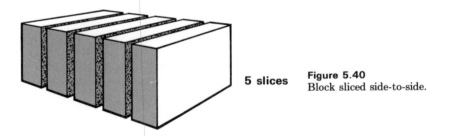

5 slices　**Figure 5.40**
Block sliced side-to-side.

Or we can slice it front-to-back in four 1 inch slabs (Figure 5.41).

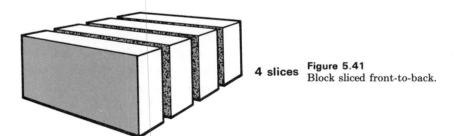

4 slices　**Figure 5.41**
Block sliced front-to-back.

Or in three 1 inch layers (Figure 5.42).

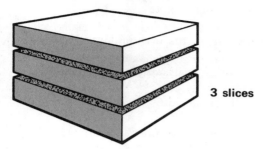

3 slices

Figure 5.42
Block sliced top-to-bottom.

Or, in all three directions (Figure 5.43).

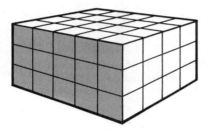

Figure 5.43
Block sliced all three ways
into small cubes.

Example 5.20: Find the volume of a rectangular box, 5 by 4 by 3.

Solution: One layer consists of 4 rows of little blocks. Each row has 5 blocks.

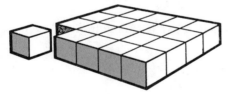

Figure 5.44
One layer of cubes.

So each layer has (4 × 5) or 20 blocks. Because there are 3 layers, there are altogether, 60 blocks:

$$3 \boxed{\times} (4 \boxed{\times} 5) \boxed{=} 60.$$

Now, each small block in the above example is a cube that measures

1 inch by 1 inch by 1 inch

and is one *cubic inch* in volume. Thus, this little cube is a unit of measure for volume.

The volume of the rectangular block in Example 5.20 is thus

60 cubic inches   (abbreviated 60 cu. in.).

Even if the measurements were not whole numbers, we can argue (as was in the case of areas) that, in general,

volume of a box = length × width × height (cubic units).

As with area, we shall insist that each measure be recorded in the same unit of length.

Let us start again with the rectangular solid measuring 5″ by 4″ by 3″. From this, we shall derive some other formulas.

We slice this solid diagonally as shown in Figure 5.45.

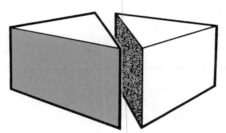

Figure 5.45
Rectangular box sliced diagonally.

Clearly, the volume has been halved. That is, the volume of each half is

$$V = \frac{1}{2} \times \ell \times w \times h$$

$$= \left(\frac{1}{2} \times \ell \times w \right) \times h$$

Notice that the factor, $\frac{1}{2} \times \ell \times w$, is the area of the triangular face on the bottom (the base.) Hence we have

volume = (area of base) × height = $B \times h$,

where B is the area of the base. In fact, the base does not have to be a triangle; this formula is equally applicable.

The triangular wedge considered is an example of a prism. A *prism* is a solid with two faces that are of the same size and shape (its bases) and with parallelograms for the other faces. Its volume, as indicated here, is given by the formula above.

Example 5.21: A hexagonal prism. Find the volume of a prism whose bases are hexagons, each with area 36.54 sq. in. and with height 13.17 in.

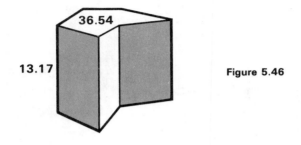

36.54

13.17

Figure 5.46

Solution:

$$\text{Volume} = 36.54 \boxed{\times} 13.17 \boxed{=} 481.2318$$

Therefore the volume is 481.2318 cubic inches.

Notice that the edges along the sides of the prism (the *lateral edges*) do not have to be perpendicular to the bases. They can be at a slant. But the height, h, is measured perpendicular to the bases.

A theorem due to Buonaventura Cavalieri (1598–1647) known as *Cavalieri's Principle* states that—

if two solids have the same altitude and parallel cross-sections have the same area, then the solids have the same volume (see Figure 5.47).

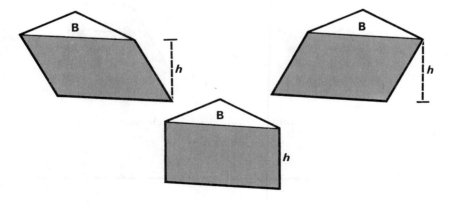

Figure 5.47

We can picture Cavalieri's principle in the following way.

Think of a stack of playing cards stacked straight up or stacked out of kilter. Either way, the two stacks occupy the same volume (see Figure 5.48).

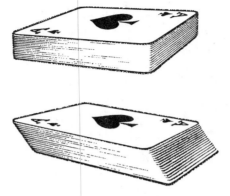

Figure 5.48
A model for Cavalieri's Principle.

The principle is even more general. The cross-sections do not have to be all the same within the stack—as long as each cross-section has a matching one in the other stack.

If the solid is such that the bases are curved as in Figure 5.49 (rather than polygons) and the sides are all parallel, then it is a *cylinder*. If the bases of the cylinder are circles, then it is a circular cylinder. The formula

$$V = B \times h$$

still holds.

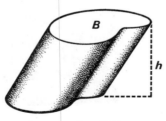
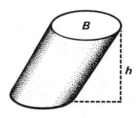

A general cylinder   **A circular cylinder**

Figure 5.49
Cylinders.

Example 5.22: Find the volume of a circular cylinder with a radius of 5 centimetres and a height of 7 centimetres.

Solution: Because $V = B \times h$, first find the area of a circle 5 cm in radius:

$$B = 5^2 \times \pi$$

$$\doteq 5 \ \boxed{\times} \ 5 \ \boxed{\times} \ 3.55 \ \boxed{\div} \ 1 \, 1 \, 3 \ \boxed{=} \ 78.54 \ (\text{cm}^2)$$

then, multiply by the height, 7. Altogether, then,

$$\text{volume} = 5 \ \boxed{\times} \ 5 \ \boxed{\times} \ 3.55 \ \boxed{\div} \ 1 \, 1 \, 3 \ \boxed{\times} \ 7 \ \boxed{=} \ 549.778 \, 76 \ (\text{cm}^3)$$

So the volume is about 549.78 cubic centimetres.

As we can see from Example 5.22, if the radius of a circular cylinder is r and its height is h, then

$$B = \pi \times r^2$$

and $$V = B \times h = \pi \times r^2 \times h.$$

In the previous section, we constructed a cone and found its area. The *volume* of a cone can be shown to be $\frac{1}{3}$ the volume of the circular cylinder having the same radius r and the same height h. So

$$V = \frac{1}{3} \times \pi \times r^2 \times h.$$

Example 5.23: Find the volume of a cone with radius 5 cm and height 7 cm.

Solution: $V = \frac{1}{3} \times \pi \times r^2 \times h.$
  Because the quantity $\pi \times r^2 \times h$ has been calculated for the circular cylinder in Example 5.22 already, we continue from that result:

$$V = 549.778 \, 76 \ \boxed{\div} \ 3 \ \boxed{=} \ 183.25958$$

So the volume is about 183.26 cm³ (cc.)

A very important solid is the sphere.

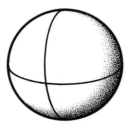

Figure 5.50

A balloon is spherical or nearly so; a bathysphere (diving bell) is spherical or nearly so. The volume of a sphere is known to be

$$V = \frac{4}{3} \times \pi \times r^3.$$

Example 5.24: How much gas is needed to inflate a balloon to a radius of 6 feet?

Solution: $V = \frac{4}{3} \times \pi \times r^3.$

Volume = $(4 \div 3) \times (355 \div 113) \times 6 \times 6 \times 6 = 904.7787$

Therefore, the volume of the balloon (which is also how much gas is required) is about 904.78 cubic feet.

Exercises 5.5

1. Find the volume of each of the following solids.
    •a. A rectangular solid 23.54 cm by 9.23 cm by 9.23 cm.
    •b. A rectangular solid 2 ft. 3 in. by 13.3 in. by 1 foot. (HINT: Express each dimension in a common unit—say, in inches.)
     c. A triangular prism whose triangular base has a base of 4.2″ and an altitude of 5.3″; the altitude of the prism is 3.1″.
     d. A hexagonal prism with $B = 53.29$ sq. cm. (cm²) and altitude 15.43 cm.
    •e. A circular cylinder with $r = 2.32″$ and $h = 6.51″$.
     f. A cone with $r = 3.22″$ and $h = 8.67″$.

2. How many cubic centimetres are there in 1 cubic inch?

3. Estimate the volume of your house in cubic feet; in cubic metres. (Think of a metre as a bit longer than a yard.)

• 4. The Earth has a radius of about 4,000 miles. What is its volume in cubic miles?

5. The Sun has a diameter of about 864,000 miles.

   •a. Find the volume of the Sun in cubic miles.

   b. How many times as large as the Earth is the Sun in volume?

6. Find the difference in volume between a 6″ cube and a 7″ cube. (A 6″ cube means a cube that is 6 inches on a side.)

• 7. Refer to Example 5.24.

   a. How much gas is needed to inflate a balloon from a radius of 6 ft. to a radius of 7 ft.? (Use the result in example 5.24 to help solve this problem.)

   b. Find the ratio of increase in volume of the balloon in part a:

$$\text{ratio} = \frac{\text{increase in volume}}{\text{volume at 6 foot radius}}.$$

8. By a procedure similar to that in Exercise 7b, find the ratio of increase in volume from a 6″ cube to a 7″ cube. Compare this answer with that in Exercise 7b.

9. Find the volume of a hemisphere (half of a sphere) whose radius is 5.13 cm.

10. A rotunda has the shape of a hemisphere just fitting on top of a cylinder. If the diameter is 64 feet and the height of the vertical wall is 23 feet, what is the volume of space in this room?

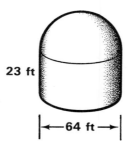

23 ft

←—64 ft —→

•11. Before pocket calculators were readily available, and when calculations had to be done by hand, an approximation for π was $3\frac{1}{7}$.

   a. Find the difference between $3\frac{1}{7}$ and the more accurate approximation, 3.1415927. ($3\frac{1}{7} > 3.1415927$.)

   b. Find the ratio of the difference $3\frac{1}{7}$ is from the more accurate approximation, 3.1415927. (This ratio is the answer to part a divided by 3.1415927.)

•12. A large can of fruit juice measures 4 in. in diameter and $6\frac{11}{16}$ in. in height. It is claimed to hold 46 fluid ounces—that is, 1 qt, 14 oz. If the measurements are accurate, is the content over or under its claim? By how many cubic centimetres? (1 qt = 960 cm³ = 32 fluid oz; so 1 oz = 30 cm³.)

13. Find the *area* of the triangular prism formed by slicing the box in Figure 5.45 along a diagonal as illustrated in Figure 5.51. The length of the diagonal is approximately 6.4 inches.

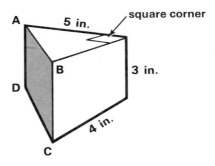

Figure 5.51

14. Find the area of the prism in Exercise 12 if it is an open box with—

a. one of the triangular faces as a base;

b. the rectangular face ABCD as an open lid.

Negative Numbers

6.1 The Old "Mirror-in-the-Wall" Routine

In this chapter, we study negative numbers. To carry out many of the operations of this chapter on your calculator, it should have a way of entering negative numbers. Sometimes this is done with a *"change-sign"* key shown as ⊠ or **CHS.** If you have such a key, you will be able to do all the basic operations in this chapter. If not, you will have to experiment to see what is possible on your machine.

Hold your arm in front of a mirror. The mirror shows you holding out your arm with the illusion of depth. For example, the image of your hand appears closer than your shoulder.

Figure 6.1
A mirror suggests negative numbers.

The mirror faithfully reports the distance of objects in its reflection. Near things appear near; far things appear far; things at a middle distance appear in between.

Let's put a mirror on the following number line.

Where shall we put the mirror? There is a great emptiness to the left of 0; let's put the mirror at 0 and see what the numbers see when they look into the mirror.

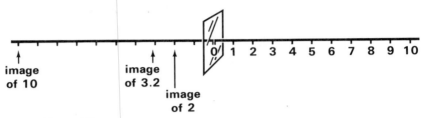

Figure 6.2
Creation of negative numbers as mirror images of counting numbers.

Where shall we show the image of 2? of 10? of 3.2? Surely, the image of 2 should be 2 units to the left of 0, the image of 10 should be 10 units to the left of 0, and the image of 3.2 should be in between—at 3.2 units to the left of zero.

By putting a mirror at zero, we are led to create an image for each number to the right of zero. It is traditional to call the numbers of these images *negative numbers,* and to denote them with a minus (−) sign preceding the number. The old numbers pictured to the right of zero on the number line are now called *positive numbers,* and are sometimes denoted with a + sign preceding the number for special emphasis. The number, 0, is neither positive nor negative.

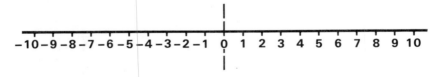

With the device of a mirror at zero, we now have a number for every point on the number line.

It is convenient to extend this notion of mirror image and to think of the mirror as "two-way." Thus, placing a minus sign before any number simply denotes its mirror image.

-2 is the mirror image of 2.
$-(-2)$ is the mirror image of -2; and so it is 2:

$$-(-2) = 2.$$

$-(2+3)$ is the mirror image of $(2+3)$; that is, of 5:

$$-(2+3) = -5.$$

$-(-(-2))$ is the mirror image of $-(-2)$, of 2; and so it is -2:

$$-(-(-2)) = -2.$$

It is worth noting that, for example,

$$-13.2378 \text{ means } -(13 + 0.2378)$$

and so -13.2378 denotes a point to the left of -13 (is smaller than -13). As 13.2378 is between 13 and 14, -13.2378 is between -13 and -14. On the number line, -13.2378 is therefore to the right of (is larger than) -14 as shown in Figure 6.3.

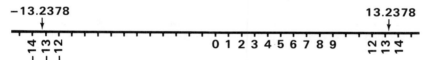

Figure 6.3
The location of negative decimal numbers.

An application of negative numbers is the temperature scale. Both Fahrenheit and Celsius (centigrade) scales have a zero point. Temperatures are reported "above zero" $(+)$ or "below zero" $(-)$.

In business, profits and losses can be interpreted in terms of positive (profit) and negative (loss) numbers. Where there is neither profit nor loss, the value is zero (breaking even).

To enter a negative number on your calculator,

Key in the number. Press ⊠ key. For example,

$$2 \boxtimes -2.$$

Many inexpensive machines enter a negative number by the following sequence:

$$-, \text{ number}, +.$$

For these machines, here is the way the number -2 is entered:

Action	Display
Turn ON	$\mathit{0.}$
Press $\boxed{-}$	$\mathit{0.}$
Key in $\mathit{2}$	$\mathit{2.}$
Press $\boxed{+}$	$\mathit{-2.}$

Exercises 6.1

1. Draw a number line. Locate each of the points listed below and its mirror image with an appropriate ($+$ or $-$) sign.

   •a. 4.8

   b. 0.5

   •c. -3

   •d. $(2.785 - 1.632)$

   e. -0.9

   •f. $5 \div 3$ (use calculator)

   g. -6.75

   h. $-(7 \times 6)$

2. For each of the following pairs of numbers, determine which is the smaller—that is, which lies to the left of the other.

   a. 78, 87

   •b. -78, -87

   •c. 2, -3

   d. 0, -5

   •e. $-(3 \div 2)$, $-(5 \div 2)$

   f. -1000, -100

3. On a cold winter day, the temperature was $-10°$ F. at 6 A.M. but had warmed to $20°$ F. by noon. How many degrees had it warmed up? Show this on the number line. Show it on your calculator:

$$20 - (-10) = \mathit{20} \boxed{+} \mathit{10} \boxed{=} \mathit{30.}$$

•4. On a summer day, the temperature at 6 A.M. was $65°$ F. If the day warmed up the same amount as the winter day in Exercise 3, how hot did it get?

5. A company showed net profits of $-\$212{,}000$ (a loss) one year, and net profits of $-\$287{,}000$ the following year. Is the company improving financially the second year, or is it doing worse?

6.2   Addition and Subtraction with Negative Numbers

Now that we have a new kind of numbers (negatives), we need to know how they operate. How do we add two numbers that are both negative? one positive and one negative? In particular, let us first ask, "What is $(-2) + (-3)$?"

In Chapter 2, we pictured addition of positive numbers on the number line in the following manner.

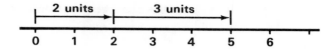

Now then, as mirror images, we should picture the addition of the negative numbers $(-2) + (-3)$ in a mirror-like fashion, thus:

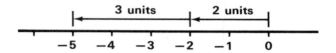

So, $(-2) + (-3) = -5$.

Let us see how this addition is done on the model of calculator that does not have the ⊠ key (or the **CHS** key.) Please refer to your instruction book if you have one of the other models.

Example 6.2: $(-2) + (-3) = ?$

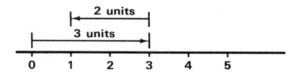

Subtraction with negative numbers is derived from comparison with subtraction of positive numbers. Here is a familiar example: $3 - 2 = 1$.

Thus, in subtracting 2 from 3 on the number line, we reach 1 by going 2 units to the left of 3. We show this by reversing the direction of the arrow. Note that the arrow to indicate subtraction is 2 points in the direction of -2. Thus, it is natural to define

$$3 - 2 = 3 + (-2) = 1.$$

That is, *adding* a *negative* number is the same as *subtracting* a *positive* one.

Now what about $2 - 3$, or, $2 + (-3)$? In Chapter 2, when we dealt only with positive numbers, there was no way to cope with this problem. With the full number line of positive and negative numbers, we can do it. With arrows, the answer can be determined by exactly the same scheme used to find $3 - 2$.

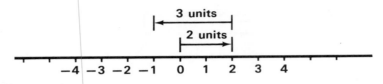

Figure 6.4
Subtraction of 3 shown by a left-pointing arrow.

Thus, $2 - 3 = -1$.

In summary, *adding* a *negative* number is the same as *subtracting* a *positive* one. To add any two numbers (positive or negative) on the number line, locate the first, then move in the direction of the sign of the second number (right for positive; left for negative), moving the number of units indicated.

In symbols:

$$A + (-B) = A - B.$$

To add more than two numbers, simply follow the same rule, moving in the directions indicated.

This rule may help if your calculator doesn't have a "change sign" key. Just use ⊟ instead of ⊞ when adding negative numbers.

Example 6.3: $2 + (-3) + 4 + (-6) = ?$

Solution: On the number line:

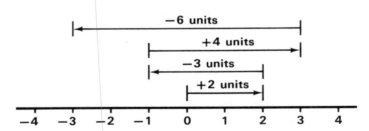

Figure 6.5
Addition of several numbers on the number line. Answer: -3.

On a calculator, it is a chain computation:

$$2 + (-3) + 4 + (-6) = ?$$

$$2 \boxplus \boxminus 3 \boxplus 4 \boxplus \boxminus 6 \boxminus = -3.$$

NOTE: Most machine algorithms in this chapter are written for models of calculators with Ⓩ or **CHS** keys. If your calculator gives erroneous answers when you follow these algorithms, consult your instruction book to find out how to enter negative numbers in your machine.

One of the most common applications of positive and negative numbers is their use with a credit–debit tally sheet, such as a personal checking account. All entries may be thought of as additions, deposits being + and payments (checks or withdrawals) being −.

Example 6.4: Suppose an account opened with a $100 deposit. What is the balance after these operations?

+100	(deposit)	
−48.55	(check for clothes)	
−36.78	(check for utilities)	
+132.64	(deposit paycheck)	
−222.00	(insurance payment)	

Solution:

100 ⊞ 48.55 **CHS**
⊞ 36.78 **CHS**
⊞ 132.64
⊞ 222 **CHS** ⊟ - 74.69

Balance: −$74.69 OVERDRAWN. (We hope this person has a reserve account to cover such overdrafts!)

Remember:

The sum $A + B$ can be computed whether or not A and B are positive numbers. The difference $A − B$ can be written as the sum $A + (−B)$.

Exercises 6.2

1. Perform the indicated additions or subtractions on your calculator. For each result in Column A, record its negative in (−A), and then show that the answer in the third column, (A*), is the same as that in (−A).

A	−A	A*
•a. $2.78 + 3.76 =$	$−(2.78 + 3.76) =$	$−2.78 − 3.76 =$
b. $5.36 − 3.89 =$	$−(5.36 − 3.89) =$	$−5.36 + 3.89 =$
•c. $1.78 − 2.15 =$	$−(1.78 − 2.15) =$	$−1.78 + 2.15 =$

2. Calculate the following (try each first without your calculator; then check).

A	$-$ A
•a. $3 - 0.5 =$	$0.5 - 3 =$
b. $0 - 0.5 =$	$0.5 - 0 =$
c. $8 - 8 =$	$-8 + 8 =$

3. Calculate.

•a. $5.46 - 3.14 + 1.41 - 6.78 =$

•b. $-5.46 + 3.14 - 1.41 + 6.78 =$

•c. $9876.3 - 3.6789 - 12.875 =$

•d. $-9876.3 + 3.6789 + 12.875 =$

e. $-6.32 - 7.82 =$

f. $6.32 + 7.82 =$

g. $-3.45 - (-1.32) =$

h. $3.45 + (-1.32) =$

4. Consult the financial pages of your paper. Select 10 stocks. Record their gains ($+$, up) or losses ($-$, down) as reported for the preceding day. If you had owned one share each of these stocks, what would have been your total gains or losses for the day? (You will find that the daily changes in prices are reported in fractions. For ease of calculations, you may wish to convert these to decimals.)

5. A miler runs four laps. The times elapsed at the completion of each lap follow:

Laps	Total elapsed time in minutes and seconds	Total elapsed time in seconds
Start	0:0	0
1	1:03	63
2	2:09.5	129.5
3	3:19.5	199.5
4	4:24	264

•a. How fast did he run each lap?

Compute: time for lap 2 $-$ time for lap 1 $=$
time for lap 3 $-$ time for lap 2 $=$
time for lap 4 $-$ time for lap 3 $=$

b. On which lap did he speed up?

•c. On which lap did he slow down?

Consider these two typical multiplication problems.

1. You earn $120 each week. What do you earn in a year?
   Answer: $120 × 52 (weeks) = $6240.

2. You buy a refrigerator on time. You make monthly payments of $15. How much do you pay in a year?
   Answer: $15 × 12 (months) = $180.

Considering the flow of money, there is a big difference in these two problems. In the first one, $120 was coming in to you each week. In the second one, $15 was being paid out by you each month. From a personal point of view, it is natural to think of money that comes in as + (positive) and money that goes out as − (negative).

Thus, in the first problem, the income to you can be thought of as +6240 dollars; in the second, the outgo may be considered as −180 dollars. We represent these two situations as follows:

$$(1) \qquad +120 \times 52 = +6240$$

$$(2) \qquad -15 \times 12 = -180$$

In both problems, time (weeks or months) was positive.
We have the rule:

> The product of a positive and a negative number is negative.

The rule may be justified, for example, by thinking of -15×12 in terms of a repeated addition of twelve (-15)'s.

$$\underbrace{(-15) + (-15) + \cdots + (-15)}_{12 \text{ times}} = -180$$

Here are some examples for your calculator making use of the ⌧ key if it has one.

Example 6.5: **(a)** $123 \times 45.6 = ?$

Solution: $123 \times 45.6 = 5608.8$

(b) $123 \times (-45.6) = ?$

Solution: $123 \times 45.6 \; \boxtimes \; = -5608.8$

Example 6.6: $-15 \times 12 = ?$

Solution: $15 \boxed{\%} \boxed{\times} 12 \boxed{=} -180.$

In line with the idea of repeated addition, you can try -15×12 by adding -15 twelve times on your calculator. In effect, we can describe this routine simply as follows:

$$15 \boxed{\%} \boxed{+} 15 \boxed{\%} \cdots \boxed{+} 15 \boxed{\%} \boxed{=} -180.$$

Example 6.7: $-123 \times (-45.6) = ?$

Solution: $123 \boxed{\%} \boxed{\times} 45.6 \boxed{\%} \boxed{=} 5608.8$

Justifying why the product of two negative numbers should be positive is not easy. We may see the reasonableness by the following pattern.

1. Multiplying a positive number by a negative number changes the sign of the positive number.
2. Multiplying any number by a negative number changes the sign of the number. Thus the product of a negative number and a negative number is positive.

Hence we have the rule:

> The product of two negative numbers is positive.

This rule is very important in using the distributive law and in other mathematical situations.

Here is an example of use of the double negative in the distributive law.

Example 6.8: $(-2) \times (4 - 3) = ?$

Solution: First,

$$(-2) \times (4 - 3) = (-2) \times 1 = -2.$$

On the other hand,

$$(-2) \times (4 - 3) = (-2) \times [4 + (-3)].$$

Now apply the distributive law:

$$(-2) \times [4 + (-3)] = [(-2) \times 4] + [(-2) \times (-3)]$$

$$= \quad -8 \quad + \quad ? \quad = -2.$$

Now we know that $-8 + \mathbf{6} = -2$; so it must be that

$$(-2) \times (-3) = +6.$$

Division works in exactly the same way as multiplication except that the $\times$ is replaced by the $\div$ sign.

Example 6.9: $8 \div (-2) = ?$

Solution: 8 ÷ 2 ⅀ = - 4.

Example 6.10: $-8 \div 2 = ?$

Solution: 8 ⅀ ÷ 2 = - 4.

Example 6.11: $-8 \div (-2) = ?$

Solution: 8 ⅀ ÷ 2 ⅀ = 4.

In summary, we have these rules for multiplication and division involving negative numbers:

The product of a positive and a negative number is negative.

The product of two negative numbers is positive.

The quotient of a positive and a negative number is negative.

The quotient of two negative numbers is positive.

If your calculator does not have a "change sign" key, then solve problems involving products and quotients and signed numbers as follows.

1. Compute as though all numbers were positive.

2. Determine the *sign* by one of the rules above.

Exercises 6.3

1. Compute the following products on your calculator. From your computation for the problem under Column A, predict the answer under Column B. Confirm on your calculator as an exercise in proper entering of negative numbers in your machine.

A	B
•a. $3 \times (-120) =$	$-3 \times 120 =$
b. $2.5 \times 18 =$	$-2.5 \times (-18) =$
•c. $-536 \times 797 =$	$536 \times (-797) =$
d. $18.75 \times (-32.5) =$	$-18.75 \times (-32.5) =$
•e. $2 \times 3 \times (-7) =$	$2 \times (-3) \times (-7) =$
f. $32.9 \div (-3) =$	$(-32.9) \div 3 =$
•g. $(24 \div -3) \times 7 =$	$(24 \div -3) \times -7 =$
h. $(3 - 6) \times 7 =$	$(6 - 3) \times 7 =$
i. $3 \times (-6 + 7) =$	$-3 \times (6 - 7) =$
•j. $2.5 \times -17.6 =$	$-2.5 \times 17.6 =$
•k. $0.5 \times -17.6 =$	$-0.5 \times 17.6 =$
l. $0.05 \times -17.6 =$	$-0.05 \times 17.6 =$
m. $0.5 \div -17.6 =$	$0.5 \div 17.6 =$
•n. $-17.6 \div 0.5 =$	$17.6 \div -0.5 =$

2. Without computing an answer, determine the *sign* ($+$ or $-$) of the answer to these problems.

•a. $1823 \times (25 - 1000)$

b. $2 \times (-3) \times 4$

•c. $2 \times (-3) \times 4 \times (-5)$

d. $(-101) \times (-102) \times (-103)$

•e. $2 \times 4 \times 6 \times (-1) \times (-3) \times (-5)$

f. $(2 \div 4) \times 6 \div [(1) \times (3) \div 5]$

g. $(2 \div 4) \times 6 \div [(-1) \times (-3) \div (-5)]$

3. Verify the distributive law by computing the problems under Columns A and B.

A	B
•a. $(12 - 6) \div 3$	$(12 \div 3) - (6 \div 3)$
b. $(6 - 12) \div 3$	$(6 \quad 3) - (12 \div 3)$
•c. $(5 + 6) \times (8 - 3)$	$[(5 + 6) \times 8] + [(5 + 6) \times (-3)]$
d. $[(3 \times 7) + (4 \times -5)] \times 1121$	$(3 \times 7 \times 1121) + (4 \times (-5) \times 1121)$
•e. $(18.4 \times 12.1) + [18.4 \times (-12)]$	$18.4 \times (12.1 - 12)$

•4. At 6 P.M. on Christmas Eve, the temperature is 18° F. A cold front comes in and the temperature changes steadily at the rate of $-3.5°$ F. per hour. What is the temperature at midnight?

•5. The stock of the Bright Star Manufacturing Company falls $0.50 per day for 4 days, and then rises $0.25 per day for 6 days. Which of the following computations describes the result of these ten days of trading?

a. $(-.50 \times 4) + (.25 \times 6)$

b. $-.25 \times 10$

c. $.25 \times 10$

d. $.75 \times 10$

e. $(-.75 \times 4) + (.75 \times 6)$

6. Don Duffer played 18 holes of golf. On the first 7 holes, he was 2 over par on each hole. On the last 11 holes, he was 1 under par on each hole. Write a mathematical expression to describe the total number of strokes he was over (or under) par.

•7. In bowling, a formula for a bowler's handicap is

$$[(175 - A) \times 3] \div 4$$

where A is the bowler's average. A team has 4 bowlers whose averages are given below.

Player 1	147
Player 2	119
Player 3	171
Player 4	175

a. Compute the total handicap for the team.

b. The handicap is added to the actual score of each game played by the team. What is the total "score" for this team after given its handicaps?

c. For what average score is no handicap allowed? (Examine the results for each of the four players.)

8. Suppose S is the sum of the scores for the 4 players in Exercise 7. Calculate $\{[(175 \times 4) - S] \times 3\} \div 4$ and compare the result with the answer to Exercise 7c.

7

Powers, Exponents, and Scientific Notation

7.1 The Old "Grains-on-the-Chessboard" Routine

Another old fable tells the story of the inventor of chess.

> The jaded ruler of an ancient land called in his scientific advisor and demanded a new game. The outcome was chess. The king was so delighted with the game, that he offered its inventor any prize. Instead of asking for the hand of the princess, the inventor made a simple request. He asked for a "few" grains of wheat! He asked that

1 grain be placed on the first square;
2 grains be placed on the second square;
4 grains be placed on the third square;
8 grains be placed on the fourth square;
16 grains be placed on the fifth square;
32 grains be placed on the sixth square;
  . . . and so on.

This goes on for the 64 squares on the chessboard. The number of grains on the next square is to be double the number on the preceding square.

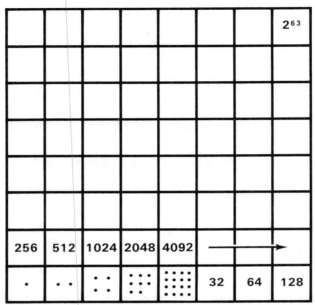

Figure 7.1
A chessboard. The prize to the inventor is wheat—one grain on the first square, two on the next, four on the next, and so on, doubling on each succeeding square. How many grains of wheat in total? How many pounds in total?

We don't know whether the inventor of chess should have chosen the beautiful princess or all the oil in the kingdom, but he surely got a lot of wheat. What do you think? Did he get a pound of wheat? 100 pounds? a ton? 100 tons? a million tons? more than all the wheat that existed?

We can try to calculate the total number of grains of wheat. Then we can estimate the weight in tons as follows. It takes 7,000 grains of wheat to weigh a pound. There are 2,000 pounds in a ton. Hence,

$$7,000 \times 2,000 = 14,000,000$$

or 14 million grains of wheat in a ton of wheat. That's a lot of grains of wheat! So, all we have to do is to compute how many grains of wheat will be on the chessboard, and then divide by 14 million to find the total weight in tons of all the grain. By the way, the world's annual output of wheat is about 400 to 500 million tons.

This problem of finding the total number of grains of wheat on the (mythical) chessboard is one of the most complicated ones we have attempted so far. But we can at least get started.

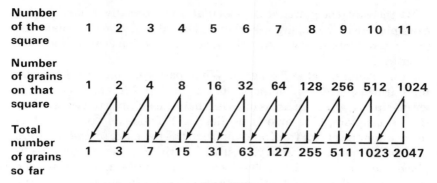

Number of the square	1	2	3	4	5	6	7	8	9	10	11
Number of grains on that square	1	2	4	8	16	32	64	128	256	512	1024
Total number of grains so far	1	3	7	15	31	63	127	255	511	1023	2047

Table 7.1

It is a good idea to begin with a table showing the number of grains on the squares of the chessboard and the total number up to a given square. (This table grows out of hand very quickly, as we shall see, but it is worthwhile to get started!)

Let's pause a moment to see how the calculator can help us. It can, especially if it has a constant (K) switch or has a constant-multiplier feature built in. Here's how it works.

Example 7.1: Doubling with the K switch.

$$\boxed{K}\ 2\ \boxed{\times}\ \boxed{\times}\ \boxed{\times}\ \cdots\ \boxed{\times}\ \boxed{=}\ \circ\ \mathit{1.3421772}$$

$$\underbrace{\qquad\qquad\qquad\qquad}_{26\ \text{times}}$$

Or, depending on how the calculator works:

$$\boxed{K}\ 2\ \boxed{\times}\ \boxed{=}\ \boxed{=}\ \boxed{=}\ \cdots\ \boxed{=}\ \boxed{=}\ \circ\ \mathit{1.3421772}$$

$$\underbrace{\qquad\qquad\qquad\qquad}_{26\ \text{times}}$$

We got an overflow after 27 doublings from 1. The false decimal point tells us that the correct entry should be

$$13421772\square,$$

and we know that digit in the units place is 8 from the fact the units digit of the last number was "4," and $4 \times 2 = 8$. Thus,

$$\underbrace{2 \times 2 \times \cdots \times 2}_{27\ \text{factors}} = 134,217,728$$

From Table 7.1, we can see that the count for the number of doublings is one less than the number of the square. Notice that the first square had 1, not 2, grains on it. So the 28th square has 134,217,728 (27 doublings) grains on it. (There were 14,000,000 grains in a ton, so that square has about 9 to 10 tons of wheat on it! The inventor's reward is growing!)

By the way, the power of the calculator is especially evident in this example. Just by tapping the ⊟ or ⊠ key 26 times, we calculate the product, 67108864, above. Just think of how laborious this would have been by hand!

Now examine Table 7.1 again. What patterns emerge? The solid lines drawn suggest one important relationship. The total number of grains up to a given square is one less than the number to be placed on the next square. (See arrows in diagram.) Thus the TOTAL doubles (except for one grain) from one square to the next. So, to find the total number of grains, it is enough just to find the number of grains to be placed on the 64th square. Then we should double that number and subtract 1 to find the total number of grains on the entire chessboard.

In fact, we are running out of room just to write down the product of the number of 2's being calculated—we're even running out of room to write down the product in decimal notation—because the final answer will use 20 places! So we need a new idea; a better way to record our numbers. Our calculations will be continued after we have some convenient notation and more skills on hand. In any event, we shall have to compute

$$\underbrace{2 \times 2 \times 2 \times \cdots \times 2}_{\text{63 factors}}$$

the number of grains of wheat to be placed on the last square.

Exercises 7.1

1. Using the constant switch, calculate these products.

    a. $3 \times 3 \times 3 \times 3 \times 3$

•b. $2.5 \times 2.5 \times 2.5 \times 2.5 \times 2.5$

    c. $3.1 \times 3.1 \times 3.1 \times 3.1 \times 3.1$

2. The constant switch can be used for multiplication (or division) by a constant. Repeat this example, and then compute the remaining problems by a similar routine.

    Examples: $2 \times 3 = ?$; $2 \times 4 = ?$; $2 \times 6 = ?$
       Solutions: Ⓚ 2 ⊠ 3 ⊟ 6
                        ⊠ 4 ⊟ 8
                        ⊠ 6 ⊟ 12

a. 33×6;	33×6.5;	33×87;	33×1
•b. 3×6;	3×189;	3×12372	
c. 1.1×5;	1.1×55;	1.1×555	

3. How many grains were there on the 20th square on the (mythical) chess-board?

4. Which was the first square to have more than—
   a. 100 pounds of wheat?
   b. one ton of wheat?

7.2   Squares and Cubes

Aside from problems like those of the chessboard, there are many others in which it is necessary to repeatedly multiply by the same factor. For example, in Chapter 5, we found the area of a square 12 inches on a side:

$$\text{area} = (12 \times 12) \text{ or } 144 \text{ sq. in.}$$

Likewise, the volume of a cube, 4 cm on a side is given by

$$\text{volume} = (4 \times 4 \times 4) \text{ or } 64 \text{ cubic centimetres.}$$

Notationally, we indicate the product of several equal numbers with an *exponent,* a number written above and to the right of the factor that is multiplied.

$$2 \times 2 \quad \text{is written} \quad 2^2; \quad \text{thus} \quad 2^2 = 4$$

$$3 \times 3 \quad \text{is written} \quad 3^2; \quad \text{thus} \quad 3^2 = 9$$

$$17 \times 17 \quad \text{is written} \quad 17^2; \quad \text{thus} \quad 17^2 = 289$$

The exponent indicates how many times the number is used as a factor. Similarly,

$$2 \times 2 \times 2 \quad \text{is written} \quad 2^3; \quad \text{thus} \quad 2^3 = 8;$$

$$3 \times 3 \times 3 \quad \text{is written} \quad 3^3; \quad \text{thus} \quad 3^3 = 27;$$

$$17 \times 17 \times 17 \quad \text{is written} \quad 17^3; \quad \text{thus} \quad 17^3 = 4913$$

We write $2^5 = 2 \times 2 \times 2 \times 2 \times 2 = 32$

and $\qquad 3^6 = 3 \times 3 \times 3 \times 3 \times 3 \times 3 = 729$

The number that is being multiplied is called the *base*. In

$$2^3 = 8,$$

3 is the exponent, 2 is the base. In

$$3^2 = 9,$$

2 is the exponent, 3 is the base. It can be seen that, just as 23 and 32 represent different numbers, 2^3 and 3^2 stand for two different numbers. (See also Exericse 7 at the end of this section.)

The value of the number is called the *power*. Thus 8 is a power of 2. Specifically,

8 is the 3rd power of 2

Likewise,

32 is the 5th power of 2;

729 is the 6th power of 3; and so on.

Moreover, in Section 7.1, we found that the number of grains to be placed on the last square of the chessboard would be

the 63rd power of 2; that is, 2^{63}.

Exercises 7.2

For these exercises, you may use your calculators as you wish.

1. Compute 10^2, 10^3, 10^4.
   - •a. What is the relationship between the number of zeroes after 1 and the exponent?
   - b. How many zeros follow 1 in 10^{23}?

2. Recall (page 41) that a "googol" is represented by a 1 followed by 100 zeros. Write "1 googol" in the form

$$1 \text{ googol} = 10^z.$$

• 3. What is 1^2, 1^3, 1^4, 1^{10}, 1^{101}?

• 4. Use your calculator to compute the following and record your answers in a table.
   $2(=2^1)$, 2^2, 2^3, 2^4, 2^5, 2^6, . . . , 2^{20}.

• 5. Construct a similar table for the powers of 3 from 3^1 to $3^{15} = 14348907$.

6. Find each of the following sums.
   a. $1 + 3$
   c. $1 + 3 + 3^2 + 3^3$
   b. $1 + 3 + 3^2$
   d. $1 + 3 + 3^2 + 3^3 + 3^4$
   e. Can you guess what would be $1 + 3 + 3^2 + \cdots + 3^{14}$?

• 7. Which is larger: 2^3 or 3^2? 3^4 or 4^3? 5^2 or 2^5?

8. For each of the following, identify the exponent and the base.
   •a. 2^{10}
   •c. 7^1
   e. 4^3
   •b. 5^3
   d. 10^3
   f. 3^4

9. Suppose a cube 4 in. on a side is painted red and then diced into 64 one-inch cubes as in Figure 5.43. How many of these will have *NO* paint on them?

10. By making guesses, determine the size of a square (the length of a side)—
    a. whose area is 64 square units;
    b. whose area is 64.0625 square units.

11. For your guesses in Exercise 10—
    a. how would you determine whether the guesses are correct?
    b. Check each of the guesses according to your suggestion in part a.

•12. Complete this table of squares of positive integers from 1 to 33.

N	N^2	N	N^2	N	N^2
1	1	12		23	
2	4	13		24	
3		14		25	
4		15		26	
5		16		27	
6		17		28	
7		18		29	
8		19		30	
9		20		31	
10		21		32	
11	121	22	484	33	1089

•13. Make a similar table for the cubes of positive integers from 1 to 33. These tables will be useful in Chapter 12.

N	N^3	N	N^3	N	N^3
1	1	12		23	
2	8	13		24	
3		14		25	
4		15	3375	26	
5		16		27	
6		17		28	
7		18		29	
8		19		30	
9		20		31	
10		21		32	
11		22		33	

7.3 Exponents

In the foregoing section, we talked about exponent—a notation for powers of numbers. For example, some powers of 2:

$$2^2 = 2 \times 2 = 4$$

$$2^3 = 2 \times 2 \times 2 = 8$$

$$2^4 = 2 \times 2 \times 2 \times 2 = 16$$

Of great importance are the powers of 10, because of the way in which they relate to our base-ten numeral system. Some powers of 10:

$10^2 = 10 \times 10 = 100$ $(10^2 = 1$ hundred$)$

$10^3 = 10 \times 10 \times 10 = 1000$ $(10^3 = 1$ thousand$)$

$10^4 = 10 \times 10 \times 10 \times 10 = 10000$ $(10^4 = 1$ ten-thousand$)$

$10^5 = 10 \times 10 \times 10 \times 10 \times 10$ $(10^5 = 1$ hundred-thousand$)$
$= 100000$

$10^6 = 10 \times 10 \times 10 \times 10 \times 10 \times 10$ $(10^6 = 1$ million$)$
$= 1000000$

Incidentally, it is useful to remember that $10^6 = 1$ million. There are a few prefixes that would be equally useful to recognize. The word for "great" in Greek is *"mega,"* and this word appears as a prefix to stand for "million." For example:

$$\text{"megaton"} = \text{a million tons,}$$

$$\text{"megabuck"} = \text{a million dollars.}$$

A "megabomb" is the equivalent of a million tons of TNT. Items in the U.S. National Budget are often listed in "megabucks"—the mark of an affluent society (or an inflationary one!). Other frequently used prefixes are listed below. (The prefixes at the left are derived from the Greek language and the ones at the right from Latin.)

deka- = ten   deci- = tenth

hecto- = hundred   centi- = hundredth

These and other prefixes used in the metric system can be found in the table on page 26, extended on page 217.

SUMMARY: For any number B and any positive integer N,

$$B^N = \underbrace{B \times B \times \cdots \times B}_{N \text{ factors}}$$

is called the Nth *power* of B. N is called the *exponent*. B is called the *base*.

There are three especially important rules for juggling exponents. Here is an example of the first one.

Example 7.2: (Multiplication) $2^3 \times 2^4 = 2^7$.
Rewrite this product showing all the factors:

$$\underbrace{2^3}_{} \quad \times \quad \underbrace{2^4}_{}$$

$$= (2 \times 2 \times 2) \times (2 \times 2 \times 2 \times 2) = 2^7$$

because there are 7 factors of 2 altogether. But look at the exponents: $3 + 4 = 7$. Thus, the total number of factors is determined by adding the two exponents.

Example 7.3: (Multiplication using exponents) $3^2 \times 3^5 = ?$

Solution:

$$3^2 \times 3^5 = (3 \times 3) \times (3 \times 3 \times 3 \times 3 \times 3) = 3^7.$$

Note that $2 + 5 = 7$.

> RULE 1M: The product of two or more numbers, each in exponential form with the *same* base, may be expressed in exponential form by *adding* exponents:
>
> $$B^N \times B^M = B^{N+M}.$$

Just because division is the reverse operation to multiplication, we can expect to find a rule for division involving *subtraction* of exponents similar to Rule 1M. By now, you may have guessed that the "M" stands for multiplication!

Begin with an easy example, by reversing Example 7.2. It will be helpful to have at your fingertips the relationships between multiplication and division summarized on page 106:

$$A \div B = C \quad \text{and} \quad B \times C = A$$

state the same fact.

Example 7.4: (Division) $2^7 \div 2^3 = ?$

Solution: $2^7 \div 2^3 = C$ and $2^3 \times C = 2^7$

state the same fact. But we know from Example 7.2 (Multiplication) that

$$2^3 \times 2^4 = 2^7;$$

so we conclude that

$$C = 2^4.$$

Now the easy part. We computed $2^4 \times 2^3$ ($=2^7$) by *adding* exponents: $4 + 3 = 7$. Just so, compute $2^7 \div 2^3$ ($=2^4$) by *subtracting* exponents: $7 - 3 = 4$.

Example 7.5: (Division) $3^7 \div 3^5 = ?$

Solution: $7 - 5 = 2$; so $3^7 \div 3^5 = 3^2$.

Check: $3^7 \stackrel{?}{=} 3^5 \times 3^2$. Yes, because $7 = 5 + 2$.

From these examples, we are led to the following rule.

RULE 1D: The quotient of two numbers, each in exponential form with the same base, may be expressed in exponential form by *subtracting* the two exponents:

$$B^N \div B^M = B^{N-M}.$$

The next rule helps us to simplify expressions like the following.

Example 7.6: $(2^3)^4 = 2^{12}$.

Solution: In $(2^3)^4$, we have a number 2^3, with the exponent 4. Thus this means,

$$(2^3)^4 = 2^3 \times 2^3 \times 2^3 \times 2^3$$

$$= (2 \times 2 \times 2) \times (2 \times 2 \times 2) \times (2 \times 2 \times 2) \times (2 \times 2 \times 2)$$

$$= 2^{12},$$

because there are 12 factors of 2. But look at the exponents: $3 \times 4 = 12$.

Example 7.7: $(5^3)^2 = ?$

Solution: Expand $(5^3)^2$ as

$$(5^3)^2 = 5^3 \times 5^3$$

$$= (5 \times 5 \times 5) \times (5 \times 5 \times 5) = 5^6.$$

Note that $3 \times 2 = 6$.

RULE 2: To raise a number already in exponential form to a power, *multiply* the exponents:

$$(B^N)^M = B^{N \times M}.$$

The next rule describes how we can simplify or expand factors in exponential form using different bases in a special situation.

Example 7.8: $(3 \times 4)^2 = 3^2 \times 4^2$.

Solution: $(3 \times 4)^2 = (3 \times 4) \times (3 \times 4)$
$$= 3 \times 3 \times 4 \times 4$$
$$= (3 \times 3) \times (4 \times 4)$$
$$= 3^2 \times 4^2,$$
by counting the factors of 3 and of 4.

Example 7.9: $(2 \times 7)^5 = ?$

Solutions:

$$(2 \times 7)^5 = (2 \times 7) \times (2 \times 7) \times (2 \times 7) \times (2 \times 7) \times (2 \times 7).$$

How many factors of 2 appear on the right? (5.)
How many factors of 7 appear on the right? (5.) Thus,

$$(2 \times 7)^5 = 2^5 \times 7^5.$$

RULE 3M: The product of two or more factors raised to a power is the product of each factor raised to that power:

$$(A \times B)^N = A^N \times B^N \text{ or } (A \times B \times C)^N = A^N \times B^N \times C^N.$$

There is a similar rule for division.

RULE 3D: The quotient of two numbers raised to a power is the quotient of each number raised to that power:

$$(A \div B)^N = A^N \div B^N.$$

Finally, it is convenient to extend the notion of exponents to include zero as an exponent. Let's examine first, the following sequence.

$$32 = 2^5$$

$$16 = 2^4$$

$$8 = 2^3$$

$$4 = 2^2$$

$$2 = 2^1$$

$$1 = 2^?$$

Notice that each number in the lefthand column is half the preceding number. Notice also, each exponent in the righthand column is one less than the preceding.

According to this pattern, it would be reasonable to expect the exponent indicated by the question mark to be 0. That is, $2^0 = 1$.

This pattern holds true for any base B, as long as $B \neq 0$. Therefore, we *define*

$$B^0 = 1,$$

for any nonzero number B. Thus,

$$2^0 = 1; \quad 10^0 = 1; \quad \left(\frac{1}{2}\right)^0 = 1;$$

and so on.

Note that

$$B^0 \times B^N = 1 \times B^N = B^N \quad \text{by the above definition,}$$

$$= B^{0+N} = B^N \quad \text{by Rule 1M.}$$

So Rule 1 holds with $M = 0$.

Similarly,

$$(B^0)^N = (1)^N = 1 \quad \text{by the above definition,}$$

$$= B^{0 \times N} = B^0 = 1 \quad \text{by Rule 2}$$

while

$$(B^N)^0 = 1$$

$$= B^{N \times 0} = B^0 = 1$$

so that Rule 2 holds with $M = 0$ or $N = 0$. Similarly, if we check, we would find that Rule 3M holds if $N = 0$.

Rule 1M permits us to replace an act of multiplication by an act of addition. This step was the first major simplification in computation; it is the system of logarithms invented by John Napier.

To see how the system works, let us build a table of powers of 2.

$2^0 = 1$	$2^8 = 2^7 \times 2 = 256$
$2^1 = 2^0 \times 2 = 2$	$2^9 = 2^8 \times 2 = 512$
$2^2 = 2^1 \times 2 = 4$	$2^{10} = 2^9 \times 2 = 1024$
$2^3 = 2^2 \times 2 = 8$	$2^{11} = 2^{10} \times 2 = 2048$
$2^4 = 2^3 \times 2 = 16$	$2^{12} = 2^{11} \times 2 = 4096$
$2^5 = 2^4 \times 2 = 32$	$2^{13} = 2^{12} \times 2 = 8192$
$2^6 = 2^5 \times 2 = 64$	$2^{14} = 2^{13} \times 2 = 16384$
$2^7 = 2^6 \times 2 = 128$	$2^{15} = 2^{14} \times 2 = 32768$

To use this table to calculate, one need only convert the numbers to exponential form.

Example 7.10: $256 \times 32 = ?$

Solution: Rewrite the factors as powers of 2.

Thus, $256 = 2^8$ and $32 = 2^5$.

$$256 \times 32 = 2^8 \times 2^5 = 2^{8+5} = 2^{13},$$

and from the table, $2^{13} = 8192$.
Therefore, we know that

$$256 \times 32 = 8192.$$

Check: $256 \;\boxed{\times}\; 32 \;\boxed{=}\; 8192.$

The table of the powers of 10 is especially easy to construct because the base-ten place-value system permits us to show an integral power of 10 by the number of zeros after the "1."
In brief,

$$10^N = \underbrace{10 \cdots 0}_{N \text{ zeros}}$$

Thus

$$\underbrace{100000000}_{} \times \underbrace{100000}_{} = \underbrace{10000000000000}_{}$$

or $\qquad 10^8 \qquad \times \quad 10^5 \quad = \qquad 10^{13}$ (count the zeros)

Now you may have noticed that there is one big catch in all of this: we are restricted in computing products to those numbers that appear in the table. Thus, if asked to compute even such a simple product as

$$13 \times 15,$$

we are stymied because neither 13 nor 15 appears in the tables given.

A table of logarithms gives a much more detailed listing for powers of 10. Such a table, for instance, might show that

$$13 = 10^{1.1139} \quad \text{and} \quad 15 = 10^{1.1761}$$

so that

$$13 \times 15 = 10^{1.1139} \times 10^{1.1761} = 10^{1.1139+1.1761}$$

$$\doteq 10^{2.29} = 195,$$

the last equality again coming from the table of logarithms. Prior to electronic calculators, most products were calculated this way. The slide rule, itself, is simply a mechanical version of a table of logarithms.

Now we are in excellent position to make more progress on the problem of the number of grains in the prize claimed by the inventor of chess and, finally, to compute the total weight of the grain in tons.

In exponential notation, we see that the number of grains to be placed on the last, or 64th, square is

$$\underbrace{2 \times 2 \times \cdots \times 2}_{\text{63 factors}} = 2^{63}$$

and so the total number of grains on the chessboard is

$$2^{64} - 1$$

Now $2^{64} = \underbrace{2 \times 2 \times \cdots \times 2}_{\text{64 factors}}$

$$= \underbrace{(2 \times 2 \times \cdots \times 2)}_{\text{32 factors}} \times \underbrace{(2 \times 2 \times \cdots \times 2)}_{\text{32 factors}} \qquad \text{(Rule 1M)}$$

But we do not yet know 2^{32}.

We *do* know 2^{27}, and after all,

$$2^{32} = \underbrace{2 \times 2 \times \cdots \times 2}_{\text{32 factors}} = \underbrace{2 \times 2 \times \cdots \times 2}_{\text{27 factors}} \times \underbrace{2 \times 2 \times 2 \times 2 \times 2}_{\text{5 factors}}$$

$$= 2^{27} \times 2^5 \qquad \text{(Rule 1M)}$$

$$= 2^{27} \times 32.$$

Now the numbers are within our grasp. By the methods of Section 4.5, we calculate

$$2^{32} = 2^{27} \times 32 = 134217728 \times 32$$

$$= 4294967296$$

Briefly,

```
    1342 | 17728
  ×      |    32
  ───────┼──────
   42944 | 00000
       5 | 67296
  ───────┼──────
   42949   67296
```

Then by these same methods, we calculate

$$2^{64} = 2^{32} \times 2^{32} = 18{,}446{,}744{,}073{,}709{,}551{,}616$$

Let's call this number c, for "chess."

Here's how that calculation can be made.

$$
\begin{array}{r|r|r|r}
 & 42949 & 67296 \\
\times & 42949 & 67296 \\
\hline
 & 45287 & 51616 \\
28902 & 95904 & 00000 \\
28902 & 95904 & 00000 \\
18446 \quad 16601 & 00000 & 00000 \\
\hline
18446 \quad 74407 & 37095 & 51616 \\
\end{array}
$$

To express the total weight of all the grains $(= c - 1)$ of wheat in tons, we must then calculate

$$(c - 1) \div 14{,}000{,}000$$

Before we do this, it will be helpful to employ *"scientific notation"* to reduce the computations to manageable size! This is discussed in the next section.

Exercises 7.3

- 1. Make a table of the positive powers of 3 from 3^1 to 3^{16}. That is, find 3^N for integers N between 1 and 16.

- 2. Using exponents and the table constructed in Exercise 1, compute:

$$81 \times 243 = ? \qquad 27 \times 6561 = ? \qquad 729^2 = ?$$

3. From the table in Exercise 1, extract a table for the positive integral powers of 9 from 9^1 to 9^8.

4. Make a table of the powers of $\frac{1}{2}(=0.5)$; find $(0.5)^N$ for the positive integral values of N between 1 and 25.

5. Refer to the table constructed in Exercise 4.

   a. Explain why $(0.5)^{24} = 0$ on an 8-digit calculator.

   b. Using exponents, compute 0.03125×0.0078125.

6. Calculate these two columns of products and powers.

   - a. $(2 \times 3)^2 =$ $2^2 \times 3^2 =$
   - b. $(2 \times 10)^3 =$ $2^3 \times 10^3 =$
   - c. $(5 \times 5)^4 =$ $5^4 \times 5^4 = 5^8 =$
   - d. $(1.3 \times 3.08)^5 =$ $(1.3)^5 \times (3.08)^5 =$

- 7. Show that the following is *NOT* a rule of exponents:

$$A^N \times B^M = (A \times B)^{N+M}.$$

(Find some values of A, B, N, and M for which it is false.)

8. Show that the following is *NOT* a rule of exponents:

$$B^N + B^M = B^{N+M}.$$

9. Use your calculator to compute each of the following, and compare results in Column A with corresponding ones in Column B.

A	B
a. $(6 \div 2)^3 =$	$6^3 \div 2^3 =$
•b. $(6 \div 3)^4 =$	$6^4 \div 3^4 =$
c. $(8 \div 2)^2 =$	$8^2 \div 2^2 =$
•d. $(12 \div 3)^3 =$	$12^3 \div 3^3 =$

10. Using the pattern from Exercise 9, guess the results for each of the following.

•a. $12^5 \div 6^5 = ?$	c. $9^4 \div 3^4 = ?$
•b. $8^7 \div 4^7 = ?$	d. $6^5 \div 2^5 = ?$

7.4   Scientific Notation

We can take an initial step toward understanding logarithms without going deeply into that theory. This first step is useful because it is the device called *scientific notation*. In fact, this notational device is used on some of the more sophisticated calculators to enter numbers and to express results. But more than that, it gives us a handy way of writing large or, as we shall see, small numbers. Furthermore, it is a great aid in estimating products and quotients.

As we know, multiplication by 10 or a power of 10 simply adds zeros to the number. Thus:

$$23 \times 10 \quad = 230 \quad = 23 \times 10$$

$$23 \times 100 \quad = 2300 \quad = 23 \times 10^2$$

$$23 \times 1000 = 23000 = 23 \times 10^3, \quad \text{and so on.}$$

It is convenient to think of this in terms of moving the decimal point:

$$23 \times 10 \quad = 23.0 \times 10 \quad = 230, \quad \text{or } 23.0$$

$$23 \times 100 = 23.00 \times 100 = 2300 \quad \text{or } 23.00$$

Similarly, division by 10 or a power of 10 simply "moves" the decimal point in the other direction:

$$23 \div 10 = 2.3 \quad \text{or} \quad 23.$$

$$23 \div 100 = .23 \quad \text{or} \quad 23.$$

Any number can be written with the decimal place at any point if compensated by multiplication or division of a power of 10. It is standard practice to choose a number between 1 and 10, and then follow it by the appropriate factor of a power of 10. (The number between 1 and 10 includes the possibility of 1 but excludes the possibility of 10.) Thus,

$$23 = 2.3 \times 10 \qquad\qquad 0.23 = 2.3 \div 10$$

$$230 = 2.3 \times 100 \qquad\qquad 0.023 = 2.3 \div 100$$

$$2300 = 2.3 \times 1000 \qquad\qquad 0.0023 = 2.3 \div 1000$$

If the power of 10 in each of the products is written in exponential form, then we have the number in scientific notation. Thus, in scientific notation,

$$23 = 2.3 \times 10^1$$

$$230 = 2.3 \times 10^2$$

$$2300 = 2.3 \times 10^3$$

Example 7.11: $23 \times 321 = ?$

Solution: First, we express each number in scientific notation.

$$23 \times 321 = 2.3 \times 10 \times 3.21 \times 10^2$$

$$= 2.3 \times 3.21 \times 10 \times 10^2$$

$$= 2.3 \times 3.21 \times 10^3 = 2.3 \times 3.21 \times 1000$$

Now 2.3×3.21 is bigger than $2.3 \times 3 = 6.9$, so we might estimate

$$2.3 \times 3.21 \doteq 7,$$

and so

$$23 \times 321 \doteq 7 \times 1000 = 7000.$$

(A check with a calculator shows that $23 \times 321 = 7383$.)

Since Example 7.11 deals with familiar numbers, the solution we give here seems to be needlessly confusing! However, in dealing with large numbers—or very small ones—scientific notation helps us make quick estimates and keeps track of decimal points. Moreover, it is used frequently in "scientific calculators."

Astronomers sometimes measure long distances in space in units of *light-years*. A light-year is the distance light travels in one year. (See Exercises 3 and 4 for Section 4.5.) Now light travels approximately 186,284 miles in one second. To determine how far light travels in one year, we need to know the number of seconds in a year. (For purposes of this example, we shall assume that a year is 365 days; actually, it is somewhat longer and thus necessitates our leap-year convention—but that's another story.)

Example 7.12: Find the number of miles in a light-year.

Solution: There are—

60 seconds in 1 minute;
60 minutes in 1 hour; hence 60×60 seconds in 1 hour;
24 hours in a day; hence $24 \times 60 \times 60$ seconds in 1 day;
365 days in 1 year; hence $365 \times 24 \times 60 \times 60$ seconds in 1 year.
To your calculator!

$$365 \;\boxed{\times}\; 24 \;\boxed{\times}\; 60 \;\boxed{\times}\; 60 \;\boxed{=}\; 31536000.$$

Thus a light year is 31536000×186284 miles
$$= \text{?? miles.}$$

If that last multiplication is attempted on a calculator with only 8- or 12-digit capacity, an overflow results and no further calculations can be done. However, we can write

$$31536000 = 3.1536 \times 10^7$$

$$186284 = 1.86284 \times 10^5$$

The product of 3.1536 and 1.86284 can be accepted and will be computed to the number of places your machine is capable of handling. Moreover, further calculations can be done. On an 8-digit machine,

$$3.1536 \;\boxed{\times}\; 1.86284 \;\boxed{=}\; 5.8746522$$

So $31536000 \times 186284 = 3.1536 \times 1.86284 \times 10^7 \times 10^5$

$$\doteq 5.8746522 \times 10^{12}$$

is the number of miles in a light year, approximately.

To complete our work with exponents, we need one other bit of juggling involving the order of performing multiplication and division. Here are some examples.

Example 7.13:

(1) $600 \div 30 = (6 \times 100) \div (3 \times 10)$
$\qquad\qquad = (6 \div 3) \times (100 \div 10)$

$\qquad$ *Check:* $\quad 600 \div 30 = 20$
$\qquad\qquad\qquad (6 \div 3) \times (100 \div 10) = 2 \times 10 = 20.$

(2) $42000 \div 140 = (42 \times 1000) \div (14 \times 10)$
$\qquad\qquad\qquad = (42 \div 14) \times (1000 \div 10)$

$\qquad$ *Check:* $\quad 42000 \div 140 = 300$
$\qquad\qquad\qquad (42 \div 14) \times (1000 \div 10) = 3 \times 100 = 300.$

(3) $400 \div 50 = (4 \times 100) \div (5 \times 10)$
$\qquad\qquad = (4 \div 5) \times (100 \div 10)$

$\qquad$ *Check:* $\quad 400 \div 50 = 8$
$\qquad\qquad\qquad (4 \div 5) \times (100 \div 10) = .8 \times 10 = 8.$

More generally,

(4) $(56 \times 39) \div (7 \times 13) = (56 \div 7) \times (39 \div 13)$

$\qquad$ *Check:* $\quad (56 \times 39) \div (7 \times 13) = 2184 \div 91 = 24$
$\qquad\qquad\qquad (56 \div 7) \times (39 \div 13) = 8 \times 3 = 24.$

Incidentally, Example 7.13(4) shows how useful this rule is. Most of us would agree that it is easier to multiply 8×3 than to divide $2184 \div 91$.

In symbols, this rule states that

$$(A \times B) \div (C \times D) = (A \div C) \times (B \div D).$$

Pay close attention to the roles of A, B, C, and D. Notice that C and D are divisors on both sides of the equation. We can write this rule in fraction form.

$$\frac{A \times B}{C \times D} = \frac{A}{C} \times \frac{B}{D}$$

As a final application, we shall calculate a good estimate for the total weight of the wheat in the prize for the inventor of chess. Recall that earlier we had found

$$c - 1 = 18,446,744,073,709,551,615$$

grains of wheat on the chessboard.

Example 7.14: What is the total weight of the grains on the chessboard?

Solution: Write

$$c - 1 = 1.844674407309551615 \times 10^{19}.$$

Approximate $c - 1$ by 1.8×10^{19}.
  Now recall that there are

$$14,000,000 = 1.4 \times 10^7$$

grains in a ton of wheat. Thus, the total weight is approximately

$$(1.8 \times 10^{19}) \div (1.4 \times 10^7) = (1.8 \div 1.4) \times (10^{19} \div 10^7)$$

$$\doteq 1.29 \times 10^{12}. \qquad \text{(Here we use Rule 1D.)}$$

So the total weight is approximately

$$1.29 \times 10^{12} \text{ tons} = 1,290,000,000,000 \text{ tons}$$

$$= 1 \text{ trillion, 290 billion tons!}$$

The world production of wheat each year is about 500 million tons or, in scientific notation, 5×10^8 tons. Comparing this with 1.29×10^{12}, we have

$$(1.29 \times 10^{12}) \div (5 \times 10^8) = \frac{1.29}{5} \times 10^4$$

$$= 0.258 \times 10^4 = 2580$$

So the total weight of grains "on" the chessboard is more than 2,500 times the present annual world production of wheat!

Exercises 7.4

1. Note that $2^{10} \doteq 10^3$.

  •a. Make an estimate for 2^{63}, the number of grains of wheat "on" the final square of the chessboard in Section 7.3. [HINT: $2^{63} = 2^{60} \times 2^3 = 8 \times (2^{10})^6$.]

  b. Use scientific notation and the relation,

$$2^{10} = 1.024 \times 10^3$$

  to obtain an estimate for the number of grains of wheat "on" the last square. [HINT: $(2^{10})^6 = (1.024 \times 10^3)^6 = (1.024)^6 \times (10^3)^6$.]

• 2. There are 30.48 centimetres in a foot. How many centimetres are there in a mile ($= 5280$ feet)? Express your answer in the form: $1. _____ \times 10^N$.

3. Consult your financial page. It will probably publish the total number of shares traded and the highest price per share for the most active stock. Compute the total value of the shares traded, based on the highest price. Give your answer in megabucks!

7.5 Negative Exponents

As we have seen, we can perform division of powers of numbers by subtracting exponents. Rule 1D states that

$$B^N \div B^M = B^{N-M}.$$

Now, we have to be a bit careful with this rule. Does it make sense when M is larger than N? Suppose we try it with

$$2^2 \div 2^5 = ?$$

Applying Rule 1D, we have

$$2^2 \div 2^5 = 2^{2-5} = 2^{-3}.$$

The symbol 2^{-3} has not yet been defined. What should it mean, if anything? Well, to be consistent, we should also have

$$2^0 \div 2^3 = 2^{-3};$$

or, because $2^0 = 1$,

$$1 \div 2^3 = 2^{-3};$$

that is,

$$1 \div 8 = 2^{-3}.$$

Or again, recalling our notation from Section 1.5, $1 \div 8 = \frac{1}{8}$. So

$$2^{-3} = 1 \div 2^3 = \frac{1}{2^3}.$$

Is it true that

$$2^2 \div 2^5 = 1 \div 2^3?$$

We can test this by our general rule for checking division: We ask

$$2^2 \overset{?}{=} 2^5 \times (1 \div 2^3).$$

Well, we note that $2^5 = 2^2 \times 2^3$, so that

$$2^5 \times (1 \div 2^3) = (2^2 \times 2^3) \times (1 \div 2^3)$$

$$= 2^2 \times (2^3 \times (1 \div 2^3))$$

$$= 2^2 \div 1 \qquad \text{(Why?)}$$

$$= 2^2$$

From all of this—and these arguments are quite general—it would be consistent with what we have done to *define*

$$2^{-3} = 1 \div 2^3 = \frac{1}{2^3}$$

and thus give meaning to *negative* exponents.

In summary, then, we shall *define*

$$B^{-N} = 1 \div B^N = \frac{1}{B^N} \quad \text{if } B \neq 0$$

It is important to know that this definition makes Rules 1, 2, and 3 true for all exponents, positive or negative.

These conventions about negative exponents are particularly useful in dealing with powers of 10. Thus,

$$10^{-1} = 1 \div 10 = \frac{1}{10} = .1$$

$$10^{-2} = 1 \div 100 = \frac{1}{100} = .01$$

$$10^{-3} = 1 \div 1000 = \frac{1}{1000} = .001$$

and so on.

Note that, if N is positive,

$$10^{-N} = \frac{1}{10^N} = \overbrace{0.0 \cdots 01}^{N \text{ places}}$$

$$(N-1) \text{ zeros}$$

This convention means that small numbers can be neatly expressed with the help of negative exponents. These topics are discussed further in Chapter 8.

Example 7.15: Express these small numbers using negative exponents.

$$0.2345 = 2.345 \times \frac{1}{10} = 2.345 \times 10^{-1}$$

$$0.02345 = 2.345 \times \frac{1}{100} = 2.345 \times 10^{-2}$$

Let's look at the pattern as the exponent of 10 decreases from 3 to -3. In the table below, each line is obtained by dividing the preceding line by 10.

$$2345 \quad = 2.345 \times 10^3$$

$$234.5 \quad = 2.345 \times 10^2$$

$$23.45 \quad = 2.345 \times 10^1$$

$$2.345 \quad = 2.345 \times 10^0$$

$$0.2345 \quad = 2.345 \times 10^{-1}$$

$$0.02345 \quad = 2.345 \times 10^{-2}$$

$$0.002345 = 2.345 \times 10^{-3}$$

Thus, any number can be written as the product of a number between 1 and 10 (excluding 10) and an integral power of 10, simply by adjusting the decimal point. This notation is called *"scientific notation"* because it is so useful in all branches of science. Moreover, the exponent of 10 gives a rough idea of the magnitude of the number, and often suffices for crude approximations in calculations.

Application: Small Numbers

Television signals travel with the speed of light. How long does it take the signal from a TV transmitter to reach a house 20 miles away?

Answer: 20 miles ÷ speed of light

$$= 20 \text{ miles} \div 186284 \text{ miles per second}$$

$$= 20 \div 186284 \text{ second}$$

$$= 20 \div (1.86284 \times 10^5) \text{ seconds}$$

$$= (2 \div 1.86284) \times (10 \div 10^5) \text{ seconds}$$

(Use your calculator on the first quotient.)

$$= 1.0736295 \times 10^{-4} \text{ seconds} \doteq .000107 \text{ seconds}.$$

Many physical constants are small. For example, the weight of an atom varies

between 1.675×10^{-24} grams for a hydrogen atom
and 3.96×10^{-22} grams for a uranium atom.

Thus, uranium is approximately (crude estimate) 100 times as heavy as hydrogen, because

$$10^{-22} \times 10^{-24} = 10^{-22-(-24)}$$

$$= 10^{-22+24} = 10^2 = 100$$

More precisely, uranium is

$$(3.96 \times 10^{-22}) \div (1.675 \times 10^{-24}) = (3.96 \div 1.675) \times 10^{-22-(-24)}$$

$$= 2.36 \times 10^2 = 236$$

times as heavy as hydrogen.

As the metric system is the preferred system of units for scientific work, the table of prefixes from Section 2.2 is reproduced here, together with the power of ten for each factor.

Prefix	Factor	Prefix	Factor
tera-	10^{12}	deci-	10^{-1}
giga-	10^9	centi-	10^{-2}
mega-	10^6	milli-	10^{-3}
kilo-	10^3	micro-	10^{-6}
hecto-	10^2	nano-	10^{-9}
deka-	10^1	pico-	10^{-12}

Exercises 7.5

1. Here are some easy problems dressed up in exponential form. (No calculators, please!)

   a. $6 \times 2^{-1} =$

   •b. $(\frac{1}{2})^{-1} =$

   c. $(2^{-1} \times 3^{-1})^{-1} =$

   •d. $3 \times 4 \times 3^{-1} =$

   •e. $6^{-2} \times (2 \times 3) =$

   f. $6 \div (2 \times 3)^{-1} =$

2. Some of these are examples of the rules of exponents. Some are not. Answer "Yes" or "No." Check with your calculator if you are in doubt.

   •a. $10^3 \times 10^7 \overset{?}{=} 10^{10}$

   b. $10^3 \times 10^{-7} \overset{?}{=} 10^{-4}$

   •c. $2^{-1} \times 3^{-1} \overset{?}{=} (2 \times 3)^{-1}$

   •d. $2^{-1} + 3^{-1} \overset{?}{=} (2 + 3)^{-1}$

   e. $5^2 - 5^{-2} \overset{?}{=} 5^0$

   f. $(2^3)^{-6} \overset{?}{=} 2^{-18}$

3. Place the appropriate sign, $>$ (greater than), $<$ (less than), or $=$ (equals), in the box between these numbers to make a correct statement.

   a. $2^3 \square 2^2$

   •b. $2^3 \square 3^2$

   •c. $2^{-3} \square 2^{-2}$

   •d. $5^{-1} \square -5$

   e. $5^{-2} \square 2^5$

   •f. $(7^{-2})^3 \square 7^{-6}$

   g. $(7^{-2})^{-3} \square 7^6$

   •h. $(\frac{1}{3})^2 \square 3^{-2}$

   i. $3^{-2} + 5^{-1} \square (3 + 5)^{-1}$

   j. $(7^3)^2 \square (7^2)^3$

•4. Henry Aaron hits a fly ball deep to centerfield. The centerfielder stands 350 feet from home plate.

   a. If sound travels 1,087 feet per second, how soon does the centerfielder hear the crack of the bat? (Less than 1 second.)

   b. How soon does the TV sound microphone standing 50 feet away from home plate hear (receive) the crack of the bat?

   c. The electronic signal of that sound now travels at the speed of light 20 miles to a home where Joey Little Leaguer watches the game. He is 10 feet from his set. How soon after the crack of Aaron's bat does Joey hear the sound? (The speed of light is approximately 186,284 miles per second.)

   d. Who hears it first—Joey or the centerfielder?

5. Follow the procedure in Example 7.15 to show that

$$24^7 \div 12^7 = (24 \div 12)^7.$$

6. Use your calculator to find

$$24^7 \div 3^7$$

the easy way.

Rational Numbers

8.1 Troublesome Routines

Exercise 7 in Section 4.12 asked for the calculator answers of the four divisions shown in Example 8.1.

Example 8.1: Divide each of these numbers by 8131:

91351785; 91351787; 91351790; 91351795

Solution:

$$9\,1\,3\,5\,1\,7\,8\,5 \; \boxdot \; 8\,1\,3\,1 \; \boxminus \; 1\,1\,2\,3\,5.$$

$$9\,1\,3\,5\,1\,7\,8\,7 \; \boxdot \; 8\,1\,3\,1 \; \boxminus \; 1\,1\,2\,3\,5.$$

$$9\,1\,3\,5\,1\,7\,9\,0 \; \boxdot \; 8\,1\,3\,1 \; \boxminus \; 1\,1\,2\,3\,5.$$

$$9\,1\,3\,5\,1\,7\,9\,3 \; \boxdot \; 8\,1\,3\,1 \; \boxminus \; 1\,1\,2\,3\,5.$$

In Section 4.8 (Example 4.24) we found that, if C is constant, then $A \div C$ increases as A increases. But in the examples above we see no increase as A increases. In each case, the calculator gives the same answer, 11235. We should be getting different answers! Is the calculator stuck?

These examples illustrate one kind of calculator error to watch out for. Another kind of calculator error gives different answers when the same answer is expected! Both errors have the same source. To understand why these problems arise and what we can do about them, we shall look at rational numbers.

The name *rational* comes from the word *ratio,* a relation used for the purpose of comparison. Here are some examples to show the way ratios are used.

1. The ratio of men students to women students in the freshman class at Bigtown State University is 3 to 4.

2. The ratio of honor cards (ace, king, queen, jack, ten) to others in a standard deck of playing cards is 5 to 8.

3. The ratio of guppies to swordtails in a fish tank is 3 to 1.

Such expressions are so common that we have learned to interpret them (correctly) in a natural way. Thus, in example 1, we expect that a male will have an easier time getting a date with a girl than he would if the ratio were reversed: 4 men to 3 women (or, 3 women to every 4 men.)

Ratios are related to fractions. Recall that

> A *fraction* is a symbol indicating division. The dividend is the *numerator,* and the divisor is the *denominator.*

For example,

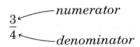

is a fraction where 3 is the numerator and 4 is the denominator. From the definition,

$$\frac{3}{4} = 3 \div 4 = 0.75$$

The ratio, 3 to 4, may be expressed as $\frac{3}{4}$.

In general, the ratio A to B may be expressed as

$$\frac{A}{B}, \text{ indicating } A \div B,$$

and a number that may be expressed as a ratio, $A \div B$, with whole numbers A and B (where $B \neq 0$) is a *rational number.* We may restate the previous examples in terms of fractions.

1. There are $\frac{3}{4}$ as many men as women in the freshman class.

2. There are $\frac{5}{8}$ as many honor cards as there are others.

3. There are $\frac{3}{1}$ as many guppies as there are swordtails.

In this last example, because $3 \div 1 = 3$, we usually say: "There are 3 times as many guppies as there are swordtails."

Now back to the troubles in performing some divisions on a calculator.

> For some problems, we get the same answer when we should be getting different ones.

> For some problems, we get different answers when we should be getting the same ones.

Let us begin looking at the first kind of trouble by tracing some typical algorithms through long division. We'll use a calculator when we can.

Example 8.2(a): $91351785 \div 8131$

Solution:

```
              1 1 2 3 5
8 1 3 1 ) 9 1 3 5 1 7 8 5
          9 1 3 5 1 7 8 5
                        0   Remainder
```

Check: 8131 ⊠ 11235 ⊟ 9135 1785.

Example 8.2(b): $91351787 \div 8131$

Solution:

```
              1 1 2 3 5
8 1 3 1 ) 9 1 3 5 1 7 8 7
          9 1 3 5 1 7 8 5
                        2   Remainder
```

Check: 8131 ⊠ 11235 ⊟ 9135 1785. So
$(8131 \times 11235) + 2 = 91351787$.

Here, the "quotients" are the same, but the remainders are different. From the display for the second problem, it is clear that the division should not have ended with the answer 11235. Rather, to 8 digits,

$$91351787 \div 8131 \doteq 11235.000$$

For a more accurate picture, we need to continue the division.

Example 8.3: 91351787 ÷ 8131; a closer look.

Solution: From the long division in Example 8.2(b), and by using the distributive law,

$$91351787 \div 8131 = (91351785 + 2) \div 8131$$
$$= (91351785 \div 8131) + (2 \div 8131)$$
$$= 11235 + (2 \div 8131).$$

Now let's see what the calculator has to say about the last term in parentheses:

$$2 \; \boxplus \; 8131 \; \boxminus \; 0.0002459$$

So $91351787 \div 8131 = 11235 + 0.0002459$

To find $11235 + 0.0002459$, we again turn to the calculator.

$$11235 \; \boxplus \; .0002459 \; \boxminus \; 11235.$$

Roadblock; we're getting nowhere fast!
But we know from Section 3.3 that

$$
\begin{array}{r}
11235.0000000 \\
+ \quad\;\; 0.0002459 \\
\hline
11235.0002459
\end{array}
$$

If we accept the answer, 11235.0002459, we can see that the calculator was not wrong after all (nor was our mathematics!). To the best of its ability, the machine gave us the first 8 digits of the answer; namely, 11235.000, suppressing the zeros after the decimal point. In other words, the calculator chopped off the answer beyond 8 digits. This chopping off process is called *"rounding off."* We can see from our work above with the remainder how we use the ideas from Section 4.13 to gain more accuracy.

Here's the second kind of trouble. We know that two-halves, three-thirds, or seven-sevenths is one whole.

A half is $\frac{1}{2}$. Two-halves is $\frac{1}{2} \times 2 = 1$.
A third is $\frac{1}{3}$. Three-thirds is $\frac{1}{3} \times 3 = 1$.
A seventh is $\frac{1}{7}$. Seven-sevenths is $\frac{1}{7} \times 7 = 1$.

Let's check these on our calculator.

$$\frac{1}{2} \times 2 = ?$$

$\boxed{1} \boxed{\div} \boxed{2} \boxed{\times} \boxed{2} \boxed{=} \; \textit{1.}$ Check!

$$\frac{1}{3} \times 3 = ?$$

$\boxed{1} \boxed{\div} \boxed{3} \boxed{\times} \boxed{3} \boxed{=} \; \textit{0.9999999}$ Trouble!

$$\frac{1}{7} \times 7 = ?$$

$\boxed{1} \boxed{\div} \boxed{7} \boxed{\times} \boxed{7} \boxed{=} \; \textit{0.9999997}$ Trouble!

We get different answers from those our common sense tells us to expect.

The source of this kind of trouble can be seen to be due to rounding off again, this time for fractions, $\frac{1}{3}$ and $\frac{1}{7}$. We can use the calculator values, 0.9999999, and 0.9999997, to tell us how far off we are from 1 in each case.

Example 8.4: What is the round-off error in $7 \times \frac{1}{7}$?

Solution: The calculator procedure,

$$\boxed{1} \boxed{\div} \boxed{7} \boxed{=} \; \textit{0.1428571}$$

corresponds to the division,

$$7 \overline{)\; 1}^{\; 0.1428571}$$

which means

$$7 \times .1428571 = 1$$

However, the calculator says

$$\boxed{7} \boxed{\times} \; \textit{0.1428571} \; \boxed{=} \; \textit{0.9999997}$$

To find the discrepancy, subtract this number from 1 (using the calculator):

$$\boxed{1} \boxed{-}\textit{.9999997} \boxed{=} \; \textit{0.0000003}$$

In other words,

$$1 = (7 \times 0.142857) + 0.0000003$$

The round-off error in $7 \times \frac{1}{7}$ is then 0.0000003.

Now suppose we asked our calculators to compute in the reverse order!

$$2 \times \frac{1}{2} = ?$$

2 ⊠ 1 ÷ 2 = $1.$ Check!

$$3 \times \frac{1}{3} = ?$$

3 ⊠ 1 ÷ 3 = $1.$ Check!

$$7 \times \frac{1}{7} = ?$$

7 ⊠ 1 ÷ 7 = $1.$ Check!

In the next section, we review these general laws for arithmetic and see how they help us offset some of the round-off errors.

Exercises 8.1

1. For each fraction, $\frac{1}{A}$, find $A \times \frac{1}{A}$ on the calculator. Which ones did not turn out to be 1?

   a. $\frac{1}{3}$ •d. $\frac{1}{4}$ g. $\frac{1}{20}$

   •b. $\frac{1}{5}$ e. $\frac{1}{9}$ •h. $\frac{1}{25}$

   •c. $\frac{1}{6}$ f. $\frac{1}{10}$ •i. $\frac{1}{30}$

•2. Observing that $91351785 \div 8131$ and $91351793 \div 8131$ both give the same answer on an 8-digit calculator, suggest seven other numbers that would give the same "quotient" on the calculator when each is divided by 8131.

3. Use the method of Example 8.3 to express each of the following to 12 digits.

   a. $91351786 \div 8131$ e. $91351791 \div 8131$

   b. $91351788 \div 8131$ f. $91351792 \div 8131$

   c. $91351789 \div 8131$ g. $91351793 \div 8131$

   d. $91351790 \div 8131$

4. Express each of the following ratios in fraction form.

   •a. The ratio of the two gears is 15 to 55.

   b. The ratio of pesos to dollars is 100 to 8.

   •c. The ratio of land area to water is 1 to 5.

   d. The ratio of dollars to doughnuts is 1 to 6.

5. Express each in ratio form.

   a. The picture is reduced to $\frac{1}{4}$ the original size.

   •b. This calculator is $\frac{2}{3}$ the price of the other.

   •c. Bob's father is twice as old as he.

   d. There are $\frac{7}{6}$ as many red stripes as white stripes in the flag.

6. Follow the routine in Example 8.4 to find the round-off error for each.

   a. $\frac{2}{7} \times 7$ •c. $\frac{4}{7} \times 7$ e. $\frac{6}{7} \times 7$

   •b. $\frac{3}{7} \times 7$ d. $\frac{5}{7} \times 7$ •f. $\frac{7}{7} \times 7$

7. List the products of Exercise 6 (including the one in Example 8.4) in order from least round-off error to the most.

8.2   Reciprocals and Equivalent Fractions

To justify procedures with fractions and their decimal equivalents, we must look into the four arithmetic operations $(+, -, \times, \div)$ on rational numbers in general. But first we need to establish, once and for all, the meaning of fractions and decimal notations to the right of the decimal point.

We shall assume that addition and multiplication of rational numbers have two important properties that whole numbers have. These are the properties that allow us to shuffle the order of the numbers involved and to regroup the numbers as needed.

In the first section, we recalled that products like $\frac{1}{2} \times 2$, $\frac{1}{3} \times 3$, and $\frac{1}{7} \times 7$ should all be equal to 1. This property is so important that a special name is given to two numbers that are so related.

> If B and C are numbers such that $B \times C = 1$, then B and C are called *reciprocals* of each other.

Each number (other than 0) has only one reciprocal. Zero has no reciprocal because it can never happen that 0 times any number is 1, and this is true because

$$0 \times A = 0 \quad \text{for any number } A.$$

Notice that reciprocals come in pairs: if B is the reciprocal of C, then C is the reciprocal of B. Can you find a number that is the reciprocal of itself? (There are two such numbers.)

Reciprocals of powers of 10's show clearly the place values to the left and right of the decimal point.

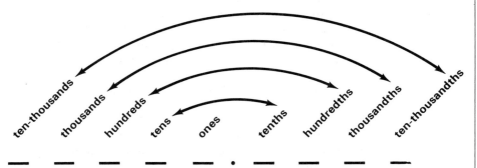

Example 8.5: Reciprocals of powers of tens.

$\dfrac{1}{10}$ is the reciprocal of 10; $\dfrac{1}{10}$ is read: "one-tenth."

$1 \div 10 = 0.1$ Hence 0.1 is read: "one-tenth."

$\dfrac{1}{100}$ is the reciprocal of 100; $\dfrac{1}{100}$ is read: "one-hundredth."

$1 \div 100 = 0.01$ Hence 0.01 is read: "one-hundredth."

$\dfrac{1}{1000}$ is the reciprocal of 1000; $\dfrac{1}{1000}$ is read: "one-thousandth."

$1 \div 1000 = 0.001$ Hence 0.001 is read: "one-thousandth."

Notice that 1000 has 3 zeros and 0.001 is in the third place to the right of the decimal point.

$$\frac{1}{1 \text{ googol}} \text{ is the reciprocal of 1 googol.}$$

$$\text{Hence } \frac{1}{1 \text{ googol}} = \underbrace{0.000 \cdots 1}_{99 \text{ zeros}}$$

The extension of place-values can thus be shown as in Figure 8.1.

Figure 8.1

So we have found how to interpret a digit "1" in any decimal place. Now, what is the meaning of

0.04?   0.002?   0.15?   0.0002459?

Part of the answer can be given right away. Of course,

0.04 is 4-hundredths or, alternately, $\dfrac{4}{100}$.

Similarly,

$$0.002 \text{ is 2-thousandths or, alternately, } \frac{2}{1000}.$$

More importantly, we shall show below that

$$\frac{4}{100} = 4 \times \frac{1}{100}, \qquad \frac{2}{1000} = 2 \times \frac{1}{1000},$$

and, in general,

$$\frac{A}{B} = A \times \frac{1}{B}.$$

Why is this so? The demonstration requires a little patience. Take the case of $\frac{4}{100}$. We wish to show that this is the same as $4 \times \frac{1}{100}$. Suppose $4 \times \frac{1}{100}$ is equal to *some* number B. That is,

$$4 \times \frac{1}{100} = B.$$

Now, $\qquad \left(4 \times \dfrac{1}{100}\right) \times 100 = B \times 100$

$$= 100 \times B \quad \text{(re-ordering)}$$

$$4 \times \left(\frac{1}{100} \times 100\right) = 100 \times B \quad \text{(regrouping)}$$

Because $\frac{1}{100} \times 100 = 1$ (they are reciprocals), then

$$4 \times 1 = 100 \times B$$

or $\qquad\qquad\qquad 100 \times B = 4$

On the other hand, remember that

$$100 \times B = 4 \quad \text{and} \quad B = 4 \div 100$$

state the same fact. Hence

$$B = \frac{4}{100}.$$

Because B was designated to be the number $4 \times \frac{1}{100}$, we have, therefore,

$$B = 4 \times \frac{1}{100} = \frac{4}{100}; \quad \text{or} \quad 0.04 = 4 \times \frac{1}{100}.$$

Similarly,
$$0.002 = \frac{2}{1000} = 2 \times \frac{1}{1000}.$$

The proof that $\frac{2}{1000} = 2 \times \frac{1}{1000}$ follows the same line of reasoning. So does the proof for

$$\frac{A}{B} = A \times \frac{1}{B}.$$

A special result of this general statement is

$$B \times \frac{A}{B} = A.$$

To see this, notice that

$$B \times \frac{A}{B} = B \times \left(A \times \frac{1}{B} \right)$$

$$= A \times \left(B \times \frac{1}{B} \right) \quad \text{(Shuffling)}$$

$$= A \times \frac{B}{B}$$

$$= A \times (B \div B)$$

$$= A \times 1 = A.$$

To explain the meaning of 0.15 or, even worse, of 0.0002459, we need to know about equivalent fractions and addition of rational numbers.

When two fractions, $\frac{A}{B}$ and $\frac{C}{D}$, are equivalent, we write

$$\frac{A}{B} = \frac{C}{D}.$$

For example,

$$\frac{2}{3} = \frac{4}{6}; \quad \text{so} \quad \frac{2}{3} \text{ and } \frac{4}{6} \text{ are equivalent.}$$

That is, this means $2 \div 3 = 4 \div 6$. Try it on the calculator.

Example 8.6: $2 \div 3 = ?$ $4 \div 6 = ?$

Solution:

$2 \div 3 = 0.6666666$ $4 \div 6 = 0.6666666$

Similarly, $\frac{2}{3} = \frac{6}{9}$. Confirm it on your calculator. Therefore, all three numbers ($\frac{2}{3}$, $\frac{4}{6}$, and $\frac{6}{9}$) are equivalent.

In fact, there is a little test we can use for equivalence.

$$\text{If } \frac{A}{B} = \frac{C}{D} \text{ then } A \times D = B \times C.$$

This is called *cross-multiplying*.

$$\frac{A}{B} \times \frac{C}{D}$$

Here is a proof.

Because $\frac{A}{B} = \frac{C}{D}$, multiplying both fractions by the same number will result in an equality. What number will kill the denominator in both fractions? $B \times D$!

We will therefore choose to multiply both by $B \times D$. So

$$(B \times D) \times \frac{A}{B} = (B \times D) \times \frac{C}{D}$$

$$D \times \left(B \times \frac{A}{B}\right) = B \times \left(D \times \frac{C}{D}\right) \quad \text{(rearranging terms)}$$

$$D \times A = B \times C \quad \text{(From the special result on page 228)}$$

or $A \times D = B \times C.$

That's all there is to the proof! It also works the other way around.

Example 8.7: Test $\frac{2}{3}, \frac{4}{6}, \frac{6}{9}$ by cross-multiplying.

Solution:

$\frac{2}{3} = \frac{4}{6}$: $2 \times 6 = 3 \times 4$?
$\qquad\qquad 12 = 12 \qquad$ Check.

$\frac{2}{3} = \frac{6}{9}$: $2 \times 9 = 3 \times 6$?
$\qquad\qquad 18 = 18 \qquad$ Check.

$\frac{4}{6} = \frac{6}{9}$: $4 \times 9 = 6 \times 6$?
$\qquad\qquad 36 = 36 \qquad$ Check.

Now recall that

$$91351785 \div 8131 \quad \text{and} \quad 91351787 \div 8131$$

both came out to be 11235 on the calculator. We can check these by cross-multiplying:

$$\frac{91351785}{8131} \diagdown\!\!\!\!\!\diagup \frac{91351787}{8131}\,?$$

By the method of Section 4.5,

$$91351785 \times 8131 = 742781363835$$

$$91351787 \times 8131 = 742781380097$$

Hence we see that the two divisions should not have resulted in the same answer! [Of course, we also know that in dividing by a fixed number (8131), the larger the numerator, the larger we expect the quotient to be.]

Here's another useful rule:

$$\frac{A \times K}{B \times K} = \frac{A}{B}.$$

That is, if the numerator and denominator of a fraction are both multiplied by the same fixed number K, the resulting fraction is equivalent to the original. We verify this rule by cross-multiplying.

The test:

$$(A \times K) \times B \overset{?}{=} (B \times K) \times A.$$

At this point, the answer is obvious!

It is on this idea, of course, that we got $\frac{4}{6}$ and $\frac{6}{9}$ as fractions equivalent to $\frac{2}{3}$:

$$\frac{2}{3} = \frac{2 \times 2}{3 \times 2} = \frac{2 \times 3}{3 \times 3}.$$

In fact,

$$\frac{2}{3} = \frac{2 \times 4}{3 \times 4} = \frac{2 \times 5}{3 \times 5} = \cdots$$

Example 8.8: $\dfrac{2}{3} = \dfrac{2 \times 8}{3 \times 8} = \dfrac{2 \times 17}{3 \times 17} = \cdots$

Solutions (*on your calculator!*):

$\dfrac{2}{3}$: $\qquad\qquad\qquad$ $2 \div 3 = 0.6666666$

$\dfrac{2 \times 8}{3 \times 8} = \dfrac{16}{24}$: $\qquad$ $16 \div 24 = 0.6666666$

$\dfrac{2 \times 17}{3 \times 17} = \dfrac{34}{51}$: $\qquad$ $34 \div 51 = 0.6666666$

This relationship,

$$\frac{A \times K}{B \times K} = \frac{A}{B},$$

is the basis for the usual technique for *cancellation*:

$$\frac{16}{24} = \frac{2 \times \cancel{8}}{3 \times \cancel{8}} = \frac{2}{3}.$$

By the same token, we can write fractions *"to higher terms."* This is used to get equal denominators for addition and subtraction of fractions. (More of this later.)

Example 8.9: Equivalent fractions.

$$\frac{1}{2} = \frac{1 \times 2}{2 \times 2} = \frac{1 \times 3}{2 \times 3} = \frac{1 \times 4}{2 \times 4} = \frac{1 \times 5}{2 \times 5} = \frac{1 \times 6}{2 \times 6} = \cdots$$

$$= \frac{2}{4} = \boxed{\frac{3}{6}} = \frac{4}{8} = \frac{5}{10} = \frac{6}{12} = \cdots$$

$$\frac{2}{3} = \frac{2 \times 2}{3 \times 2} = \frac{2 \times 3}{3 \times 3} = \frac{2 \times 4}{3 \times 4} = \frac{2 \times 5}{3 \times 5} = \frac{2 \times 6}{3 \times 6} = \cdots$$

$$= \boxed{\frac{4}{6}} = \frac{6}{9} = \frac{8}{12} = \frac{10}{15} = \frac{12}{18} = \cdots$$

(Notice that the fractions circled have a common denominator; 6 is the *least* common denominator for $\frac{1}{2}$ and $\frac{2}{3}$.)

Exercises 8.2

1. For each of the following, find $A \div B$, then $B \times (A \div B)$, on the calculator.

  •a. $\frac{24}{8}$ •d. $\frac{9}{18}$ •g. $\frac{3}{128}$

  b. $\frac{18}{2}$ e. $\frac{7}{8}$ h. $\frac{213}{32}$

  c. $\frac{18}{9}$ •f. $\frac{13}{16}$ •i. $\frac{0}{5}$

2. Give the reciprocal of each in fraction form, and use your calculator to show that the product of the number and its reciprocal is 1.

  a. 5   •d. 125   •g. 50

  •b. 20   e. 64   h. 320

  c. 25   f. 80   •i. 10000

3. Use your calculator to find each quotient.

  a. $1 \div 0.2$ •d. $1 \div 0.008$ g. $1 \div 0.02$

  •b. $1 \div 0.05$ •e. $1 \div 0.015625$ h. $1 \div 0.003125$

  c. $1 \div 0.04$ •f. $1 \div 0.0125$ i. $1 \div 0.0001$

4. Compare each divisor in Exercise 3 with the corresponding part in Exercise 2. Explain what you notice and why this observation is to be expected.

• 5. Write "1-millionth" in the following forms:

  a. as a fraction;

  b. as a decimal numeral;

  c. in scientific notation.

6. Write each in fractional form:

  •a. $3 \times \frac{1}{8} =$ d. $\frac{1}{64} \times 13 =$ •g. $5 \times \frac{1}{5} =$

  b. $7 \times \frac{1}{4} =$ •e. $\frac{1}{160} \times 79 =$ h. $20 \times \frac{1}{20} =$

  c. $\frac{1}{2} \times 5 =$ •f. $17 \times \frac{1}{32} =$ i. $\frac{1}{4} \times 24 =$

7. Write each fraction, $\dfrac{A}{B}$, as a product with $\dfrac{1}{B}$ as one of the factors.

  a. $\frac{5}{8} =$ •d. $\frac{4}{8} =$ •g. $\frac{16}{16} =$

  •b. $\frac{13}{4} =$ e. $\frac{8}{4} =$ h. $\frac{25}{25} =$

  c. $\frac{7}{32} =$ •f. $\frac{4}{1} =$ i. $\frac{10000}{10000} =$

• 8. Using your calculator, show that the following are equivalent fractions.
$$\frac{3}{4}, \quad \frac{6}{8}, \quad \frac{9}{12}, \quad \frac{15}{20}, \quad \frac{18}{24}.$$

9. Show how each of the fractions in Exercise 8 may be obtained from $\frac{3}{4}$. For example.

$$\frac{12}{16} = \frac{3 \times 4}{4 \times 4} = \frac{3}{4}.$$

10. Use cross-multiplying to show that each of these pairs is equivalent.

  •a. $\frac{3}{4}$, $\frac{6}{8}$ c. $\frac{6}{8}$, $\frac{15}{20}$ e. $\frac{12}{16}$, $\frac{15}{20}$

  b. $\frac{3}{4}$, $\frac{12}{16}$ •d. $\frac{6}{8}$, $\frac{18}{24}$ •f. $\frac{15}{20}$, $\frac{18}{24}$

11. Express each in the form $\dfrac{A \times K}{B \times K}$, and cancel to reduce to lowest terms.

  a. $\frac{6}{8}$ •b. $\frac{24}{42}$ c. $\frac{36}{48}$

12. Use $K = 2, 3, 4,$ and 5 to express each fraction to higher terms. Use the calculator to help get the new numerators and denominators.

  •a. $\frac{2}{5}$ •d. $\frac{5}{16}$ g. $\frac{3}{25}$

  b. $\frac{7}{8}$ e. $\frac{3}{8}$ h. $\frac{1}{20}$

  c. $\frac{1}{4}$ •f. $\frac{13}{4}$ i. $\frac{1}{1}$

13. Follow Example 8.9 to express each fraction to higher terms, then circle the fractions having the least common denominator for each given pair of fractions.

  •a. $\frac{3}{4}$, $\frac{2}{5}$ •c. $\frac{3}{5}$, $\frac{2}{3}$ e. $\frac{3}{4}$, $\frac{1}{2}$

  b. $\frac{2}{3}$, $\frac{5}{8}$ d. $\frac{2}{3}$, $\frac{2}{5}$ f. $\frac{5}{8}$, $\frac{2}{3}$

8.3   Multiplication of Rational Numbers

Here are some of the things we have already found out about rational numbers.

$$\frac{A}{B} = A \div B = A \times \frac{1}{B};$$

$$B \times \frac{A}{B} = A; \quad \text{or, by } \textit{"cancellation,"} \; \cancel{B} \times \frac{A}{\cancel{B}} = A;$$

in particular, $A \times \dfrac{1}{A} = 1,$

and $\dfrac{A}{B} = \dfrac{A \times K}{B \times K}.$

Now, what about multiplication of two rational numbers in general?

$$\frac{A}{B} \times \frac{C}{D} = ?$$

For this, we need a fact about multiplying reciprocals; namely,

$$\frac{1}{B} \times \frac{1}{D} = \frac{1}{B \times D}.$$

Example 8.10(a): Does $\dfrac{1}{2} \times \dfrac{1}{4} = \dfrac{1}{2 \times 4} = \dfrac{1}{8}$?

Solution:

$1 \div 2 \times 1 \div 4 = 0.125$

$1 \div 8 = 0.125$

So $\dfrac{1}{2} \times \dfrac{1}{4} = \dfrac{1}{2 \times 4} = \dfrac{1}{8}$. Yes!

Example 8.10(b): Does $\dfrac{1}{4} \times \dfrac{1}{5} = \dfrac{1}{4 \times 5} = \dfrac{1}{20}$?

Solution:

$1 \div 4 \times 1 \div 5 = 0.05$

$1 \div 20 = 0.05$

So $\dfrac{1}{4} \times \dfrac{1}{5} = \dfrac{1}{4 \times 5} = \dfrac{1}{20}$. Yes!

To show that

$$\frac{1}{B} \times \frac{1}{D} = \frac{1}{B \times D},$$

notice that $\dfrac{1}{B \times D}$ is the reciprocal of $B \times D$:

$$(B \times D) \times \left(\frac{1}{B \times D} \right) = 1$$

Observe also that

$$(B \times D) \times \left(\frac{1}{B} \times \frac{1}{D} \right) = \left(B \times \frac{1}{B} \right) \times \left(D \times \frac{1}{D} \right) \quad \text{(Rearranging)}$$

$$= 1 \times 1$$

$$= 1$$

Therefore, $\dfrac{1}{B} \times \dfrac{1}{D}$ is the reciprocal of $B \times D$. That is,

$$\frac{1}{B} \times \frac{1}{D} = \frac{1}{B \times D}.$$

As each number ($\neq 0$) has one and only one reciprocal, we are led to the conclusion that

$$\frac{1}{B \times D} = \frac{1}{B} \times \frac{1}{D}.$$

At this point, we can proceed to show that

$$\frac{A}{B} \times \frac{C}{D} = \frac{A \times C}{B \times D} \quad \text{for } B \neq 0 \text{ and } D \neq 0$$

In Section 7.4 this rule was stated and used in the calculations, but it was not explained why it should work. First, let us verify the rule on the calculator for two cases.

Example 8.11:

$$\frac{2}{5} \times \frac{3}{4} \stackrel{?}{=} \frac{2 \times 3}{5 \times 4}$$

$$2 \div 5 \times 3 \div 4 = 0.3$$

$$6 \div 20 = 0.3 \leftarrow\!$$

$$\frac{3}{5} \times \frac{3}{8} \stackrel{?}{=} \frac{3 \times 3}{5 \times 8}$$

$$3 \div 5 \times 3 \div 8 = 0.225$$

$$9 \div 40 = 0.225 \leftarrow\!$$

We provide an argument to show that this procedure is as it should be in general. From this, we shall also find an improved and more efficient way to carry out the multiplication on the calculator.

By the meaning of fractions, we have

$$\frac{A}{B} \times \frac{C}{D} = \left(A \times \frac{1}{B}\right) \times \left(C \times \frac{1}{D}\right)$$

$$= (A \times C) \times \left(\frac{1}{B} \times \frac{1}{D}\right). \quad \text{(Rearranging terms)}$$

Recall that

$$\frac{1}{B} \times \frac{1}{D} = \frac{1}{B \times D}.$$

So

$$\frac{A}{B} \times \frac{C}{D} = (A \times C) \times \left(\frac{1}{B \times D}\right)$$

$$= (A \times C) \div (B \times D) \quad \text{(Meaning of fractions)}$$

$$= \frac{A \times C}{B \times D}.$$

Thus we have the usual rule:

> To multiply two fractions, multiply the two numerators, then multiply the two denominators. Form the fraction with the product of numerators as its numerator, and the product of denominators as its denominator.

For example,

$$\frac{5}{8} \times \frac{3}{4} = \frac{5 \times 3}{8 \times 4} = \frac{15}{32} \,,$$

and

$$\frac{6}{35} \times \frac{20}{27} = \frac{6 \times 20}{35 \times 27} = \frac{120}{945} \,.$$

The answer is usually reduced to lowest terms by the method of Section 8.2.

Example 8.12: Reduce $\dfrac{120}{945}$ to lowest terms.

Solution: $\dfrac{120}{945} = \dfrac{8 \times 15}{63 \times 15} = \dfrac{8}{63}\,.$

In Chapter 11 we shall discuss how to split numbers such as 120 and 945 into factors in order to proceed with cancelling. Notice though that the calculator treats $\frac{120}{945}$ and $\frac{8}{63}$ as equal decimals. Indeed they are equal fractions.

Example 8.13:

$\dfrac{120}{945} = ?$ $120 \boxplus 945 \boxminus 0.1269841$

$\dfrac{8}{63} = ?$ $8 \boxplus 63 \boxminus 0.1269841$

Hence the calculator makes no distinction between

$$\frac{120}{945} \quad \text{and} \quad \frac{8}{63}\,.$$

Indeed, we can check these two fractions for equivalence:

$$\frac{120}{945} \overset{?}{=} \frac{8}{63} \qquad 120 \boxtimes 63 \boxminus 7560.$$

$$945 \boxtimes 8 \boxminus 7560.$$

From the statement,

$$\frac{A}{B} \times \frac{C}{D} = \frac{A \times C}{B \times D},$$

we conclude that we can process the multiplication in two different ways.

1. Divide A by B to get $\frac{A}{B}$; divide C by D to get $\frac{C}{D}$; multiply the two quotients.
2. Multiply A by C to get $A \times C$; multiply B by D to get $B \times D$; divide the first product by the second.

There are advantages to both forms.

Example 8.14: $\frac{14}{3} \times \frac{12}{35}$ vs $\frac{14 \times 12}{3 \times 35}$.

Solution:

$\frac{14}{3} \times \frac{12}{35}$: $14 \div 3 \times 12 \div 35 = 1.5999999$

$\frac{14 \times 12}{3 \times 35}$: $3 \times 35 = 105.$ (Record or store)

$14 \times 12 = 168.$

$168 \div 105 = 1.6$

In the above example, it is clear that proceeding by

$$\frac{A}{B} \times \frac{C}{D}$$

introduces errors due to rounding off; whereas, by

$$\frac{A \times C}{B \times D}$$

the result is *exact*.

Notice that the product of the denominators was calculated first. This was done in order to save a step in recalling. Here, we recalled 105 and divided it into 168 right away. Otherwise, we would have to store both 168 and 105, then recall both 168 and 105 (in that order) in an algebraic machine.

When is it an advantage to proceed by $\frac{A}{B} \times \frac{C}{D}$?

Example 8.15: $\dfrac{12345678}{98765432} \times \dfrac{34567891}{76543219}$.

Solution:

$\dfrac{A}{B} \times \dfrac{C}{D}$: $12345678 \boxdot 98765432 \boxtimes 34567891$

$\boxdot 76543219 \boxminus 0.0564515$

$\dfrac{A \times C}{B \times D}$: $12345678 \boxtimes 98765432 \boxminus$ OVERFLOW

By the methods of Section 4.5:

		1234	5678	
	×	3456	7891	
		4480	5098	
	973	7494		
	1962	3168		
426	4704			
426	7640	5142	5098	

Similarly,

		9876	5432	
	×	7654	3219	
		1748	5608	
	3179	0844		
	4157	6528		
7559	0904			
7559	8240	9120	5608	

Finally, by Section 7.4:

$426764051425098 \div 7559824091205608$
$$\doteq (4.2676405 \times 10^{14}) \div (7.5598240 \times 10^{15})$$
$$= (4.2676405 \div 7.559824) \times (10^{14} \div 10^{15})$$
$$\doteq 0.5645158 \times 10^{-1}$$
$$= 0.05645158$$

(Recall that in Section 7.4 we had depended on switching the operations in calculations such as these.)

Comparing the methods in Example 8.15, we thus see that, for large numbers, it is easier to proceed by two separate divisions first:

$$\frac{A}{B} \times \frac{C}{D}.$$

Observe that

$$\frac{A}{B} \times \frac{C}{D} = (A \times C) \times \left(\frac{1}{B} \times \frac{1}{D} \right).$$

That is,
$$\frac{A}{B} \times \frac{C}{D} = \left((A \times C) \times \frac{1}{B}\right) \times \frac{1}{D}$$
$$= [(A \times C) \div B] \div D.$$

So, what we can do is to multiply the numerators and divide by each denominator, one after the other.

Example 8.16: Find $\frac{2}{5} \times \frac{3}{4}$ by this new procedure and compare with the solutions in Example 8.11.

Solution:

$$\frac{2}{5} \times \frac{3}{4} = ((2 \times 3) \div 5) \div 4$$

$$= 2 \,\boxed{\times}\, 3 \,\boxed{\div}\, 5 \,\boxed{\div}\, 4 \,\boxed{=}\, 0.3$$

In Example 8.11, the first solution was processed thus:

$$2 \,\boxed{\div}\, 5 \,\boxed{\times}\, 3 \,\boxed{\div}\, 4 \,\boxed{=}\, 0.3$$

The only difference is the order of processing $\div 5$ and $\times 3$.
  An alternate solution was shown as

$$6 \,\boxed{\div}\, 20 \,\boxed{=}\, 0.3$$

What was not displayed in Example 8.11 was that 6 and 20 must be obtained first as results of 2×3 and 5×4. Therefore, in reality, this alternate solution involves more steps than either of the others.

Example 8.14 is equally easy to process by this new procedure.

Example 8.17: $\dfrac{14}{3} \times \dfrac{12}{35} = ?$

Solution: $14 \,\boxed{\times}\, 12 \,\boxed{\div}\, 3 \,\boxed{\div}\, 35 \,\boxed{=}\, 1.6$

Although we use the calculator to carry out our computations, occasionally we may have to do some hand calculation. Then, it is convenient to get the numbers down to "bite size."

Example 8.18: Find the product (by hand); then check using the calculator:

$$\frac{6}{49} \times \frac{35}{48}.$$

Solution:

$$\frac{6}{49} \times \frac{35}{48} = \frac{6 \times 35}{49 \times 48} = \frac{\cancel{6} \times \cancel{7} \times 5}{\cancel{7} \times 7 \times \cancel{6} \times 8}$$

$$= \frac{5}{7 \times 8} = \frac{5}{56}.$$

Check: $6 \;\boxed{\times}\; 35 \;\boxed{\div}\; 49 \;\boxed{\div}\; 48 \;\boxed{=}\; 0.0892857$
$5 \;\boxed{\div}\; 7 \;\boxed{\div}\; 8 \;\boxed{=}\; 0.0892857$

Application: A crate of oranges sells for $6. How much would $\frac{3}{4}$ of a crate cost at that price?

Answer: We want to find $\frac{3}{4}$ of 6; or, in other words, $\frac{3}{4} \times 6$. The integer 6 is equivalent to $\frac{6}{1}$ because $6 \div 1 = 6$. So

$$\frac{3}{4} \times \frac{6}{1} = \frac{3 \times 6}{4 \times 1} = \frac{3 \times (3 \times \cancel{2})}{2 \times \cancel{2}} = \frac{3 \times 3}{2}.$$

$$3 \;\boxed{\times}\; 3 \;\boxed{\div}\; 2 \;\boxed{=}\; 4.5$$

Thus, $\frac{3}{4}$ of a crate should cost $4.50.

Exercises 8.3

1. Rearrange the numbers so that you can use reciprocals to make the computations easy. Then find the obvious replacement by a number for the letter in each problem.

•a. $5 \times (3 \times \frac{1}{5}) = A$ •e. $25 \times \left(4 \times \dfrac{1}{F}\right) = 4$

•b. $8 \times (5 \times \frac{1}{8}) = B$ f. $200 \times (x \times \frac{1}{200}) = 15$

c. $32 \times (17 \times \frac{1}{32}) = C$ g. $256 \times \left(346 \times \dfrac{1}{y}\right) = 346$

d. $16 \times (13 \times \frac{1}{16}) = E$ h. $Z \times (43 \times \frac{1}{129}) = 43$

2. Use "cancellation" to find each product; check by the calculator.

•a. $5 \times \frac{4}{5} =$ •d. $\frac{234}{1000} \times 1000 =$

b. $7 \times \frac{8}{7} =$ e. $234 \times \frac{432}{234} =$

c. $\frac{9}{11} \times 11 =$ •f. $1 \times \frac{3}{1} =$

3. Complete the chain of multiplication.

$$24 \times \frac{7}{6} = (4 \times 6) \times \frac{7}{6} =$$

4. Find each product.

•a. $\frac{3}{4}$ of 12 =   •d. $\frac{3}{5}$ of 25 =   •g. $\frac{1}{3}$ of 24 =

b. $\frac{4}{3}$ of 12 =   •e. $\frac{3}{4}$ of 20 =   h. $\frac{7}{8}$ of 32 =

c. $\frac{4}{5}$ of 20 =   f. $\frac{3}{7}$ of 35 =   i. $\frac{8}{9}$ of 36 =

• 5. Using the hint in Exercise 4, parts d–f, find the missing number.

$$\frac{3}{8} \text{ of } \square = 15.$$

6. Find these products by hand, using as much cancellation as you can. Use the calculator to check your answers.

•a. $\frac{21}{4} \times \frac{12}{7} =$ •d. $\frac{42}{5} \times \frac{65}{6} =$ g. $\frac{14}{3} \times \frac{18}{7} =$

b. $\frac{18}{5} \times \frac{35}{6} =$ e. $\frac{20}{3} \times \frac{27}{4} =$ •h. $\frac{3}{4} \times \frac{4}{3} =$

•c. $\frac{36}{7} \times \frac{77}{9} =$ f. $\frac{25}{12} \times \frac{24}{5} =$ i. $\frac{7}{8} \times \frac{8}{7} =$

7. With the help of the calculator and using the idea that

$$\frac{A}{B} \times \frac{C}{D} = (A \times C) \times \left(\frac{1}{B} \times \frac{1}{D}\right),$$

find each product.

a. $\frac{3}{4} \times \frac{7}{8} =$ d. $\frac{9}{5} \times \frac{11}{10} =$ •g. $\frac{6}{7} \times \frac{14}{3} =$

•b. $\frac{3}{8} \times \frac{7}{4} =$ •e. $\frac{5}{16} \times \frac{1}{4} =$ h. $\frac{6}{3} \times \frac{14}{7} =$

c. $\frac{9}{10} \times \frac{11}{5} =$ f. $\frac{3}{16} \times \frac{5}{4} =$ i. $\frac{25}{7} \times \frac{49}{5} =$

8. Use your calculator to compare the results of $\frac{1}{B} \times \frac{1}{D}$ with $\frac{1}{B \times D}$ for each of the following.

a. $\frac{1}{2} \times \frac{1}{8} =$ d. $\frac{1}{4} \times \frac{1}{4} =$ g. $\frac{1}{3} \times \frac{1}{7} =$

b. $\frac{1}{5} \times \frac{1}{20} =$ e. $\frac{1}{8} \times \frac{1}{4} =$ h. $\frac{1}{3} \times \frac{1}{3} =$

•c. $\frac{1}{4} \times \frac{1}{20} =$ •f. $\frac{1}{25} \times \frac{1}{5} =$ •i. $\frac{1}{3} \times \frac{1}{6} =$

9. Explain why it is true that $\frac{A}{B} \times \frac{C}{D} = \frac{A}{D} \times \frac{C}{B}$.

10. Use the "numerator times numerator, denominator times denominator" rule to compute each part of Exercise 7 on the calculator.

•11. Use the idea of $[(A \times C) \div B] \div D$ to calculate each part of Exercise 7 on the calculator.

12. Show how the product of two whole numbers, A and C, is covered by the rule for multiplying two rational numbers.

13. Give the reciprocals of each fraction below. (HINT: See Exercise 6.)

•a. $\frac{3}{4}$ d. $\frac{32}{15}$ g. $\frac{16}{5}$

b. $\frac{6}{7}$ •e. $\frac{7}{16}$ •h. $\frac{6}{23}$

•c. $\frac{5}{8}$ f. $\frac{4}{13}$ i. $\frac{4}{4}$

8.4 Addition and Subtraction of Rational Numbers

Addition and subtraction of rational numbers in decimal form are quite simple on the calculator (provided, of course, that the calculator can accommodate the number of digits needed). Review Section 3.3.

Example 8.19: $0.427 + 0.832 = ?$

Solution: $0.427 \boxplus 0.832 \boxminus 1.259$

To see why we should get such an answer, first recall that place values for decimals work just as they do for whole numbers.

A digit in the thousands place is 10 times as much as that same digit in the hundreds place:

$$2000 = 200 \times 10$$

2 in thousands place   2 in hundreds place

A digit in the hundreds place is 10 times as much as that same digit in the tens place:

$$400 = 40 \times 10$$

4 in hundreds place   4 in tens place

And so on.

Similarly,

$$0.2 = 0.02 \times 10$$

2 in tenths place   2 in hundredths place

Example 8.20: Meaning of .15

0.15 means $(1 \times \frac{1}{10}) + (5 \times \frac{1}{100})$.

That is, $\qquad\qquad 0.15 = \frac{1}{10} + \frac{5}{100}$.

Now, how do we add two fractions? A basic rule is this:

$$\frac{A}{C} + \frac{B}{C} = \frac{A + B}{C}.$$

In words:

> Two fractions with the *same* denominator (a common denominator) are added by adding the numerators. The denominator remains the same.

Example 8.21: $\frac{3}{7} + \frac{2}{7} = ?$

Solution: $\dfrac{3}{7} + \dfrac{2}{7} = \dfrac{3 + 2}{7} = \dfrac{5}{7}.$

If the denominators are different, we look for equivalent fractions having a common denominator.

Example 8.22: $\frac{2}{3} + \frac{1}{5} = ?$

Solution:

$$\frac{2}{3} = \frac{2 \times 5}{3 \times 5} = \frac{10}{15} \quad \text{and} \quad \frac{1}{5} = \frac{1 \times 3}{5 \times 3} = \frac{3}{15}.$$

Now, both fractions are expressed in fifteenths; so

$$\frac{2}{3} + \frac{1}{5} = \frac{10}{15} + \frac{3}{15} = \frac{10 + 3}{15}$$

$$= \frac{13}{15}.$$

In decimals, this job becomes very easy. To see why we can rely on the easy algorithm, we take two steps. First, we review the meaning of a decimal. Then we perform addition based on the meaning.

Example 8.23: $0.1 + 0.05 = \frac{1}{10} + \frac{5}{100} = ?$

Solution:

$$\frac{1}{10} = \frac{1 \times 10}{10 \times 10} = \frac{10}{100} \quad \text{and} \quad \frac{5}{100} = \frac{5}{100}.$$

Now both fractions are expressed in hundredths; so

$$\frac{1}{10} + \frac{5}{100} = \frac{10}{100} + \frac{5}{100} = \frac{10 + 5}{100} = \frac{15}{100}.$$

That is, $\quad \dfrac{1}{10} + \dfrac{5}{100} = \dfrac{15}{100}.$

From Example 8.10, we have

$$\frac{1}{10} + \frac{5}{100} = 0.15$$

Hence, $\qquad 0.1 + .05 = 0.15$

Check on the calculator:

$$0.1 \; \boxplus \; 0.05 \; \boxminus \; 0.15$$

Here, we observe that, because

$$\frac{1}{10} + \frac{5}{100} = 0.15 \quad \text{and} \quad \frac{1}{10} + \frac{5}{100} = \frac{15}{100},$$

it follows that

$$0.15 = \frac{15}{100}.$$

From this, we see the place-value interpretation of a decimal. Also, we have the answer to the question we asked on page 226:

The last digit of 0.15 is in the hundredths place;

$$0 \quad . \quad \overset{\text{tenths}}{1} \quad \overset{\text{hundredths}}{5}$$

so 0.15 is 15 hundredths, or $\frac{15}{100}$.

Now, what is the meaning of 0.2459?

The last digit of .2459 is in the ten-thousandths place:

$$0 \quad . \quad \overset{\text{tenths}}{2} \quad \overset{\text{hundredths}}{4} \quad \overset{\text{thousandths}}{5} \quad \overset{\text{ten-thousandths}}{9}$$

so 0.2459 is 2459 ten-thousandths, or $\frac{2459}{10000}$.

Here are the details:

$$0.2459 \text{ means } \left(2 \times \frac{1}{10}\right) + \left(4 \times \frac{1}{100}\right) + \left(5 \times \frac{1}{1000}\right) + \left(9 \times \frac{1}{10000}\right)$$

$$= \frac{2}{10} + \frac{4}{100} + \frac{5}{1000} + \frac{9}{10000}$$

or,

$$= \frac{2 \times 1000}{10 \times 1000} + \frac{4 \times 100}{100 \times 100} + \frac{5 \times 10}{1000 \times 10} + \frac{9}{10000}$$

$$= \frac{2000}{10000} + \frac{400}{10000} + \frac{50}{10000} + \frac{9}{10000}$$

$$= \frac{2000 + 400 + 50 + 9}{10000}$$

$$= \frac{2459}{10000}.$$

Now let us again examine Example 8.19.

Example 8.24: $0.427 + 0.832 = ?$

$$0.427 = 427 \times \frac{1}{1000} \quad \text{and} \quad 0.832 = 832 \times \frac{1}{1000}.$$

So $0.427 + 0.832 = \dfrac{427}{1000} + \dfrac{832}{1000}$

$$= \left(427 \times \frac{1}{1000}\right) + \left(832 \times \frac{1}{1000}\right)$$

$$= (427 + 832) \times \frac{1}{1000} \quad \text{(By the distributive law)}$$

$$= 1259 \times \frac{1}{1000} = \frac{1259}{1000}.$$

But $\dfrac{1259}{1000} = (1000 + 259) \times \dfrac{1}{1000}$

$$= \left(1000 \times \frac{1}{1000}\right) + \left(259 \times \frac{1}{1000}\right)$$

$$= \quad 1 \quad + \quad 0.259,$$

which we write as 1.259.

It is the idea of adding fractions with a common denominator that allows us to write the numbers vertically, aligned by the decimal points, and add by the columns.

For example,

$$0.427 + 0.832 = \frac{427}{1000} + \frac{832}{1000} = \frac{1259}{1000}.$$

Written vertically,

$$
\begin{array}{r}
0.427 \\
+0.832 \\
\hline
1259 \quad \text{thousandths}
\end{array}
$$

As 1259 thousandths = 1.259 (Example 8.24), the decimal point in the answer falls right in line with the decimal points of the numbers we are adding:

$$
\begin{array}{r}
.427 \\
+.832 \\
\hline
1.259
\end{array}
$$

The "1" to the left of the decimal point is "carried over" from the 0.4 + 0.8.

Example 8.25: $0.4 + 0.8 = ?$

Solution:

$$0.4 + 0.8 = \frac{4}{10} + \frac{8}{10} = \frac{12}{10}$$

$$= \frac{10 + 2}{10} = 1 + \frac{2}{10} = 1.2$$

Check: $0.4 \boxplus 0.8 \boxminus 1.2$

Changing Example 8.24 a bit, we show also how to do these problems by scientific notation.

Example 8.26: $0.427 + 0.832$ by scientific notation.

Solution:

$$0.427 = 4.27 \times 10^{-1} \quad \text{and} \quad 0.832 = 8.32 \times 10^{-1}$$

$$(4.27 \times 10^{-1}) + (8.32 \times 10^{-1}) = (4.27 + 8.32) \times 10^{-1}$$

$$= 12.59 \times 10^{-1} = 12.59 \div 10$$

$$= 1.259 \quad \text{(Recall shifting decimal points, page 209)}$$

In this example, observe that both numbers have the same power of ten as factors, in order to make use of the distributive law.

The numbers we had above were easy to work with and were chosen just to illustrate the method without having the arithmetic get in the way. The strength and usefulness of this procedure is seen more clearly in the following type of problems.

Example 8.27: $0.00000005238 + 0.0000000006294 = ?$

Solution: By scientific notation,

$$0.00000005238 = 5.238 \times 10^{-8};$$
$$0.0000000006294 = 6.294 \times 10^{-10}.$$

We cannot use the distributive law right away because there is no common factor shown. But notice the alignment:

$$0.00000005238$$
$$0.0000000006294$$

We can use $\quad 5.238 \times 10^{-8} \quad$ with 0.06294×10^{-8}
or $\quad 52.38 \times 10^{-9} \quad$ with 0.6294×10^{-9}
or $\quad 523.8 \times 10^{-10}$ with 6.294×10^{-10}

and so on. Using 10^{-10} as the common factor,

$$5\ 2\ 3.8\ \boxplus\ 6.2\ 9\ 4\ \boxminus\ 5\ 3\ 0.0\ 9\ 4$$

Answer: $530.094 \times 10^{-10} \quad$ or 5.30094×10^{-8}
or 0.0000000530094

The ideas for subtraction follow those for addition. We can go on directly to parallel examples.

Example 8.28: $0.832 - 0.427 = ?$

Solution:

$$0.832 = 832 \times 10^{-3} \quad \text{and} \quad 0.427 = 427 \times 10^{-3}$$
$$0.832 - 0.427 = (832 \times 10^{-3}) - (427 \times 10^{-3})$$
$$= (832 - 427) \times 10^{-3} = 405 \times 10^{-3}$$
$$= 0.405$$

Example 8.29: $0.00000005238 - 0.0000000006294 = ?$

Solution: By scientific notation (Example 8.27):

$$.00000005238 \uparrow = 523.8 \times 10^{-10}$$

$$.0000000006294 \uparrow = 6.294 \times 10^{-10}$$

$$(523.8 - 6.294) \times 10^{-10} = 517.506 \times 10^{-10}$$

$$= 5.17506 \times 10^{-8}$$

$$= 0.0000000517506$$

Example 8.30: The mass of a hydrogen atom is about 1.675×10^{-24} grams and that of a uranium atom is 3.96×10^{-22} grams. What is the difference in mass of these atoms?

Solution:

$$\text{Uranium:} \quad 3.96 \times 10^{-22} = 3.96 \times (10^2 \times 10^{-2}) \times 10^{-22}$$
$$= (3.96 \times 10^2) \times 10^{-24}$$
$$= 396 \times 10^{-24}$$

Hydrogen: $\qquad\qquad 1.675 \times 10^{-24}$

Difference: $\quad (396 - 1.675) \times 10^{-24} = 394.325 \times 10^{-24}$ grams.

It is interesting to return for another look at the number 0.2459 (see page 244) from another point of view. There, we "converted" each digit in terms of ten-thousandths:

$$0.2 \quad = \frac{2000}{10000} = 0.2000$$

$$0.04 \quad = \frac{400}{10000} = 0.0400$$

$$0.005 \quad = \frac{50}{10000} = 0.0050$$

$$0.0009 = \frac{9}{10000} = 0.0009$$

Clearly, when we add, we can ignore the zeros we had "annexed":

$$
\begin{array}{ll}
0.2000 & 0.2 \\
0.0400 & 0.04 \\
0.0050 \quad \rightarrow & 0.005 \\
\underline{+0.0009} & \underline{+0.0009} \\
0.2459 & 0.2459
\end{array}
$$

The "annexed" zeros were simply the results of writing the number for each digit in a common denominator. If you have a calculator that can be set to show a fixed number of decimal places, try the following experiment.

Example 8.31:

	Set "decimals"	Read
Enter 0.02	2 places	0.02
	4 places	0.0200
	6 places	0.020000

Clearly this is the neat way that decimal notation has of annexing zeros in the numerator and denominator of fractions:

$$\frac{2}{100} = \frac{200}{10000} = \frac{20000}{1000000} = \cdots$$

simply becomes

$$0.02 = 0.0200 = 0.020000 = \cdots$$

We can "add" as many zeros as we please!

Adding and subtracting decimals are also subject to overflow problems as for integers and we overcome these problems the same way (Section 3.4).

Example 8.32: Find the sum $316.553 + 0.0325 + 90158.7243$

Solution: Note that we cannot even enter the third number into the machine. It has 9 digits! Lining up the numbers vertically, we see that a natural place to split the numbers might be at the decimal point:

```
        3 1 6 . 5 5 3
            . 0 3 2 5
+9 0 1 5 8 . 7 2 4 3
```

Add the numbers to the right of the decimal point:

$$0.553 \;\boxplus\; 0.0325 \;\boxplus\; 0.7243 \;\boxminus\; 1.3098$$

Record the numbers to the right of the decimal point in the answer:

```
        3 1 6 . 5 5 3
            . 0 3 2 5
+9 0 1 5 8 . 7 2 4 3
          1 . 3 0 9 8
```

Carry the "1" and add to the left-hand side numbers:

$$316 \;\boxplus\; 90158 \;\boxplus\; 1 \;\boxminus\; 90475.$$

Combine the two parts:

```
        3 1 6 . 5 5 3
            . 0 3 2 5
+9 0 1 5 8 . 7 2 4 3
  9 0 4 7 5 . 3 0 9 8
```

Subtraction is treated likewise.

Example 8.33: 89657.3862 − 343.465 = ?

Solution: Lining up and splitting at the decimal point, we find that we must borrow:

$$
\begin{array}{r}
89657.3862 \\
-\quad 343.465 \\
\hline
\end{array}
$$

$$
\begin{array}{r}
8\ 9\ 6\ 5\ 6\overset{7}{7}.\overset{1}{|}3\ 8\ 6\ 2 \\
-\qquad 3\ 4\ 3\ .\ |\ 4\ 6\ 5 \\
\hline
\end{array}
$$

Then proceed:

$$1.3862 \boxminus .465 \boxminus 0.9212$$

$$89656 \boxminus 343 \boxminus 89313.$$

So $\qquad$ 89657.3862 − 343.465 = 89313.9212

A different splitting saves borrowing:

$$
\begin{array}{r}
8\ 9\ 6\ 5\ |\ 7\ .\ 3\ 8\ 6\ 2 \\
-\qquad 3\ 4\ |\ 3\ .\ 4\ 6\ 5 \\
\hline
8\ 9\ 3\ 1\ |\ 3\ .\ 9\ 2\ 1\ 2
\end{array}
$$

$$8965 \boxminus 34 \boxminus 8931. \qquad 7.3862 \boxminus 3.465 \boxminus 3.9212$$

Exercises 8.4

1. In expanded notation, $0.23 = (2 \times \frac{1}{10}) + (3 \times \frac{1}{100})$. Write each of the following in expanded form.

   •a. 23.45   c. 5012.009

   b. 0.000617   •d. 835.2007

•2. Write each expanded form in Exercise 1 using scientific notation for the terms.

3. Write each in decimal notation.

   a. 2 ten-thousands   •c. 2 ten-thousandths

   •b. 20 thousands   d. 20 thousandths

4. State which of the numbers is the larger, and from the decimal notation, tell how many times as large it is as the smaller.

   •a. 2 ten-thousands or 20 thousands;

   b. 2 ten-thousandths or 20 thousandths.

5. Express each of Exercise 3, parts c and d as a common fraction.

6. Perform the following operations.

- •a. $(2.537 \times 10^{-12}) + (3.462 \times 10^{-14})$
- b. $(2.537 \times 10^{-12}) - (3.462 \times 10^{-14})$
- •c. $(7.284 \times 10^{-9}) + (2.84 \times 10^{-13})$
- •d. $(7.284 \times 10^{-9}) - (2.84 \times 10^{-13})$
- e. $(3.567 \times 10^{-9}) - (56.7 \times 10^{-10})$

•7. Write out the following sum in decimal notation:

$$(4.78 \times 10^7) + (2.83 \times 10^{-8}).$$

8. Compute on the calculator:

- •a. $\frac{117}{500} + \frac{3}{4} + \frac{33}{100} + \frac{1}{200} + \frac{1403}{2500} =$
- b. $5266.8862 + 33068.796 + 796.53531 =$
- •c. $82921.976102 - 5431.6992671 =$
- d. $829219.76102 - 543.16992671 =$

9. In Example 8.4, we indicated that the round-off error for $\frac{1}{7}$ to 7-place decimals was 0.0000003; show how this came about by subtraction if $\frac{1}{7}$ was taken to be 0.1428571 (see page 223).

8.5 A Different "Big-Ones-from-Little-Ones" Routine

In Chapter 3, we got "big ones from little ones" by multiplying positive integers (natural numbers.) In Section 8.1, multiplication produced "little ones" from "big ones" when we used rational numbers less than 1 for factors.

Now, dividing a natural number by another natural number gives a smaller number. What about dividing one rational number by another? Try these examples on the calculator and compare the results with the dividend (the number you are dividing into.)

Example 8.34: Dividing by rational numbers.

$5 \div \frac{1}{2}$:	$5 \div (\boxed{1} \boxed{\div} \boxed{2})$	
	$\rightarrow \boxed{5} \boxed{\div} \boxed{0.5} \boxed{=} \boxed{10.}$	$10 > 5.$
$\frac{1}{2} \div 5$:	$\boxed{0.5} \boxed{\div} \boxed{5} \boxed{=} \boxed{0.1}$	$0.1 < 0.5$
$7 \div 3.2$:	$\boxed{7} \boxed{\div} \boxed{3.2} \boxed{=} \boxed{2.1875}$	$2.1875 < 7.$
$0.6 \div 0.003$:	$\boxed{0.6} \boxed{\div} \boxed{0.003} \boxed{=} \boxed{200.}$	$200 > 0.6$

Looking at these examples, we see that, when we divide by 5 or 3.2, the answer is smaller than the dividend. Moreover, when we divide by $\frac{1}{2}$ or 0.003, the answer is larger than the dividend.

In general, when we divide by a (rational) number that is less than 1, we get a larger number; when we divide by a number that is greater than 1, we get a smaller number.

This can be explained by an important relationship between multiplication and division. Recall that

if $B \neq 0$, then $A \div B = C$ and $B \times C = A$ state the same fact.

This property is equally true if A, B, and C are rational numbers. To remind us that we are now thinking of A, B, and C as rational numbers, let us rewrite the numbers in this form:

$$A = \frac{a}{b}, \quad B = \frac{c}{d}, \quad C = \frac{e}{f}.$$

Then,

$$\frac{a}{b} \div \frac{c}{d} = \frac{e}{f} \quad \text{and} \quad \frac{c}{d} \times \frac{e}{f} = \frac{a}{b} \qquad (*)$$

state the same fact.

Now we do a bit of juggling with the multiplication statement:

$$\frac{c}{d} \times \frac{e}{f} = \frac{a}{b}.$$

Because these members are equal, multiplying both by the same number would still result in an equation.

We will choose to multiply by $\frac{d}{c}$. (How is $\frac{d}{c}$ related to $\frac{c}{d}$? See Section 8.2.) It will be clear from what follows why we happen to choose to multiply by this number.

$$\left(\frac{c}{d} \times \frac{e}{f}\right) \times \frac{d}{c} = \left(\frac{a}{b}\right) \times \frac{d}{c}.$$

Notice the canceling on the left-hand side, leaving simply $\frac{e}{f}$. So

$$\frac{e}{f} = \frac{a}{b} \times \frac{d}{c}.$$

But,

$$\frac{e}{f} = \frac{a}{b} \div \frac{c}{d}, \qquad \text{(See ($*$) above)}$$

so we arrive at the conclusion:

$$\frac{a}{b} \div \frac{c}{d} = \frac{a}{b} \times \frac{d}{c}.$$

Thus we have the following rule.

> To divide by a fraction, invert and multiply.

Notice the relationship between the divisor $\frac{c}{d}$ and the inverted fraction $\frac{d}{c}$:

$$\frac{c}{d} \times \frac{d}{c} = \frac{\cancel{c} \times \cancel{d}}{\cancel{d} \times \cancel{c}} = 1. \quad \text{They are reciprocals!}$$

Therefore, the above rule can be reworded.

> To divide a number by $\frac{c}{d}$, multiply the number by the reciprocal, $\frac{d}{c}$.

Let's examine some numbers and their reciprocals.

Example 8.35: Comparing reciprocals.

$$\frac{2}{3} < 1; \quad \frac{3}{2} > 1. \qquad \frac{20}{7} > 1; \quad \frac{7}{20} < 1.$$

$$\frac{3}{4} < 1; \quad \frac{4}{3} > 1. \qquad \frac{4}{4} = 1; \quad \frac{4}{4} = 1.$$

$$\frac{8}{5} > 1; \quad \frac{5}{8} < 1. \qquad \frac{7}{7} = 1; \quad \frac{7}{7} = 1.$$

When is a fraction $\frac{a}{b}$ less than 1? Exactly when the numerator is less than the denominator ($a < b$). The reciprocal of $\frac{a}{b}$ is $\frac{b}{a}$; so if $a < b$, the numerator of the reciprocal is greater than its denominator, which means $\frac{b}{a}$ is greater than 1.

Thus, as in Example 8.35,

if a fraction is less than 1, its reciprocal is greater than 1;

if a fraction is greater than 1, its reciprocal is less than 1.

Now dividing by $\frac{a}{b}$ is the same as multiplying by $\frac{b}{a}$. If $\frac{a}{b} < 1$, then dividing by $\frac{a}{b}$ means multiplying by $\frac{b}{a}$, which is greater than 1. Therefore the number (the dividend) is increased. (Likewise, dividing by a number that is greater than 1, decreases the number.)

Sometimes we have to help the calculator a bit to get better answers.

Example 8.36: $0.000005428 \div 43987 = ?$

Solution: We cannot even begin to key in 0.000005428 in an 8-digit calculator. To get around the calculator's limitations, consider scientific notation:

$$0.000005428 = 5.428 \times 10^{-6}$$

$$43987 = 4.3987 \times 10^4$$

$$0.000005428 \div 43987 = (5.428 \times 10^{-6}) \div (4.3987 \times 10^4)$$

$$= (5.428 \div 4.3987) \times (10^{-6} \div 10^4)$$

$$5.428 \; \boxed{\div} \; 4.3987 \; \boxed{=} \; 1.2340009$$

So $0.000005428 \div 43987 = 1.2340009 \times 10^{-10}$ $(10^{-6} \div 10^4 = 10^{-10})$

$$= 0.00000000012340009$$

(correct to 17 places)

Using this procedure, we can now push Example 8.4 further:

$$
\begin{array}{r}
0.1428571 \\
7\overline{)1.0000000} \\
0.9999997 \\
\hline
0.0000003
\end{array}
$$
$\quad (= 7 \times .1428571)$

Continue: $0.0000003 \div 7 = (3 \times 10^{-7}) \div 7$

$$= (3 \div 7) \times 10^{-7}$$

$$
\begin{array}{r}
0.4285714 \\
7\overline{)3.0000000} \\
2.9999998 \\
\hline
0.0000002
\end{array}
$$
$\quad (= 7 \times 0.4285714)$

$\quad (= 2 \times 10^{-7})$

and $(3 \div 7) \times 10^{-7} = (0.4285714 \times 10^{-7}) + (2 \div 7) \times (10^{-7} \times 10^{-7})$

$$= 0.0000004285714 + \left(\frac{2}{7} \times 10^{-14}\right)$$

So $\quad 1 \div 7 = 0.1428571 + 0.0000004285714 + \left(\frac{2}{7} \times 10^{-14}\right)$

$$= 0.14285714285714 + \left(\frac{2}{7} \times 10^{-14}\right)$$

By now, the pattern for getting greater accuracy is clear.

Exercises 8.5

1. Use your calculator to divide each to 8 places.
   - a. $5.9 \div 0.295$
   - •b. $24.8 \div 3.1$
   - •c. $0.038469312 \div 0.9024$
   - •d. $1263 \div 42.1$
   - e. $0.039852 \div 0.00123$
   - f. $12.4 \div 300$

2. Look at the answers to Exercise 1.
   - •a. How can you tell whether or not there are some more remainders not shown on the calculator?
   - b. Check each answer in Exercise 1 to find out if the answer is exact.

3. Use the calculator to compute each in two ways: (1) by direct division; (2) by using reciprocals.
   - •a. $\frac{45}{32} \div \frac{2}{3}$
   - b. $\frac{2}{3} \div \frac{32}{45}$
   - •c. $\frac{9}{2} \div \frac{16}{2}$
   - •d. $4.5 \div \frac{9}{2}$
   - e. $63 \div \frac{7}{10}$
   - f. $\frac{2}{3} \div \frac{3}{7}$

4. Divide.
   - •a. $542800 \div 0.0043987$
   - b. $7.6295 \div 3052685$
   - •c. $0.000000007425 \div 0.000000225$

•5. A human body has about 5,600 millilitres of blood. Normally, there are about 0.0000165 grams of growth hormone in the blood. How many grams per millilitre of blood are there of hormone? (This is the *concentration* of hormone; divide the number of grams by the number of millilitres.)

Percent and Interest

9.1 What Is Percent?

Much numerical data is expressed in *"percents"* (or "percentages") shown by the symbol %. Here are some examples taken from one issue of a newspaper.

1. . . . *The President's recommendation that an 8 percent cost of living adjustment in Social Security payments . . . be held to 5 percent . . . The administration said the difference between an 8 percent and a 5 percent increase was $2.2 billion.*

2. *Facing a deficit of $2.5 million the . . . school board voted to cut . . . salaries by 15%.*

3. *The County Agent . . . estimates that this year's Perique tobacco crop will be down 40 to 50 percent . . .*

4. *Limited Offer! 20% Discount During Grand Opening!*

5. *. . . Drive in. Save 10% . . .*

6. *They (IUDs) are roughly 95 percent effective in preventing pregnancy.*

7. *Excursion fares are 25 percent lower than . . . last year.*

8. *The Secretary of Housing and Urban Development appearing on CBS's interview program . . . said mortgage interest rates might drop to 8.5% next year. They have been as high as 10.5% in recent months and now range between 9% and 9½%.*

9. *The San Fernando Valley, which represents about 40% of the city's population, is a key to the mayor's political future . . .*

Let's face it; in reading a newspaper, we frequently stumble over "percent."

What is a "percent"? What does the symbol % mean? First note that, in the examples, the term expresses the notion of *change* or *comparison* of two numbers. The words used are "adjustment, cut, down, discount, save, effective, lower rate, and represents." In practice, the actual numbers involved are not always spelled out; they are sometimes large or even variable. Thus, in example 1, the total Social Security benefits are large and not stated (although it is possible to get an estimate by computation). In example 5, because customers' bills will vary, the amount of the discount depends on the bill. The rate of discount is 10%. Thus we may expect to use percents when we cannot specify the precise numbers involved, or when we want to express the rate at which one quantity varies with respect to another.

One of the most common usages of percent is in connection with a state sales tax. For example, California levies a 6% (six percent) sales tax on each item purchased in a store. (In most states, food items are exempt from the tax, but this is irrelevant to our discussion of percent.) In stores, the clerk determines the tax from a *"sales-tax table."*

An equivalent phrase is that the tax is figured at "6 cents on the dollar." Thus a purchase of $1 is taxed 6 cents; the total bill is $1.06. A purchase of $2 is taxed twice as much, 12 cents. The total bill is computed thus:

$$\left(\begin{array}{c}\text{Amount}\\\text{of purchase}\\\text{in dollars)}\end{array}\right) \times \left(\begin{array}{c}\text{Rate of sales tax}\\\text{on each dollar}\\\text{(in dollars)}\end{array}\right) = \left(\begin{array}{c}\text{Total}\\\text{sales tax}\\\text{(in dollars)}\end{array}\right);$$

$$2 \quad \boxed{\times} \quad 0.06 \quad \boxed{=} \; 0.12 \quad (= 12 \text{ cents}).$$

Similarly, a purchase of 75 cents or $\frac{75}{100}$ dollars is computed as

$$75 \; \boxed{\div} \; 100 \quad \boxed{\times} \quad 0.06 \quad \boxed{=} \quad 0.045$$

or $\quad\quad 0.75 \quad \boxed{\times} \quad 0.06 \quad \boxed{=} \quad 0.045 \quad (= 4\frac{1}{2} \text{ cents}).$

Because we cannot pay a half cent, this result is rounded off. This rounding-off is made according to the tax table referred to above, issued by the state (Board of Equalization) in which some taxes are rounded down and some are rounded up. In the long run, overestimates are offset by underestimates.

A purchase of 93 cents or 0.93 dollars is handled the same way:

$$0.93 \; \boxed{\times} \; 0.06 \; \boxed{=} \; 0.0558 \quad (= 5.58 \text{ cents}),$$

and in this case, because 5.58 > 5.5 and so is closer to 6 than to 5, the tax is rounded up to 6 cents.

The tax table determines (as approved by law) just how to round off the parts of a cent to exact penny amounts. If you examine the tax table that a clerk refers to at a store, you will notice that it involves a bit more than the state sales tax. It includes also any local tax that might apply. For example, in San Francisco, an additional $\frac{1}{2}$% sales tax is collected for the city, so the state calculates what it considers to be fair for $6\frac{1}{2}$% in total. The display below shows a small part of this table as an example. Later, when we learn more about fractions of percents, we shall pick a few examples to calculate in order to compare the actual results with some of the entries in the abbreviated chart.

State of California $6\frac{1}{2}$% Sales Tax Reimbursement

Transaction	Tax	Transaction	Tax
0.01–0.10	0.00	1.47–1.61	0.10
0.11–0.20	0.01	1.62–1.76	0.11
0.21–0.35	0.02	1.77–1.92	0.12
0.36–0.51	0.03	1.93–2.07	0.13
0.52–0.67	0.04	2.08–2.23	0.14
0.68–0.83	0.05	2.24–2.38	0.15
0.84–0.99	0.06	2.39–2.53	0.16
1.00–1.15	0.07	2.54–2.69	0.17
1.16–1.30	0.08	2.70–2.84	0.18
1.31–1.46	0.09	2.85–2.99	0.19

This tax example contains a good clue as to the meaning of percent. Note that 6 percent means the same as 6 per hundred, or $\frac{6}{100}$, or 0.06. Thus the "cent" in "percent" has exactly the same meaning as "cent" does with reference to "dollars and cents."

To help us fix in our minds the meaning of the symbol %, we might imagine that we were living at a time when the percent idea was just beginning to catch hold—when bookkeepers, accountants, and scientists, for instance, began to rely on its use. Imagine also, you were laboring away at one of these professions and needed to express percents again and again. First, you faithfully write out the word *percent* each time as needed. Next you might feel justified in abbreviating the word to *Pct*, or *Pc*, or $\frac{1}{100}$, and maybe make do with some corruption of $\frac{1}{100}$ like /00. Maybe this last symbol appears out of balance to you, and you finally turn to a more symmetrical and aesthetic %.

We invented this account of the way the symbol % evolved to emphasize the connection with the factor $\frac{1}{100}$. Hence,

$$9\% \text{ means } 9 \times \frac{1}{100} = \frac{9}{100}.$$

Example 9.1: Use the calculator to express $9\frac{1}{4}\%$ as a decimal.

Solution: Because $9\frac{1}{4}$ means $9 + \frac{1}{4}$, and $\frac{1}{4} = 0.25$,

$$9\tfrac{1}{4}\% = 9.25\%.$$

By agreement (definition),

$$9.25\% = 9.25 \times \frac{1}{100} = 9.25 \div 100$$

On your calculator:

$$9.25 \;\boxplus\; 100 \;\boxminus\; 0.0925$$

So $9\frac{1}{4}\% = 0.0925$.

The digits, 9, 2, and 5 in *9.25%* appear in the result, 0.0925, in *exactly the same order*. This is the rule for expressing a percent as a decimal:

$$9.25 \rightarrow .0925$$

The decimal point is shifted 2 places to the *left*.
    To remember in which direction the decimal point is shifted, remember that

$$1\% \text{ means } \frac{1}{100}, \text{ or } 0.01$$

Hence, from 1., shift two places to the left to get

$$.01$$

by filling in the vacant places with zeros.
    Similarly,

$$9\% \text{ means } .09$$

and

$$9.25\% \text{ means } .0925$$

Remember that 0.0925 and .0925 are just two different ways of writing the same number. The optional zero to the *left* of the decimal point merely calls attention to the existence of the decimal point—it is a reminder that the number is between 0 and 1. In fact, you could write as many zeros as you wish to the *left* of the decimal point—00.0925 and 000.0925 also mean the same thing, although numbers are rarely written this way. However, it is very important to remember that 0.925 and 0.0925 are two *different* numbers; zeros immediately to the *right* of the decimal point *do* have meaning.

•1. Use your calculator to determine which of the following have "7" in the thousandths place.

a. 7%

c. $8\frac{3}{4}\%$

e. $\frac{7}{1000}$

b. 0.7%

d. 77%

f. 7000

2. Refer to the newspaper items listed at the beginning of Section 9.1. Express in decimal form each of the numbers there given in percents. For example, the first item refers to a 5 percent increase in Social Security payments; in decimal form,

$$5\% = \frac{5}{100} = 0.05$$

•3. A stereo record is on sale for $2.40.

a. There is a sales tax of 6% of the purchase price. Find the amount of the tax on this purchase and the total you must pay for the record (including tax).

b. At a nearby store the same record is on sale for $2.43. Find the sales tax at 6% and the total price of the purchase at that store.

c. What is the difference in total price for the purchase at the two stores?

d. Suppose that the sales tax were $6\frac{1}{2}\%$. Use the tax table given in this section to find the tax and total purchase prices at the two stores. In this case, what is the difference in total price for the purchase at the two stores?

e. Comment on the effects of "rounding off" on the transactions described here.

9.2 The Percent Equation: Converting Percents into Decimals and Decimals into Percents

In Section 9.1, we noted that

$$9\% \text{ means } 9 \times \frac{1}{100}.$$

We know from Section 8.2 that

$$9 \times \frac{1}{100} = \frac{9}{100},$$

and so

$$9\% = \frac{9}{100}.$$

Similarly,

$$25\% = 25 \times \frac{1}{100} = \frac{25}{100}.$$

In general, we can state a *percent equation:*

$$a\% = a \times \frac{1}{100} = \frac{a}{100} = a \div 100$$

Now, we shall use a customary practice in algebra. When no sign of operation is indicated, multiplication is intended. Thus, if a and b are numbers,

$$ab \text{ means } a \times b.$$

For example,

$$3a \text{ means } 3 \times a.$$

By this agreement,

$$a\left(\frac{1}{100}\right) \text{ means } a \times \frac{1}{100}.$$

Therefore, in line with our previous comments about % and "/100," it is helpful to think of the symbol % as the number $\frac{1}{100}$. Thus,

$$a\% \text{ means } a\left(\frac{1}{100}\right), \text{ which means } a \times \frac{1}{100} \text{ or } a \div 100$$

In the statement,

$$9\% = \frac{9}{100},$$

we say that 9% has been *converted into the fraction form,* $\frac{9}{100}$.

As discussed in Chapter 8, the result for dividing a given number by 100 can be obtained by moving the decimal point two places to the left:

$$9\% = .09$$

In this case, we say that 9% has been *converted into the decimal form,* .09. Thus,

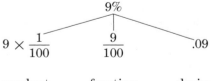

9%		
$9 \times \dfrac{1}{100}$	$\dfrac{9}{100}$	.09
product form	fraction form	decimal form

Some calculators have a % key. This key changes a percent into a decimal. For example, enter 9, press % key, read 0.09. Thus, the % key simply divides the entered number by 100.

In mathematics, whenever we perform some operation, it is important to know whether we can reverse that operation. That is, if we know how to get from A to B, can we find our way back from B to A?

For this reason, because we started with 9% and expressed it as a fraction, $\frac{9}{100}$ (or 0.09), it makes sense to ask

Can a fraction or decimal be written as a percent? Yes, indeed!

Because 9% = 0.09, it must be that 0.09 = 9%, for such is the nature of an equality:

$$\text{If } a = b, \text{ then } b = a.$$

In practice, how do we accomplish this reversal? One way is to note that

$$0.09 \text{ means } \frac{9}{100}, \text{ hence it means 9\%.}$$

This is simple enough. For more complicated expressions, other procedures may be used. Once each routine is firmly settled, you can choose whichever method (algorithm) seems best for any particular problem.

Observe that

$$0.09 = (0.09) \times 1$$

$$= (0.09) \times \frac{100}{100}$$

$$= (0.09 \times 100) \times \frac{1}{100}$$

$$= 9 \times \frac{1}{100} = 9\%.$$

Example 9.2: 0.0925 = ? %

Solution:

$$0.0925 = (0.0925) \times \tfrac{100}{100}$$
$$= (0.0925 \times 100) \times \tfrac{1}{100}$$
$$= 9.25 \times \tfrac{1}{100} \quad \text{(Multiplying by 100 shifts the decimal point 2 places to the right.)}$$

$$= 9.25\%$$

In this way, every decimal can be expressed as a percent.

Example 9.3: $7 = ?\%$

Solution:

$$7 = 7 \times 1 - 7 \times \tfrac{100}{100}$$
$$= (7 \times 100) \times \tfrac{1}{100}$$
$$= 700 \times \tfrac{1}{100} = 700\%.$$

So, $7 = 700\%$.

In general, if we wish to express some number b as a percent, we follow this same procedure:

$$b = b \times 1 = b \times \frac{100}{100}$$

$$= (b \times 100) \times \frac{1}{100}$$

$$= (100 \times b)\%.$$

In other words, to express a number b as a percent, we multiply it by 100 and attach the symbol % to form $(100b)\%$.

Note that, if $b > 1$, then the percent is greater than 100.

To convert a fraction into percent, we have two choices, both of which use the above rule.

Example 9.4: Convert $\tfrac{1}{4}$ to percent.

First Choice: Going by the definition above, we have

$$\frac{1}{4} = \left(\frac{1}{4} \times 100\right)\% = \frac{100}{4}\%$$

$$= 25\%.$$

Second Choice: Express as a decimal.

$$\frac{1}{4} = 0.25, \quad \text{so} \quad (0.25 \times 100)\% = 25\%.$$

Example 9.5: Convert $\tfrac{1}{3}$ to percent.

$$\frac{1}{3} = \left(\frac{1}{3} \times 100\right)\% = 100 \times \frac{1}{3}\%$$

$$= (100 \div 3)\% = 33\frac{1}{3}\%.$$

Second Choice: Express $\frac{1}{3}$ as a decimal.

$$/ \; \boxed{\div} \; 3 \; \boxed{=} \; 0.3333333$$

So
$$\frac{1}{3} = (0.3333333 \times 100)\%$$

$$= 33.333333\% \quad \text{or} \quad 33\frac{1}{3}\%.$$

Exercises 9.2

•1. For each of the following percents, make all three conversions required. In the case of the fraction form, reduce your answer when possible.

	Percent	Product Form	Fraction Form	Decimal Form
a.	11%	$11 \times \frac{1}{100}$		
b.	20%		$\frac{1}{5}$	
c.	8%			0.08
d.	75%			
e.	5%			
f.	100%			
g.	820%			
h.	7.5%			
i.	$\frac{7}{5}\%$			
j.	$\frac{1}{4}\%$			
k.	12.5%			
l.	23%			
m.	$2\frac{1}{2}\%$			
n.	$6\frac{3}{4}\%$			
o.	0.5%			
p.	0.02%			
q.	0.007%			

2. Convert each of the following to percent.

a. 0.06

b. 0.006

c. 0.6

d. 6

e. 0.20

f. 23

g. 3.7

h. 0.2525

i. 0.752

j. 0.0001

3. List all the numbers in Exercise 2 whose percents are—

a. less than 100;

b. greater than 100.

•4. Convert to percent, using both methods in Example 9.4.

a. $\frac{1}{2}$

b. $\frac{1}{5}$

c. $\frac{1}{10}$

d. $\frac{1}{8}$

e. $\frac{3}{8}$

f. $\frac{3}{4}$

g. $1\frac{1}{2}$

h. $6\frac{1}{4}$

•5. It is estimated that in a few years $\frac{3}{20}$ of all oil used by the United States will come from Arab countries. What percent will that be?

6. Mr. Mons Zalott complains that he spends only $\frac{1}{40}$ of his income on pleasure trips. What percent of his income is spent in this manner?

•7. A Housing Authority recommends that no family pays more than 250% of its yearly income for a house. How many times the family income does this represent?

9.3 Comparing Parts of a Whole

Frequently we need to examine how various parts of a quantity compare. Percents are a convenient way of doing this. By using percents, large numerical quantities are expressed as smaller numbers, and so are understood more easily. Here is just such an example.

Example 9.6: A breakdown of Goliath Intergalactic Industries' first-quarter sales looks like this:

Division	Sales (in dollars)
Clear Air Division	146,319,450
Clean Water Division	341,412,050
Solar Energy Division	243,865,750
Recycling Division	195,092,600
Weapons Division	48,773,150
Total Sales	975,463,000

Consider the amount of sales in the weapons division. How much of Goliath's sales is in weapons? Nearly 49 million dollars! Is this portion of Goliath sales large enough to affect the public image of the company or its stock price?

A more accurate picture of the activities of the company in Example 9.6 can be obtained by comparing the weapons division's sales compare with those of the other divisions. At this point, percent becomes useful.

Example 9.7: Compare the sales of the divisions of Goliath Intergalactic Industries (Example 9.6), using percent.

Solution: For the Clear Air Division,

$$\frac{146,319,450}{975,463,000} = \boxed{1\;46\;31\;945} \div \boxed{9\;75\;46\;300} = \boxed{0.15}$$

So the Clear Air Division contributes $\frac{15}{100}$ or 15% to the total sales picture. (Note that we were able to perform this division on our 8-digit calculator because we could "drop a zero" from each number—that is, divide both top and bottom of the fraction by ten before proceeding with the division.)

A small number like 15% is certainly much easier to comprehend than an awesome number like

$$\frac{146,319,450}{975,463,000}.$$

By making similar calculations for the rest of the company's divisions, we obtain the following table.

Division	Percent of Total Sales
Clear Air Division	15%
Clean Water Division	35%
Solar Energy Division	25%
Recycling Division	20%
Weapons Division	5%
Total	100%

In Example 9.7 where percent is used, it is easier to see what is happening. For each hundred dollars of total sales, the Clear Air Division contributes 15 dollars; the Clean Water Division contributes 35 dollars, and so on.

To make an even greater visual impact on the public, the Goliath Company might present the above information in the form of a circle graph (commonly called a *"pie"* diagram).

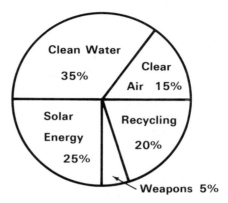

Figure 9.1
Sales of Intergalactic Industries.

The area of each wedge of the "pie" corresponds to a portion of the company's business as labeled. A diagram is only an approximation. More accurate information is given as a percent in each portion. The company has made its point that weapons play only a small part in its operations.

Example 9.8: You want to buy a TV set, and you are told that the store requires a 25% down payment on the $500 selling price. How much are you expected to deposit as down payment?

First Solution: The first thing to do here is to change the question in such a way that we know how to handle it. Because $25\% = \frac{1}{4}$, the question becomes

$$\text{What is } \frac{1}{4} \text{ of } \$500?$$

This is easy to answer; we have

$$1 \div 4 \times 500 = 125.$$

So the down payment is $125.

Second Solution: Because 25% = 0.25, we can process this problem directly on the calculator as follows:

$$0.25 \;\boxed{\times}\; 500 \;\boxed{=}\; 125.$$

Example 9.9: A salesman earns a 30% commission on his sales. Suppose he needs to earn $20,000 a year from his work in order to survive. How many dollars of total sales must he come up with?

Solution: We know that, if his sales are N dollars for the year, then he receives

$$30\% \text{ of } N = \frac{30}{100} \times N \text{ dollars.}$$

Now all he has to do is find out what the number N should be so that

$$\frac{30}{100} \times N = 20{,}000.$$

From Chapter 8, we know that

$$\frac{30}{100} \times N = 20{,}000 \quad \text{and} \quad N = 20{,}000 \div \frac{30}{100}$$

state the same fact. As $\frac{100}{30}$ is the reciprocal of $\frac{30}{100}$, we have

$$N = 20{,}000 \times \frac{100}{30}$$

$$= (20{,}000 \times 100) \div 30$$

$$20000 \;\boxed{\times}\; 100 \;\boxed{\div}\; 30 \;\boxed{=}\; 66666.666$$

or

$$20000 \;\boxed{\div}.3 \;\boxed{=}\; 66666.666$$

Rounding up to the nearest dollar, the answer is that the salesman must sell about $66,667 worth of merchandise in order to earn his $20,000 (at the commission rate of 30%). How would you verify this answer?

Example 9.10: A martini recipe calls for 18% vermouth and 82% gin. (Note that $18 + 82 = 100$, so all the ingredients have been accounted for.) A bon vivant wishes to make a larger pitcher of martini. He has only 6 ounces of vermouth left at his bar, but he has lots of gin. If he uses this recipe, how many ounces of martini would he be able to make? If each glass holds 4 ounces, how many glasses can he serve? How much gin will he use?

Solution: We have to find the number N (the total) so that

$$18\% \text{ of } N = 6 \qquad \text{or} \qquad .18 \times N = 6$$

As $.18 \times N = 6$ and $N = 6 \div .18$ state the same fact, we have

$$N = 6 \div .18$$

$$\boxed{6} \boxed{\div} \boxed{.18} \boxed{=} \;33.333333$$

He can make $33\frac{1}{3}$ ounces (a bit more than 1 quart) of martini. To find the amount of gin that must be added, we have only to subtract the vermouth from the total:

$$33.333333 \boxed{-} \boxed{6} \boxed{=} \;27.333333$$

or $27\frac{1}{3}$ ounces of gin. (In practice, one can just pour the 6 ounces of vermouth first, then pour enough gin to fill up to $33\frac{1}{3}$ ounces.)

Finally, to find the number of glasses of martini that the bon vivant has, we have only to divide the total amount by the number of ounces in each glass:

$$33\tfrac{1}{3} \div 4 = [(\boxed{1} \boxed{\div} \boxed{3}) \boxed{+} \boxed{33}] \boxed{\div} \boxed{4} \boxed{=} \;8.3333332$$

So the bon vivant can serve 8 glasses (with a bit left over which he polishes off while working this problem!)

Example 9.11: You are buying a coffeemaker that sells for \$35.80, and the store wants \$10.74 as a down payment. What percent of the price must be given as a down payment?

Solution: We are asked: What percent of \$35.80 is \$10.74? Or,

$$\text{For what number } R \text{ is } R \times 35.80 = 10.74?$$

Because

$$R \times 35.80 = 10.74 \qquad \text{and} \qquad R = 10.74 \div 35.80$$

state the same fact, we divide 10.74 by 35.80 to find R:

$$R = \boxed{10.74} \boxed{\div} \boxed{35.80} \boxed{=} \;0.3$$

Thus a 30% down payment is required.

In the preceding examples, you may have observed the pattern of relationship among three things: a basic amount W, another amount X being related to (or compared with) W, and a percent p. The pattern lies in this relationship:

$$p \times W = X.$$

For example, when we say that 20 is 25 percent of 80, we mean that

$$0.25 \times 80 = 20$$

(Remember that "25 percent" means "25 per hundred" or "$\frac{25}{100}$" or "0.25.") Because

$$p \times W = X \qquad \text{and} \qquad p = X \div W$$

state the same fact, and

$$p \times W = X \qquad \text{and} \qquad W = X \div p$$

state the same fact, we can express each quantity in terms of the other two:

$$X = p \times W; \qquad p = \frac{X}{W}; \qquad W = \frac{X}{p}.$$

With practice, any of these equations is as handy as the others in solving problems. Thus, the percent (p) is given by the fraction $\frac{X}{W}$.

Exercises 9.3

- 1. Calculate each of the following.
    - a. 15% of 6421
    - b. $6\frac{1}{2}$% of 246.23
    - c. 18.37% of $94\frac{1}{2}$
    - d. 0.037% of $\frac{83}{17}$
    - e. Find N if 9% of N is 43.47.
    - f. 15.3% of N is 732.717. What is N?
    - g. 0.03% of what number is 0.02295?
    - h. What percent of 420 is 63?
    - i. What percent of 8465 is 11004.5?
    - j. 0.000964 = ?% of 0.357?

2. A pair of shorts weighs 12 ounces and contains 30% cotton. How many ounces of cotton are in the shorts?

• 3. If a quart of milk costs 40 cents and 2% of this amount goes to political contributions, how much money from each bottle of milk sold goes to political contributions?

• 4. Find the amount of money contributed in 10 days if 20,000,000 quarts of milk are sold per day (based on the information in Exercise 3).

• 5. If a quart of milk weighs 2 pounds (= 32 ounces) and contains 4% butterfat, how many ounces of butterfat are there in a quart?

6. If a potato chip is 12% vegetable oil, how many ounces of oil are consumed in eating a 9-ounce bag of potato chips?

7. A teacher spends 22% of her time in lesson preparation. How much time in a normal 8-hour workday is spent in lesson preparation? (Give your answer to the nearest minute.) How much time is spent during a normal 5-day week?

• 8. A sport fishing boat catches 39 bonitos, 234 calico bass, and 493 rock cod. What percent of the total number of fish caught were bonito?

9. A baseball player has 87 hits of which 14 are homeruns. What percent of his hits were homeruns?

•10. A salesman is paid a commission of 24% of his sales. If he sells $94,658 worth of goods in one year, what would he be paid in commission that year?

11. A survey questioned the opinion of taxicab users in order to appraise the effectiveness of the taxi service. If out of 20,000 questionnaires distributed, only 1230 were answered, what percent of the questionnaires were returned?

•12. One year, for single taxpayers, the Federal Income Tax Computation Table prescribes that if you itemize your deductions and if your taxable income (after all deductions) is between $16,000 and $18,000, you pay a flat amount of $3,830 plus 34% of the excess of your income over $16,000. If you are single and your taxable income that year is $17,180, what is the amount of tax you must pay?

13. A company that makes automobiles decides to reduce the weight of its small and large models for the coming year. Currently the small model weighs 2,530 pounds, while the larger model weighs 4,320 pounds. The small model is to be reduced in weight by 3.7% and the large by 8.5%.

a. How much will the new small model weigh?

b. How much will the new large model weigh?

c. How much more (in percent) will the new large model weigh than the new small model?

14. Superstar Moo Van Pick made 247 free throws out of 260 attempts.

    a. What percent of free throws did he make?

    b. If Moo maintains this percent, how many free throws would you expect him to make out of his next 180 attempts?

15. An ore contains 8.7% copper. How many pounds of ore will be required to produce 75,000 pounds of copper?

•16. While traveling, you see a sign saying, "Next 5 miles, 6% grade. Trucks use low gear." What is the number of feet of drop from the beginning of the descent to the end of this stretch? (A mile is 5280 feet.)

17. Example 9.1 (Section 9.1) states that the difference between an 8% and a 5% income was $2.2 billion. On this basis, about how much is the total amount in Social Security payments? (HINT: First find the percent difference between 8% and 5%.)

9.4   Percent Change and Percents Greater than 100

In the previous section, all of the examples and exercises involved percents of 100 or less. The reason for this is simple enough; in dealing with "part to whole relationships," a part can never be more than 100% of the whole. You cannot eat 120% of a particular pie; nor can a sweater be 200% wool.

On the other hand, there are many circumstances where percents greater than 100 are not only meaningful, but quite useful as well. If 100% of the price of something is so much, then (by comparison) something else that costs twice as much is 200% of the price of the first item.

Percents greater than 100 often occur when rates of change are to be calculated such as change in price (as above), profit, productivity, speed, weight, and so on. Here are some examples where we compute "percent change."

Example 9.12: The price of a stock was $30 per share five days ago; today it is selling at $45. What is the percent change in value?

Solution: The change in price is $15.

Starting with a value of $30, this price moved up $15. The fraction, $\frac{15}{30}$, tells us that the value changed at the rate of $15 for each $30 invested. In terms of percents,

$$\frac{15}{30} = 0.5 = 50\%.$$

Thus the stock has increased in value by 50%.

Example 9.13: A bamboo shoot is 13 inches tall. Twenty-four hours later, the shoot is 52 inches tall. What is the percent change in height?

Solution: The change in height (in inches) is

$$52 - 13 = 39$$

So the number, $\frac{39}{13}$, tells how the change is compared with the original height: it is 3 times. In terms of percents, the number 3 is 300%:

$$3 = 3 \times \frac{100}{100} = (3 \times 100) \times \frac{1}{100}$$

$$= 300\%.$$

NOTE: A common mistake is to think that, because

$$52 = 4 \times 13,$$

the change is 400%. The present height, 52, does *not* represent the *change* in height. The change is

$$52 - 13, \quad \text{or} \quad 39$$

The percent change is

$$\frac{39}{13} \times 100\% = 300\%.$$

From these examples, we should be able to find an algorithm to compute percent change in any situation. Here it is.

We start with a basic amount B, and end up with a final amount F. To find how much has been increased (or decreased), we subtract:

$$\text{change} = F - B.$$

Then the increase (decrease) is compared with the basic (or initial) amount:

$$\frac{F - B}{B}.$$

Finally, this fraction is expressed in percent:

$$\left(\frac{F - B}{B}\right) \times 100\% = \left(\frac{F - B}{B} \times 100\right)\%.$$

Observe that if $B < F$, we have an increase;

if $B = F$, there is no change;

if $B > F$, we have a decrease.

The decrease is signaled by a negative number in subtraction, because $B > F$ means $F - B$ is less than 0. The following is an example of percent decrease.

Example 9.14: Man-Mountain Gordo, who weighed 268 pounds, decided to go on a diet. Ten months later, he weighs 183 pounds. What is the percent change in his weight?

Solution: Here, B (his initial weight) is 268, and F is 183. So

$$\frac{183 - 268}{268} = (\boxed{1\,8\,3} \; \boxed{-} \; \boxed{2\,6\,8}) \; \boxed{\div} \; \boxed{2\,6\,8} \; \boxed{=} \; \boxed{-0.3\,1\,7\,1\,6\,4\,1}$$

Rounding to the nearest tenth of a percent, we find that there had been a 31.7% change in weight. The negative sign indicates a reduction.

A common mistake is to think that an $x\%$ increase is exactly offset by an $x\%$ decrease. Look at the next example.

Example 9.15: A pair of slacks was made of some material that was expected to shrink by 10%. If the manufacturer decided to make a pair of 40″ slacks 10% longer, how long would they measure after shrinkage?

Solution:

$$10\% \text{ of } 40 = \boxed{0.1\,0} \; \boxed{\times} \; \boxed{4\,0} \; \boxed{=} \; \boxed{4}.$$

So the length of the unwashed slacks would be

$$\boxed{4\,0} \; \boxed{+} \; \boxed{4} \; \boxed{=} \; \boxed{4\,4} \text{ (inches)}.$$

On shrinkage,

$$10\% \text{ of } 44 = 4.4$$

and

$$\boxed{4\,4} \; \boxed{-} \; \boxed{4.4} \; \boxed{=} \; \boxed{3\,9.6}$$

That is, the slacks would be almost half an inch too short:

$$\boxed{4\,0} \; \boxed{-} \; \boxed{3\,9.6} \; \boxed{=} \; \boxed{0.4} \doteq 0.5$$

Exercises 9.4

1. The stock of Rolli-Koster, Inc., has had its ups and downs. The following table summarizes the facts.

Year	Average Price per Share (in dollars)	Price Change	Percent Change
1970	10		
1971	27		
1972	85		
1973	280		
1974	112		
1975	8		

Calculate the values and complete the table. For the last column, round off your answers to the nearest whole percent.

2. The price of coffee was 40 cents a pound 10 years ago. If the current price of coffee is $4.25 a pound, what percent increase in price does this represent?

•3. A manufacturer knows that, if he is to break even, his sales on a certain product have to increase 300% over last year's sales. If sales of that product last year were $87,500, what should the sales be this year in order to break even?

•4. The circle graph at the right shows. the racial makeup of a high school.

a. If the school has 240 Chicano students, what is the total number of students at the school?

b. How many students are Black?

c. How many are White?

d. How many students are Oriental?

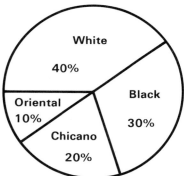

5. The enrollment at the school in Exercise 4 increases by 30% and the racial makeup remains the same.

a. What will the total number of Chicano students be?

b. What will the number of Black students be?

•6. A chain of pizza parlors has 35 stores in Megalopolis.

   a. If the chain owns about 60% of all pizza parlors in that city, what is the total number of pizza parlors in Megalopolis?

   b. If the chain adds 5 more parlors next year and no other new parlors open, what percent will the chain own now?

   c. What is the percent increase of this chain's share of the pizza market in that city?

7. We know that a substance can increase more than 100%. Can it decrease more than 100%? Why?

8. A football coach says to his team, "I want you all to give 120% effort today!" You can give two possible meanings to this statement. Do they both make sense? Explain.

9. Suppose the order of operations in Example 9.15 were changed. That is, starting with 40 inches, consider first, a 10% reduction. Follow this with a 10% increase. How does this result compare with the result in the example?

9.5 Discounts and Markups

Discount, the magic word that perks up the ears of any money-conscious person, plays a dominant role in the business world. Discounts are offered in various forms and vary with quantity, time span, form of payment, and numerous other conditions. Discounts are usually expressed in terms of percents. First, let us examine quantity discounts.

Example 9.16: Wine is sold by the bottle or by the case (twelve bottles). The quoted price is usually for a bottle. When a *whole case* is purchased, a 10% discount is allowed for the case lot. The following table shows the prices of some typical domestic wines in the market.

Wine	Price per Bottle	Case	10% Discount	Price per Case
Burgundy	$1.75	$21.00	$2.10	$18.90
Chablis	1.60	19.20	1.92	17.28
Cabernet	2.30	27.60	2.76	24.84
Champagne	2.60	31.20	3.12	28.08

Comments:

1. The discount of 10% in the example above is very easily calculated because $10\% = \frac{1}{10}$, and so

$$N \times 10\% = N \div 10$$

Hence the discount can easily be computed by moving the decimal point one place to the left.

2. The price per case can be computed by either of these methods:

$$21.00 - (0.10 \times 21.00) = 21.00 - 2.10$$
$$= 18.90$$

or by noting that the price is

$$21.00 - (21.00 \times 10\%) = (21 \times 100\%) - (21 \times 10\%)$$
$$= 21 \times (100\% - 10\%)$$
$$= 21 \times 90\%$$
$$= 21 \times 0.90 = 18.90$$

This last observation is useful for computing discounted prices on your calculator:

Discounted price = Price $\times$ (100 − Rate of discount)%.

Thus, the cost of a case of Cabernet is

$$\$27.60 \times (100 - 10)\% = \$27.60 \times 90\% = \$24.84$$

Another form of discount is called *"cash discount."* Here a wholesaler (distributor, manufacturer) will offer a discount if the amount due is paid within a certain period of time (usually 30 days). If the amount due is paid within the required period, it is considered a *"cash transaction."*

Example 9.17: A merchant receives a bill for $27,480 from her wholesaler for goods purchased. The wholesaler offers a 4.85%, 30-day cash discount on this bill. The merchant plans to pay the bill within the 30 days, thus obtaining the cash discount. What amount will she pay the wholesaler?

Solution: The amount of discount is determined by finding 4.85% of $27,480. Thus we have a discount of

$$(4.85 \boxplus 100) \boxtimes 27480 \boxminus 1332.78$$

so the merchant deducts this amount from the total bill before paying. The amount due on the discounted bill is

$$27480 \boxminus 1332.78 \boxminus 26147.22$$

Note that you can perform the first operation in a single step if you simply "move the decimal point" mentally to convert 4.85% into 0.0485, and punch the decimal form of the percent directly into the calculator. However, the two-step operation may be less likely to produce errors.

Comments:

1. Observe that the algorithm used in Example 9.17 is

    Bill − (Discount rate × Bill) = Discounted bill.

2. Just as for the quantity discounts, we can arrive at the same answer by calculating

    Bill × (100 − Discount rate)% = Discounted bill.

    Notice that

$$27{,}480 \times (100 - 4.85)\% = 27{,}480 \times 95.15\%$$

$$= 27480 \boxtimes 0.9515$$

$$\boxminus 26147.22$$

which is the same result that we obtained before.

3. The "discounted bill" is also known as the *"selling price"*; that is, it is the price at which the article is being sold.

4. Notice that 4% = 0.04, so you know that 4.85% should be close to 0.04; in fact, 4.85% = 0.0485. Such mental checks may help you be sure to get the decimal point in the right place.

The following example deals with the concept of percent markup under two interpretations.

Example 9.18: A merchant bought an item at a wholesale price of $100. If the merchant put a sales price tag for $150 on that item, what would you figure the percent markup to be?

Solution: Certainly, you might reasonably base the percent markup on the wholesale price, because the article was being marked up from that price. The percent change is then given by

$$\left(\frac{150 - 100}{100} \times 100\right)\% = \left(\frac{50}{100} \times 100\right)\% = 50\%.$$

(The "100" in the denominator is the wholesale price of $100 on which the markup is based.) Because we arrived at this concept of the markup through our common sense, let us label it the C-markup (for "common sense").

Example 9.19:   The following table shows the C-markups for a few items with wholesale prices of $100 in each case.

Wholesale Price	Sale Price	Percent C-Markup
$100	$200	100%
100	500	400%
100	1,000	900%
100	10,000	9,900%

The concept of markup is commonly computed differently in practice. To identify the markup used by merchants, let us use the label "the M-markup."

For the M-markup, the merchant computes a percent change in price, but uses the sales price as the basis instead of the wholesale price.

Example 9.20:   Compute the M-markup for an article with wholesale price $100 and sales price $150.

Solution: The merchant would say that the price has changed from $150 to $100, by an amount of $50. So the percent change is

$$\left(\frac{50}{150} \times 100\right)\% = \left(\frac{1}{3} \times 100\right)\% = 33\frac{1}{3}\%.$$

When studying profits from a "markup" point of view, be sure you know whether the wholesale or the sales price is used as the basis.

1. What is the largest percent discount possible? Explain.

2. For the given regular prices and percent discounts listed below, find the selling price of each item.

	Regular Price	Percent Discount
•a.	$ 5.50	12%
b.	138	35%
•c.	25	$40\frac{1}{2}$%
•d.	2.98	20%
e.	234	36%
•f.	43.60	$38\frac{3}{4}$%

3. Verify each of the percent markups in the last column of the table for Example 9.19.

4. Compute each percent M-markup for the items in Example 9.19.

5. Examine the percent M-markups in Exercise 4. Why is it that the percent M-markup must be less than 100?

•6. What percent of the regular price are you paying if you are getting a discount of—

a. 5%?   e. 90%

b. $2\frac{1}{2}$%?   f. 85%?

c. 15%?   g. $33\frac{1}{3}$%?

d. $18\frac{3}{4}$%?   h. 100%?

9.6 Percents in Action: Interest

Just as you can rent an apartment, a car, or a camper, you can also rent money. In this case, the rent that you pay for using the money is called *interest*.

How is the interest (rent) for a loan of money determined? To see what is going on, let us first consider a particularly simple situation.

Example 9.21: Suppose you need to borrow $1,000 for home improvements. If your credit is good, the bank might tell you that you can have the amount you need at an annual interest rate of 15%. This means that, one year from now, you must pay back the borrowed amount plus 15% of that amount. What must you pay back?

Solution: $1,000 + (15\% \text{ of } 1,000) = 1,000 + (0.15 \times 1,000).$

$$(0.15 \boxed{\times} 1000) \boxed{+} 1000 \boxed{=} 1150.$$

The $150 (15% of $1,000) is the "rent" for the loan and is the *simple interest* for one year. If $1,000 were borrowed for 2 years, then you would expect to pay twice as much interest:

$$(0.15 \times 1000) \times 2 = 300.$$

Again by simple interest, if you borrowed $1,000 (at 15%) for only half a year, then you might expect to pay half as much interest:

$$(0.15 \times 1000) \times (1 \div 2) = 75.$$

As in Example 9.21, the total amount to be repaid at the end of the period is

the amount borrowed + the interest owed.

So, for a loan of $1,000 at 15% for half a year (6 months,) the amount due is

$$\$1,000 + \$75 = \$1,075$$

Example 9.22: Find the interest and amount due on a loan of $8,475 at $9\frac{3}{4}\%$ for 18 months. (The interest rate, $9\frac{3}{4}\%$, is understood to mean $9\frac{3}{4}\%$ per year.)

Solution: Because $9\frac{3}{4}\% = 9.75\% = 0.0975$, and because 18 months means $\frac{18}{12}$ year or $\frac{3}{2}$ year, to find the interest we calculate

$$8475 \times 0.0975 \times (3 \div 2) = 1239.4687$$

Rounding off to the nearest cent,

$$\text{interest} = \$1,239.47$$

Therefore,

$$\text{amount due} = 8475 + 1239.47 = 9714.47$$

These problems occur frequently in everyday transactions, and special names have been given to label the important parts involved. The amount of the loan is called the *principal,* the yearly rate of interest is often referred to by the Latin description, interest *per annum,* and the time period is simply called the *period.*

Example 9.23: Find the interest and amount due for each.

	Principal	Interest Rate per Annum	Period
a.	$1,000	5%	1 year
b.	800	4%	6 months
c.	375	$6\frac{1}{2}\%$	1 year
d.	56	12%	1 year 3 months

Solutions:

a. $1000 \boxed{\times} 0.05 \boxed{\times} 1 \boxed{=} 50.$ interest
$50. \boxed{+} 1000 \boxed{=} 1050.$ amount due

b. $800 \boxed{\times} 0.04 \boxed{\times} (6 \boxed{\div} 12) \boxed{=} 16.$ interest
$16. \boxed{+} 800. \boxed{=} 816.$ amount due

c. $375 \boxed{\times} 0.065 \boxed{\times} 1 \boxed{=} 24.375$ interest ($24.38)
$24.375 \boxed{+} 375 \boxed{=} 399.375$ amount due ($399.38)

d. $56 \boxed{\times} 0.12 \boxed{\times} (15 \boxed{\div} 12) \boxed{=} 8.40$ interest (1 yr 3 mo.
$= \frac{5}{4}$ year)
$8.40 \boxed{+} 56 \boxed{=} 64.40$ amount due

If P represents the principal,
 r represents the rate per annum,
 t represents the period in terms of years, and
 i represents the amount of simple interest, then

$$i = P \times r \times t.$$

The total amount due is

$$A = P + i, \quad \text{or} \quad A = P + (P \times r \times t)$$
$$= P \times [1 + (r \times t)]$$
$$= P(1 + rt).$$

Of course, if the rate is stated as $r\%$ per annum, as is sometimes done, then because $r\%$ means $\dfrac{r}{100}$,

$$i = P \times \frac{r}{100} \times t.$$

The procedure just described applies also for interest earned on savings deposits at the bank.

In banking practice, interest earned is added to the principal on deposit. The next time interest is calculated, it is based on the total you have on deposit by now. Interest earned on such an accumulating amount is called *compound interest.*

Example 9.24: If $800 is deposited in a savings account earning interest at 4% compounded annually, how much is on deposit after 3 years?

Solution:

Interest earned during 1st year: $800 \times 0.04 = 32.$
Amount on deposit at end of 1st year: $800 + 32 = 832.$

Interest earned during 2nd year: $832 \times 0.04 = 33.28$
Amount on deposit at end of 2nd year: $832 + 33.28 = 865.28$

Interest earned during 3rd year: $865.28 \times 0.04 = 34.6112$
$(= 34.61)$
Amount on deposit at end of 3rd year: $865.28 + 34.61 = 899.89$

In our next example, we use some simple figures in order to illustrate a pattern.

Example 9.25: Suppose you opened a savings account with $1,000 at a bank that gave a steady rate of 10% interest per annum. Suppose further, you left this amount untouched (no withdrawals, no additional deposits) for four years. How much interest will you earn in that time?

Solution: The interest at the end of each year is shown in the following table.

Time	Amount on Deposit	Interest after 1 Year at 10%	Total Amount at End of Year
Start	$1,000.00	$100.00	$1,100.00
1 year later	1,100.00	110.00	1,210.00
2 years later	1,210.00	121.00	1,331.00
3 years later	1,331.00	133.10	1,464.10
4 years later	1,464.10		

So the amount on deposit after 4 years is $1,464.10.

Let us analyze the entries in the last column a bit. At the end of 1 year, we have

$$\text{Amount} = 1{,}000 + (1{,}000 \times 0.1) \qquad (10\% = 0.10 = 0.1)$$

$$= 1{,}000 \times (1 + 0.1) \qquad \text{(distributive law)}$$

$$= 1{,}000 \times 1.1$$

Let's use the symbol A_1 to represent the amount at the end of 1 year. At the beginning of the second year, we have this amount as the new principal. So, at the end of the second year, we have

$$\text{Amount} = A_1 + (A_1 \times 0.1)$$

$$= (1000 \times 1.1) + [(1000 \times 1.1) \times 0.1]$$

$$= (1000 \times 1.1) \times (1 + 0.1) \qquad \text{(distributive law)}$$

$$= (1000 \times 1.1) \times (1.1)$$

$$= 1000 \times (1.1 \times 1.1)$$

$$= 1000 \times (1.1)^2$$

Now use the symbol A_2 to represent this amount at the end of 2 years. Continuing, we have, at the end of the third year

$$\text{Amount} = A_2 + (A_2 \times 0.1)$$

$$= [1000 \times (1.1)^2] + [1000 \times (1.1)^2 \times 0.1]$$

$$= [1000 \times (1.1)^2] \times (1 + 0.1)$$

$$= [1000 \times (1.1)^2] \times (0.1)$$

$$= 1000 \times [(1.1)^2 \times (1.1)]$$

$$= 1000 \times (1.1)^3 = A_3$$

The pattern is clear. At the end of the fourth year, we have then

$$A_4 = 1000 \times (1.1)^4.$$

In fact, we can check the formula against the entries in the last column:

$$A_1 = 1000 \times (1.1)^1 = 1,100$$

$$A_2 = 1000 \times (1.1)^2 = 1000 \times 1.21 = 1,210$$

$$A_3 = 1000 \times (1.1)^3 = 1000 \times 1.331 = 1,331$$

$$A_4 = 1000 \times (1.1)^4 = 1000 \times 1.4641 = 1,464.10$$

From the expression for A_1, where

$$A_1 = P \times 1.1 \quad \text{and} \quad P = 1,000$$

We can interpret the amount at the end of 1 year as 110% ($= 1.1$) of the principal. Furthermore, each year the amount is 110% of the amount for the preceding year. Hence, we have 1.1 multiplied repeatedly. Therefore at the end of n years,

$$A_n = 1000 \times (1.1)^n.$$

Sometimes, interest is compounded at the end of every 6 months. It is then said to be *compounded semi-annually.* More often, interest is compounded *quarterly* (every 3 months.)

Example 9.26: Find the interest at the end of 1 year on $800 at 4% per annum compounded quarterly.

Solution:

Interest during quarter	+ Old amount	= Amount of deposit
$(800 \times 0.04 \times 0.25) +$	800 $=$	$808.$
$(808 \times 0.04 \times 0.25) +$	808 $=$	816.08
$(816.08 \times 0.04 \times 0.25) +$	816.08 $=$	824.24
$(824.24 \times 0.04 \times 0.25) +$	824.24 $=$	832.48

Thus in one year, the total interest earned was

$$832.48 - 800 = 32.48$$

Whereas the simple interest would have been only

$$800 \times 0.04 = 32$$

The 48 cents may seem a small gain, but later examples will show how quickly money builds under compound interest, especially when compounded daily.

Observe that, for the interest at the end of the first quarter, we have

$$800 \times 0.04 \times \frac{1}{4} = 800 \times \left(0.04 \times \frac{1}{4}\right).$$

Here, we can say that, at the end of each pay period, the interest is paid at $\frac{1}{4}$ the annual rate. In general, if interest at $r\%$ per annum is paid m times a year, then each pay period is calculated at $\frac{r}{m}\%$.

On the other hand, in these situations we have increased the number of pay periods. Each year, there are m periods. In n years, there are $n \times m$ pay periods.

Example 9.27: Suppose you have $1,000 in a savings account with interest at 5% per annum compounded monthly. What is the interest and amount on this "untouched" $1,000 at the end of 3 months?

Solution: At 5% per annum, the interest each month is

$$1,000 \times 0.05 \times \frac{1}{12}$$

or

$$1,000 \times \frac{0.05}{12}.$$

That is, a yearly rate of 5% corresponds to a monthly rate of $\frac{5}{12}\%$.
The amount at the end of the first month is

$$A_1 = 1,000 + \left(1,000 \times \frac{0.05}{12}\right)$$

$$= 1,000 \times \left(1 + \frac{0.05}{12}\right).$$

For the second month, we base the interest on what you have at the end of the first month, A_1. The interest is

$$A_1 \times \frac{0.05}{12}$$

and

$$A_2 = A_1 + \left(A_1 \times \frac{0.05}{12}\right)$$

$$= 1,000 \times \left(1 + \frac{0.05}{12}\right) \times \left(1 + \frac{0.05}{12}\right)$$

$$= 1,000 \times \left(1 + \frac{0.05}{12}\right)^2.$$

Continuing as before,

$$A_3 = 1,000 \times \left(1 + \frac{0.05}{12}\right)^3.$$

Because

$$\frac{0.05}{12} \doteq 0.0041667,$$

$$1 + \frac{0.05}{12} \doteq 1.0041667$$

Using this approximation,

$$A_1 = 1,000 \times (1.0041667)^1$$
$$\doteq \$1,004.17$$
$$A_2 = 1,000 \times (1.0041667)^2$$
$$= 1,000 \times 1.0083508$$
$$\doteq \$1,008.35$$
$$A_3 = 1,000 \times (1.0041667)^3$$
$$= 1,000 \times 1.0125523$$
$$\doteq \$1,012.55$$

The compound interest is $\$1,012.55 - \$1,000 = \$12.55$

The computations for the above example are especially easy if the calculator has a constant (K) key. The constant key saves the trouble of having to enter the number 1.0041667 repeatedly. Check your calculator instructions on how to do these repeated multiplications.

Even without a constant key, we might make our calculations a bit more simple by observing that

$$1 + \frac{0.05}{12} = \frac{12}{12} + \frac{0.05}{12}$$

$$= \frac{12.05}{12}$$

Example 9.28: Use $\dfrac{12.05}{12}$ to find A_1, A_2, A_3, for Example 9.27.

Solution:

$$A_1 = 1000 \;\boxed{\times}\; 12.05 \;\boxed{\div}\; 12 \;\boxed{=}\; 1004.1666$$

Don't clear your calculator!

$$A_2 = A_1 \;\boxed{\times}\; 12.05 \;\boxed{\div}\; 12 \;\boxed{=}\; 1008.3505$$

Don't clear your calculator!

$$A_3 = A_2 \;\boxed{\times}\; 12.05 \;\boxed{\div}\; 12 \;\boxed{=}\; 1012.5519$$

Although each of the answers in Example 9.28 agrees with the corresponding one in Example 9.27, the advantage in the method here is that it is likely to have less error due to rounding-off than in repeatedly multiplying by the approximation, 1.0041667.

From the above, we can generalize our procedure for interest compounded m times a year. If an amount P is deposited in an account for a period of n years at an annual rate of interest of $r\%$ compounded m times a year (therefore $n \times m$ payments), the amount A in the account at the end of n years is

$$A = P \times \left(1 + \frac{r}{m}\%\right)^{n \times m}$$

or

$$A = P \times \left(1 + \frac{r}{100 \times m}\right)^{n \times m}$$

Example 9.29: You have $500 on deposit in a credit union for 6 years at an annual rate of $7\frac{1}{4}\%$ compounded quarterly. How much will you have at the end of 6 years?

Solution: Here, $P = 500$, $r = 7.25$, $n = 6$, $m = 4$. Putting these values in the formula and using a constant key,

$$A = 500 \times \left(1 + \frac{0.0725}{4}\right)^{24}$$

$$\doteq 769.47842$$

$$\doteq \$769.48$$

Interest is $769.48 - \$500.00 = \$269.48$

By contrast, the simple interest for 6 years at $7\frac{1}{4}\%$ would have been

$$\$500 \times 0.0725 \times 6 = \$217.50$$

So compound interest in this case earned $51.98 more in the 6 years. (Notice that the rate in this calculation is $\frac{r}{4}\%$ or $\frac{0.0725}{4}$, and the exponent is $24 = 6 \times 4$.)

Example 9.30: Suppose that in 10 years your daughter will be going to college. How much should you deposit now in a savings and loan earning $7\frac{3}{4}\%$ interest, compounded quarterly, so that 10 years from now it will amount to $5,000 toward her education?

Solution: We still use the general equation, except in a slightly different way. We know, in this case,

$$A = 5{,}000, \quad r = 7\frac{3}{4} = 7.75, \quad m = 4, \quad n = 10$$

We need to find the amount of the principal, P. So we have

$$5{,}000 = P \times \left(1 + \frac{0.0775}{4}\right)^{4\times10}$$

or
$$5{,}000 = P \times (1.019375)^{40}$$

Therefore,
$$P = 5000 \div (1.019375)^{40}$$

To complete Example 9.30, we need to find $(1.019375)^{40}$. Repeated multiplication by 1.019375 is tedious, even on a calculator; there also is great possibility of miscounting the number of multiplications we have performed. Instead, we turn to some of the laws of exponents from Chapter 7 to help in this job.

Recall that

$$B^N \times B^M = B^{N+M} \quad \text{(page 202)}$$

and
$$(B^N)^M = B^{N\times M} \quad \text{(page 203)}$$

What we want is $(1.019375)^{40}$. This task may be broken down into smaller steps as follow:

$$(1.019375)^2 = 1.0391254$$

$$(1.019375)^4 = [(1.019375)^2]^2 = 1.0797816$$

$$(1.019375)^8 = [(1.019375)^4]^2 = 1.1659283$$

$$(1.019375)^{16} = [(1.019375)^8]^2 = 1.3593888$$

$$(1.019375)^{32} = [(1.019375)^{16}]^2 = 1.8479379$$

This may look tedious, but actually so far we have only kept squaring five times in a row. One more step, and we shall have the 40th power.

Notice that

$$40 = 32 + 8$$

So
$$B^{40} = B^{32} \times B^8.$$

We already have

$$B^{32} = 1.8479379$$

and $$B^8 = 1.1659283$$

Hence $B^{40} = $ *1.8479379* $\times$ *1.1659283* $=$ *2.1545631*

To complete Example 9.30,

$$P = 5{,}000 \div (1.019375)^{40}$$

$$= 5000 \div 2.1545631 = 2320.6561$$

or about \$2,320.66. (From $B^{40} = 2.1545631$, note that your principal would have more than doubled in the 10 years.)

Exercises 9.6

1. Compute the simple interest charged and the amount due for the following loans.

	Amount of Loan	Yearly Interest Rate	Period of Loan
•a.	\$2,700	6%	2 years
b.	\$585	$8\frac{3}{4}\%$	6 months
•c.	\$12,850	$9\frac{1}{2}\%$	45 days*
•d.	\$6,493	12%	$1\frac{1}{4}$ years
•e.	\$187	11%	2 months
f.	\$385.75	8.5%	275 days*

(* Consider 1 day to be $\frac{1}{365}$ year.)

•2. Bill loans his brother \$13 for one month and charges him \$1.75 interest. What is Bill's annual simple interest rate on this transaction?

3. You borrow \$175. After six months you are asked to pay back \$187.65. What simple interest rate (per annum) were you charged?

•4. Sue borrowed some money one year and three months ago. Today, she paid off the loan and her interest charges came to \$64.75. If the bank's simple interest rate was $9\frac{1}{4}\%$ per year, what was the original amount of her loan?

5. Find the amount of compound interest and final amount accumulated for each of the following.

	Amount Deposited	Period	Annual Interest Rate	Compounded
•a.	$2,500	5 years	$6\frac{1}{2}\%$	Monthly
b.	$750,220	15 years	$10\frac{1}{2}\%$	Yearly
•c.	$750,220	15 years	$10\frac{1}{2}\%$	Semi-annually
•d.	$750,220	15 years	$10\frac{1}{2}\%$	Quarterly
•e.	$750,220	15 years	$10\frac{1}{2}\%$	Monthly
f.	$750,220	15 years	$10\frac{1}{2}\%$	Daily
g.	$176	2 yrs, 6 mos	8%	Semi-annually
h.	$1,228.75	9 yrs, 5 mos	$7\frac{1}{2}\%$	Monthly
i.	$220,000	18 yrs, 9 mos	$5\frac{3}{4}\%$	Quarterly

6. In 1776, a citizen of Philadelphia (everybody called him "Ben") decided that the people of Philadelphia should have a nice $200-a-head bicentennial party in 1976. Ben deposited ten dollars in Ye Olde First National for this purpose, at 8% compounded monthly.

a. How much money will there be for the party?

b. How many persons can attend the party?

c. If Ben had deposited $1,000 in 1776, should the bicentennial mayor of Philadelphia hold the party at his home?

9.7 Effective Rates

Let us compare the earnings of one dollar left on deposit at 12% compounded annually with a dollar on which the interest is compounded quarterly. After one year, the amounts in the bank are

$$1 \times (1 + 0.12) = \$1.12000 \qquad \text{(compounded annually)}$$

$$\text{and} \quad 1 \times \left(1 + \frac{0.12}{4}\right)^4 \doteq \$1.12551 \qquad \text{(compounded quarterly)}$$

Because we started with one dollar, in the quarterly compounded case it is as if the bank were giving you an interest rate of 12.551% compounded annually. Does the following verify our claim?

$$1 \times (1 + 0.12551) = \$1.12551$$

Actually, the effect of one dollar at 12% compounded quarterly, produces the same amount that one dollar would produce at 12.551%, compounded annually. For this reason, we say that the *effective rate* of 12% compounded quarterly is equal to a 12.551% annual rate of interest.

How do we convert various compound interest rates into effective annual rates? We need only look at what happens to one dollar because what happens to any other amount is relative to the dollar.

If one dollar is deposited at $r\%$ compounded m times a year, then after one year, the amount is

$$\left(1 + \frac{r\%}{m}\right)^m,$$

and the interest is given by how much the dollar has grown: namely,

$$\left(1 + \frac{r\%}{m}\right)^m - 1$$

In percent increase, we would have

$$\left[\left(1 + \frac{r\%}{m}\right)^m - 1\right] \times 100\%,$$

which we can get easily by shifting the decimal point 2 places to the right and attaching the % symbol. This is, in fact the effective rate mentioned above.

The following example shows another application of this concept.

Example 9.31: Some stores charge you $1\frac{1}{2}\%$ interest on your unpaid balance each month. What is the effective rate of interest per annum that these stores really charge?

Solution: First note that a $1\frac{1}{2}\%$ rate monthly is the same as

$$1\frac{1}{2}\% \times 12 = 18\%$$

annual rate. As this is compounded monthly, we have

$$\left(1 + \frac{0.18}{12}\right)^{12} = (1.015)^{12}$$

$$= 1.1956182$$

Thus, the effective rate is more than 19.56% (or nearly 20%).

Exercises 9.7

1. Compute the effective rates for the following investment plans. Round off your answer to 4-place decimals.

   •a. 8% compounded semi-annually.

   b. 8% compounded monthly.

   c. 8% compounded daily.

   •d. 10% compounded quarterly.

   e. $7\frac{3}{4}$% compounded daily.

   •f. 22% compounded monthly.

•2. Is it better to invest your money at 5.5% compounded semi-annually or at 5.25% compounded daily?

3. One bank gives interest at $5\frac{1}{2}$% compounded semi-annually, while another gives 5.49% compounded daily.

   a. Which is the better place to invest your money?

   b. What will be the interest earned on one million dollars at each of these banks after one year?

   c. By how much do the amounts in part (b) differ?

9.8 Installment Loans

Often, arrangements are made to pay off a loan or a purchase by making equal payments at regular intervals. The loan is then said to be *amortized,* and the payments are the *installments*. They are also referred to as *installment plans*. Here are some typical examples of installment loans.

Example 9.32: A bank agrees to lend you $2,500 at 12% interest with the understanding that you are to pay the loan in quarterly installments of $672.57 for one year.

Example 9.33: A store sells you a $500 TV set with no down payment, under the condition that you pay $50 at the end of each month for the next twelve months.

Example 9.34: A savings and loan institution is willing to lend you $30,000 to build a home. The agreement is that you pay back in monthly installments of $251 (interest included) for 20 years.

An important point to observe in these kinds of problems is that with each installment the loan is reduced. So you should pay interest only on the *unpaid balance;* that is, on the portion of the loan that has not been paid. We can keep a schedule of payments to follow Example 9.32 in order to see how this works. For this example, remember $\dfrac{r\%}{m} = \dfrac{12\%}{4} = 0.03$.

Example 9.35:

Payment Period	Unpaid Balance	Interest on Balance	Amount Owed	Installment Amount Paid at End of Period	Unpaid Balance at End of Period
1	$2500.00	$75.00	$2575.00	$672.57	$1902.43
2	$1902.43	$57.07	$1959.50	$672.57	$1286.93
3	$1286.93	$38.61	$1325.54	$672.57	$652.97
4	$652.97	$19.59	$672.56	$672.56	—0—

Notice that the last payment is 1¢ less than the others; otherwise there would have been an overpayment. Such differences in the last payment, of course, are due to rounding-off errors in the calculations.

From Example 9.35, we see that the bank was quite accurate in requiring equal installments of $672.57. Now, how did the bank arrive at this figure? It can be proved mathematically, that the following formula will give the correct installment (up to rounding-off errors). For this formula,

P represents the principal or the amount of the loan;

$r\%$ represents the annual interest rate;

m represents the number of installments made each year;

n represents the number of years to clear the debt;

I represents the amount of each installment.

Then
$$I = \frac{P \times K^{n\times m} \times (K-1)}{K^{n\times m} - 1},$$

where
$$K = 1 + \frac{r\%}{m}.$$

The formula *looks* complicated, but it is not difficult once the pattern is established for the procedure. To illustrate, let us again use the information in Example 9.32 (pretending that the amount for each installment is unknown at this time).

Example 9.36: Use the formula to calculate the amount of installment.

Solution: Here, $P = \$2{,}500$, $r = 12$, $m = 4$, $n = 1$.

So
$$K = 1 + 0.03 = 1.03$$

and
$$I = \frac{2500 \times (1.03)^4 \times (0.03)}{(1.03)^4 - 1}.$$

On the calculator, this may be processed as follows.

$$(1.03)^4 = \boxed{\textit{1.03}} \; \boxed{\times} \; \boxed{\textit{1.03}} \; \boxed{\times} \; \boxed{\textit{1.03}} \; \boxed{\times} \; \boxed{\textit{1.03}} \; \boxed{=} \; \boxed{\textit{1.1255088}}$$

Don't clear the calculator, but use this result in the next step.

$$(\boxed{\textit{1.1255088}} \; \boxed{\times} \; \boxed{\textit{2500}} \; \boxed{\times} \; \boxed{\textit{0.03}}) \div (\boxed{\textit{1.1255088}} \; \boxed{-} \; \boxed{\textit{1}})$$

$$\rightarrow \boxed{\textit{84.41316}} \; \boxed{\div} \; \boxed{\textit{0.1255088}} \; \boxed{=} \; \boxed{\textit{672.56766}}$$

The amount of each installment is about $672.57. (The amount is actually a fraction of a cent less than that, explaining why the last payment is only $672.56—because each of the earlier payments has been slightly more than required.)

To get the formula used in Example 9.36 calls for quite a bit of arithmetic juggling, involving what is called a geometric progression. Just to see how $1 + \dfrac{r\%}{m}$ comes into the picture repeatedly, we show below just a few steps to indicate the reasonableness in this pattern. We will not prove the formula here.

Suppose we start out with a loan of P dollars. At the end of the first period, the interest on P would be

$$P \times \frac{r\%}{m}.$$

So altogether we owe

$$P + \left(P \times \frac{r\%}{m}\right).$$

By the distributive law, this becomes

$$P \times \left(1 + \frac{r\%}{m}\right).$$

Now, the next thing we do is pay an installment, I dollars, in order to reduce the debt. The unpaid balance is then

$$P \times \left(1 \times \frac{r\%}{m}\right) - I.$$

The interest for the next period is based on this unpaid balance. So we might label it P' to indicate that this is the new principal. Then, the next interest would be

$$\left[P \times \left(1 + \frac{r\%}{m}\right) - I\right] \times \frac{r\%}{m}$$

or
$$P' \times \frac{r\%}{m}$$

and altogether we now owe

$$P' + \left(P' \times \frac{r\%}{m}\right)$$

which by the distributive law is

$$P' \times \left(1 + \frac{r\%}{m}\right).$$

We reduce the debt further by an installment, I, so that

$$P' \times \left(1 + \frac{r\%}{m}\right) - I$$

is then our new principal, which we might designate P''.
The next interest is

$$P'' \times \frac{r\%}{m}$$

and the next new principal is

$$P'' \times \left(1 + \frac{r\%}{m}\right) - I.$$

At this point, if you *must* see this mind-blowing chain we have so far,

$$\left\{\left[P \times \left(1 + \frac{r\%}{m}\right) - I\right] \times \left(1 + \frac{r\%}{m}\right) - I\right\} \times \left(1 + \frac{r\%}{m}\right) - I.$$

We can go on. But even this pattern is enough to indicate how $\left(1 + \dfrac{r\%}{m}\right)$ enters the picture repeatedly. Moreover, the factor $\dfrac{r\%}{m}$, which we designated as $(K - 1)$ can be seen clearly in the above expression.

Other examples of these kinds of problems are shown below in the exercises. However, let us carry out in detail the schedule of payments in one last illustration.

Example 9.37: Suppose a bank lends you \$3,000 for home improvements. The bank charges $8\frac{1}{2}\%$ for such loans and wants the money paid back in semi-annual installments over a three year period. What should your installments be? Verify the correctness of your solution by completing the schedule of payments.

Solution: Referring to the formula on page 295 we find K to be

$$K = 1 + \frac{0.085}{2} = 1.0425$$

So
$$I = \frac{3000 \times (1.0425)^6 \times (0.0425)}{(1.0425)^6 - 1} = \$576.95193$$

Your schedule should look like this.

Payment Period	Unpaid Balance	Interest	Installment	Unpaid Balance at End of Period
1	\$3000.00	\$127.50	\$576.95	\$2550.55
2	\$2550.55	\$108.40	\$576.95	\$2082.00
3	\$2082.00	\$88.49	\$576.95	\$1593.54
4	\$1593.54	\$67.76	\$576.95	\$1084.32
5	\$1084.32	\$46.08	\$576.95	\$553.43
6	\$553.43	\$23.52	\$576.97	—0—

(In this schedule, we have omitted the column headed "amount owed," which is simply the sum of the entry in the second and that in the third column.)

Observe again that the last installment had a slight adjustment in order to clean up the debt; this time, you are expected to pay 2 cents more than the installment stated.

1. For each of the following loans, find the installment and make a schedule of payment.

	Amount of Loan	Interest Rate	Payment Period	Duration of Loan
a.	$4,700	9%	quarterly	2 years
•b.	$850	$7\frac{1}{2}$%	semi-annually	4 years
•c.	$10,450	$12\frac{3}{4}$%	quarterly	3 years
d.	$440,670	10.25%	semi-annually	5 years

2. In Example 9.33, the store did not declare the interest rate being charged.

   a. Assume that the store interest rate is 20%; what should your install-ments be?

   b. Compare your answer with the installments asked by the store. Are they charging more than 20%?

   c. Now assume that they charge 30%. Again compute your installments.

   d. Do the same for 35%.

   e. From your results, can you estimate what rate of interest the store is charging?

•3. To build a home, you borrow $40,000 at an interest rate of $8\frac{1}{2}$% per annum for 30 years. What are your monthly payments?

•4. Work Exercise 3 again, but assume that the interest rate is $8\frac{1}{4}$% instead of $8\frac{1}{2}$%.

5. Work Exercise 3 again, but assume that the loan is for 20 years.

•6. An automobile dealer accepted a down payment of $500 on a new car listed at $4,500. He offered the customer "6% interest" on the balance to be paid in equal monthly installments for 18 months. His calculations were:

$$\$4,500 - \$500 = \$4,000 \quad \text{amount of loan}$$

$$\$4,000 \times 0.06 \times \frac{3}{2} = \$360 \quad \text{total interest}$$

$$\$4,000 + \$360 = \$4,360 \quad \text{total amount due}$$

$$\$4,360 \div 18 = \$242.22 \quad \text{amount each installment}$$

   a. What is wrong with the dealer's claim?

   b. If the dealer's rate were really 6% per annum, what should each installment be?

   c. How much did the dealer overcharge the customer?

7. Mr. Livit Upp borrowed $40,000 at 9% interest per annum to be paid off in payments of $5,000 a year. The interest is calculated on the unpaid balance. Suppose there is a clause in the agreement saying that, if Mr. Upp wants to pay off his loan before the end of 5 years, he must pay an additional "penalty" of 15% of the unpaid balance. If he wants to pay off his debt at the end of 3 years—

a. what is the amount of his last "installment"?

b. altogether, how much interest must he pay?

•8. Janese borrows $3,000 and agrees to pay off the loan in monthly payments of $50 for 48 months. In addition, she agrees to pay off the balance of the loan on the 48th payment. If the interest rate was 8.5%, what was the amount of her 48th payment?

10

Rates, Ratios, and Similarity

10.1 What Is a Rate?

Every time we drive a car we think about rates. How fast are we driving? What is the speed limit? How slow is the traffic moving? We speak of "miles per hour" as the rate, or speed, at which we travel. When we say "50 miles per hour," we mean that at this speed, we would travel 50 miles in one hour. This rate compares distance traveled with time traveled.

Another common example of a rate is *rent,* the amount of money charged for the use of something. "The penthouse apartment rents for $550 a month."

A rate compares two quantities by forming the ratio (fraction):

$$\text{Rate} = (\text{Quantity 1}) \div (\text{Quantity 2}).$$

This number tells how many units of Quantity 1 are used for (per) each unit of Quantity 2.

Here are some examples—some real, some fanciful.

- Speed (miles per hour): $\dfrac{\text{Number of miles traveled}}{\text{Number of hours traveled}}$.

- Fuel consumption (miles per gallon): $\dfrac{\text{Number of miles traveled}}{\text{Number of gallons used}}$.

- Interest (percent: dollars per hundred):
$$\dfrac{\text{Number of dollars earned}}{\text{Number of 100 dollars deposited}}.$$

- Peanut and beer consumption (bags of peanuts per bottle of beer):
$$\dfrac{\text{Number of bags of peanuts sold}}{\text{Number of bottles of beer sold}}.$$

- Sardine packing (sardines per can): $\dfrac{\text{Number of sardines caught}}{\text{Number of cans produced}}$.

- Typing efficiency (words per minute): $\dfrac{\text{Number of words typed}}{\text{Number of minutes typed}}$.

- Scholastic performance (grade point average):
$$\dfrac{\text{Number of grade points earned}}{\text{Number of credits earned}}.$$

- Cost (unit price: cents per ounce): $\dfrac{\text{Cost in cents of a box}}{\text{Number of ounces in a box}}$.

In these illustrations the word *"per"* (for each) plays a prominent role. A rate expresses a quantity of something *for each* unit of something else. In other words, it is a ratio between two quantities.

Interest on savings or loans may be given as a ratio. Because $5\% = \frac{5}{100}$, a 5% interest rate means 5 dollars for each 100 dollars (percent) on deposit (or on loan). Here, the unit is a hundred dollars.

Equivalently, because $\frac{5}{100} = \frac{0.05}{1}$, a 5% interest rate may also be considered as 0.05 dollars for each dollar on deposit. In this case, the unit is a dollar.

If a rate may be based on different units, why do we bother to base it on any particular unit at all? Consider the following problem.

Example 10.1: Jackie got $16 interest a year from the First Bicentennial Bank; her sister, Patty, got $24 per year from the Second Centennial Bank. Which bank paid a higher rate of interest?

Solution: This question cannot be answered correctly, for the answer depends on the amount of money each has on deposit. If both Patty and Jackie started out with the same amount, clearly Patty's bank did better. However, without such additional information, we cannot make a valid comparison.

Suppose that Jackie had $320 and Patty had $600 in their own banks at the start. The rates are

for Jackie: $\boxed{16}$ ÷ $\boxed{320}$ = $\boxed{0.05}$ = 5%

for Patty: $\boxed{24}$ ÷ $\boxed{600}$ = $\boxed{0.04}$ = 4%.

We conclude that Jackie's bank paid at a higher rate of interest.

Exercises 10.1

•1. Suggest five examples of rates and their measurements.

•2. To determine how economical it is to operate a particular car, you can use a rate: miles per gallon of fuel, or gallons of fuel per mile.

   a. What are these rates if a car goes 100 miles on 5 gallons of gas?

   b. Explain how the numbers obtained for each of these rates settle the question, "How economical?" For example, tell which measure indicates more economy with a smaller number and which with a larger number.

3. A baseball batter's standing is often given by his batting average. If Ty Cobb has a batting average of .300, he is supposed to have 300 hits for each 1,000 times at bat.

   a. Why is it technically incorrect to call the rate (.300) the batter's *"percentage,"* as is sometimes done?

   b. What is his true percentage?

   c. Suppose Ty actually got 615 hits out of 2046 times at bat. What would be his batting average then?

•4. Explain how there can be more than one way of using rates to describe the ratio between two quantities. (See Exercise 2 above.)

10.2   The Old "Economy-Package" Routine

"Buy the Economy Size and Save!!!" Have you ever tried to decide whether you *do* save? Figure 10.1 and Table 10.1 present some actual examples.

Figure 10.1
Economy sizes and regulars.

Table 10.1 Which Size Is the Best Buy?

Item	Large Size	Small Size
Detergent	$4.39	$2.19
Milk	69¢	36¢
Cereal	69¢	51¢
Salad dressing	$1.38	69¢
Film for camera	$2.80	$2.00

For each item, how shall we decide which size is the more economical? We approach the problem by finding the cost rate of each size—how much we must pay for a given (unit) amount. This cost rate is the *unit price.*

For detergent, the appropriate unit might be an ounce (or gram), so the unit price would be the price for each ounce (or gram). The unit price of film might be based on the number of exposures in each roll.

Now, we see that we didn't record enough information in Table 10.1. Back to the store for more data!

(left margin) 10 Rates, Ratios, and Similarity

Detergent	
family size (10 lb., 11 oz.)	$4.39
economy size (5 lb., 4 oz.)	$2.19

Milk	
$\frac{1}{2}$ gallon	69¢
1 quart	36¢

Cereal	
large size (15 oz.)	69¢
small size (10 oz.)	51¢

Salad dressing	
large size (16 oz.)	$1.38
regular size (8 oz.)	69¢

Film	
black & white (36 exposures)	$1.55
(20 exposures)	$1.05
color (36 exposures)	$2.80
(20 exposures)	$2.00

Now we're ready to compare prices. Let's start with cereal.

Example 10.2: Cereal: comparison of large and small sizes.

Solution:

Large size: 69¢ for 15 ounces.
To find the price of 1 ounce, divide:

$$\boxed{69} \text{ (cents)} \;\boxed{\div}\; \boxed{15} \text{ (ounces)} \;\boxed{=}\; \boxed{4.6}$$

Unit price = 4.6 cents per ounce.

Small size: 51¢ for 10 ounces.
To find the price of 1 ounce, divide:

$$\boxed{51} \text{ (cents)} \;\boxed{\div}\; \boxed{10} \text{ (ounces)} \;\boxed{=}\; \boxed{5.1}$$

Unit price = 5.1 cents per ounce.

The unit price is higher for the small box.

The determination of the unit price for photographic film is equally easy. We show the procedure for the black and white rolls.

Example 10.3: Films: comparison of 36 and 20 exposure rolls.

Solution:

$1.55 for 36 shots:

$$/55 \text{ (cents)} \boxed{\div} 36 \text{ (shots)} \boxed{=} 4.3055555$$

or about 4.3¢/exposure.

$1.05 for 20 shots:

$$/05 \text{ (cents)} \boxed{\div} 20 \text{ (shots)} \boxed{=} 5.25$$

or 5.25¢/exposure.

Here, too, it is more economical to buy the larger package.

Example 10.4: Milk: comparison of half-gallon and quart containers.

Solution: We need to know the size relationship between the half-gallon and the quart containers. There are 4 quarts in 1 gallon, so there are 2 quarts in a half-gallon. Let's use the quart as the unit.

Large size: 69¢ for $\frac{1}{2}$ gallon means 69¢ for 2 quarts:

$$69 \text{ (cents)} \boxed{\div} 2 \text{ (quarts)} \boxed{=} 34.5 \text{ (¢/qt.)}$$

Small size: 36¢ for 1 quart:

$$36 \text{ (cents)} \boxed{\div} / \text{ (quart)} \boxed{=} 36. \text{ (¢/qt.)}$$

Again, in this case, it is more economical to buy the large size.

Example 10.5: Salad dressing: comparison of large and regular size bottles.

Solution:

$1.38 for 16 oz.:

$$/38 \text{ (cents)} \boxed{\div} /6 \text{ (ounces)} \boxed{=} 8.625 \text{ (¢/oz.)}$$

69¢ for 8 oz.:

$$69 \text{ (cents)} \boxed{\div} 8 \text{ (ounces)} \boxed{=} 8.625 \text{ (¢/oz.)}$$

For this item, the unit prices are equal! So unless you need salad dressing in large quantity, it may be more advantageous to buy the smaller size (for freshness or convenience in storing).

Example 10.6: Detergent: comparison of family and economy-size boxes.

Solution: First we need to convert each weight to a single unit. We choose to express all weight in ounces. Remember:

$$1 \text{ lb.} = 16 \text{ oz.}$$

10 lb., 11 oz. = (*10* ⊠ *16*) ⊞ *11* ▱ *171*. (ounces)
5 lb., 4 oz. = (*5* ⊠ *16*) ⊞ *4* ▱ *84*. (ounces).

Family size: $4.39 for 171 oz.: divide

439 (cents) ⊡ *171* (ounces) ▱ *2.56 72514* (¢/oz.)

Economy size: $2.19 for 84 oz.: divide:

219 (cents) ⊡ *84* (ounces) ▱ *2.60 71428* (¢/oz.)

Again, the larger size is the more economical, though by not much. The difference in unit price is

2.60 71428 ⊟ *2.56 72514* ▱ *0.0398914* (¢/oz.)

   Notice that, in converting the mixed units (pounds and ounces) to ounces, the number of pounds had to be multiplied by 16 because one pound is equivalent to 16 ounces.

   10 lb., 11 oz. means $[(10 \times 16) + 11]$ ounces.

Alternately, these units could be converted to pounds by dividing the number of ounces by 16:

   10 lb., 11 oz. means $[10 + (11 \div 16)]$ pounds.

On your calculator it is preferable to do the division first:

(*11* ⊡ *16*) ⊞ *10* ▱ *10.6875*

So 10 lb., 11 oz. means 10.6875 lb.
   In either case, we have to remember the conversion:

$$1 \text{ lb.} = 16 \text{ oz.}$$

Often, the weight of items like detergent is labeled in pounds and ounces, and equivalently in metric units. The metric unit of weight is a gram. In the above case, the sizes were also labeled

family size: 4,848 grams   economy size: 2,381 grams.

Using these weights, unit prices are simply obtained:

$$439 \boxdot 4848 \boxdot 0.09055528 \quad \text{or about } 0.09055\text{¢/gm.}$$

$$219 \boxdot 2381 \boxdot 0.09198\text{¢} \quad \text{or about } 0.09198\text{¢/gm.}$$

Each result was obtained easily because no mixed units occurred in the labeled weights. The conclusion is the same: the family size is the more economical.

In deciding which package to buy, there are other considerations. Time is one. The occasional picture-taker, for example, might not be able to use up the 36 exposures in the larger roll before the expiration date, or the wait may not be worth the savings. Perishable products may spoil before large packages can be used up.

Example 10.7: A 15-oz. box of cereal costs 69¢ and a 10-oz. box of the same cereal costs 51¢. In the past, the family was able to use no more than $\frac{3}{4}$ of the large box before it became stale. Now compare the unit price of these packages, based on the amount consumed.

Solution: If only $\frac{3}{4}$ of the large box is used, then the number of ounces used is

$$(3 \boxdot 4) \boxtimes 15 \boxdot 11.25$$

So, in effect, the family pays 69¢ for 11.25 oz. To find the unit price,

$$69 \boxdot 11.25 \boxdot 6.1333333 \text{ (¢/oz.)}$$

Compared with the small-size unit price, 5.1 ¢/oz. (Example 10.2), this makes the large size more expensive than the smaller size.

Exercises 10.2

1. Use the table of prices on page 305 to compare the unit prices for a roll of color film with 20 exposures and one with 36 exposures.

•2. A roll of film for color slides sells for $3.50 for 36 exposures and $2.50 for 20 exposures.

a. Suppose you bought one of the smaller rolls. If you just took one picture and then have this developed, the price per shot would be $2.50 (not including the cost of having the roll developed). If you have the roll developed after taking two pictures, then the unit price is $2.50 ÷ 2, or $1.25. Complete the table below for the unit price of different numbers of pictures shot from the roll.

Number of Shots	Unit Price
1	$2.50
2	1.25
3	——
⋮	⋮
20	——

b. Using the above information, at least how many pictures must you take of the 36-exposure roll to come out ahead of the smaller roll. (HINT: Figure the unit price for the 20-shot roll and find out how many of these $3.50 will buy.)

3. A king-size (50 oz.) dishwashing detergent sells for $1.43, and a family size (65 oz.) of the same detergent sells for $1.83.

a. Find the unit price for each package.

b. Find the difference in unit prices for the detergent.

4. A pint of half-and-half (half cream, half milk—used for coffee) sells for 37¢ and a quart of the same brand sells for 65¢.

a. Find the unit price for each. (1 qt. = 2 pt. = 32 oz.)

b. A family takes 7 days to use a quart of the half-and-half. If the product is spoiled after 6 days, is it still a bargain to buy the large carton?

•5. A brand of yogurt costs $1.17 for a 2-lb. carton and 35¢ for a $\frac{1}{2}$-lb. cup.

a. How many $\frac{1}{2}$ lb. cups would you have to have to equal one of the 2 lb. cartons?

b. Compare the price of buying 2 pounds of yogurt in the $\frac{1}{2}$-lb. cup against the price of the larger carton. How much would you be saving buying the more economical package?

10.3 The Old "mpg–km/ℓ" Routine

In Section 10.1, one of the examples of a rate was the number of "miles per gallon" (mpg). This is an important consideration in judging economy of an automobile.

Miles per gallon is just that—the number of miles you can drive on one gallon of gas.

Example 10.8: Let's say you drove 590 miles on 15.5 gallons of gas. **(a)** How many mpg did you average?

Solution: 590 miles on 15.5 gallons: divide,

$$5\,90 \;\boxed{\div}\; 1\,5.5 \;\boxed{=}\; 38.0645\,16 \;\doteq\; 38.06 \text{ mpg.}$$

(b) How many gallons of gas did it take to go one mile; that is, what was the gallons per mile (gpm) rate?

Solution: 15.5 gallons in 590 miles: divide,

$$1\,5.5 \;\boxed{\div}\; 5\,90 \;\boxed{=}\; 0.0262\,71\,1 \;\doteq\; 0.026 \text{ gpm}$$

The rate *gpm* answers the same question as the rate *mpg*. However, there are several reasons for thinking in terms of miles per gallon.

First, unless your car is horribly inefficient, the number of gpm will be less than 1; in fact, it should be a rather small fraction. People have an easier time comparing, say,

$$20 \text{ mpg} \quad \text{with} \quad 38 \text{ mpg}$$

than comparing

$$\frac{1}{20} \text{ gpm} \quad \text{with} \quad \frac{1}{38} \text{ gpm.}$$

Secondly, a driver wants to know "How far can I go on five gallons of gas?" because that is often the way gas is purchased. If it is known that the car averages 38 mpg, then in using 5 gallons, he will go

$$5 \;\boxed{\times}\; 38 \;\boxed{=}\; 1\,90 \text{ (miles).}$$

On the other hand, to answer the same question knowing that the car averages $\frac{1}{38} = 0.0263157$ gpm, requires the awkward calculation:

$$5 \;\boxed{\div}\; 0.0263\,15\,7 \;\boxed{=}\; 1\,90.0006\,4 \text{ or about 190 miles.}$$

Example 10.9: On his vacation, Kevin kept a record of his gas purchase and mileage on his trip. He filled up his tank on July 1 with 18.8 gallons of gas at the start of the trip. As usual, he filled his tank whenever he noticed that his tank was getting low.

Here is his record. The 19.5 gallons bought on July 7 was at the end of the trip when he got home. (The odometer is the gauge in his car that gives the mileage.) Round off the mpg to 1-place decimals.

Date	Odometer Reading	Number of Gallons
Start July 1	26,539.8	18.8
July 1	26,989.6	17.1
July 2	27,521.3	19.4
July 3	27,937.1	16.7
July 4	28,470.7	18.3
July 5	28,906.4	15.9
July 6	29,401.5	17.6
July 7	29,975.4	19.5

The 18.8 gallons bought on July 1 was simply to fill his tank so that we can determine the total number of gallons from that point on. Thus, the total number of gallons was

$$/7./\ \boxplus\ /9.4\ \boxplus\ /6.7\ \boxplus\ /8.3\ \boxplus\ /5.9\ \boxplus\ /7.6\ \boxplus\ /9.5\ \boxminus\ /24.5 \text{ (gals.)}$$

Now, using these 124.5 gallons, Kevin traveled

$$29975.4\ \boxminus\ 26539.8\ \boxminus\ 3435.6 \text{ (miles).}$$

To find the mpg, divide:

$$3435.6\ \boxplus\ /24.5\ \boxminus\ 27.59518 \doteq 27.6 \text{ mpg.}$$

So Kevin got about 27.6 mpg on his trip.

In discussing detergent prices earlier, along with pounds and ounces, we mentioned the gram as a unit of weight. The gram belongs to the metric system. This system is now used in most countries in the world.

In the English system using tons, pounds, and ounces, the conversion from one unit of weight to another requires multiplying or dividing by

16   (16 ounces in a pound), or

2,000   (2,000 pounds in a ton).

Similarly, converting from one unit of length (inches, feet, miles, yards, or rods) to another, requires multiplying or dividing by

$$12 \quad \text{(12 inches in a foot)},$$
$$5{,}280 \quad \text{(5,280 feet in a mile)},$$
$$3 \quad \text{(3 feet in a yard), or}$$
$$16\tfrac{1}{2} \quad (16\tfrac{1}{2} \text{ feet in a rod)}.$$

By contrast, conversion in the metric system requires merely multiplying or dividing by powers of 10:

$$100 \text{ or } 10^2 \quad \text{centimetres in a metre,}$$
$$1{,}000 \text{ or } 10^3 \quad \text{millimetres in a metre,}$$
$$1{,}000 \text{ or } 10^3 \quad \text{metres in a kilometre, or}$$
$$100 \text{ or } 10^2 \quad \text{centigrams in a gram.}$$

Thus, conversion within the metric system is a simple matter of shifting the decimal point.

When this country adopts the metric system, distances will be reported in kilometres instead of miles, and capacity will be reported in litres instead of gallons. How many mpg is equivalent to 1 km/ℓ?

Example 10.10: Find the number of mpg in 1 km/ℓ if

$$1 \text{ km} \doteq .6213712 \text{ mi.} \quad \text{and} \quad 1 \ell \doteq .2641721 \text{ gal.}$$

Solution: Replace 1 km by its equivalent in miles and 1 ℓ by its equivalent in gallons:

$$1 \text{ km}/\ell = 1 \text{ km} \div 1 \ell$$

$$\doteq \boxed{0.6213712} \text{ (miles)} \boxed{\div} \boxed{0.2641721} \text{ (gal.)}$$

$$\boxed{=} \boxed{2.3521454} \text{ (mpg)}.$$

Because both of the equivalents are approximations in this example, the result is slightly different from the official conversion factor. Officially, the conversion factor is

$$1 \text{ km}/\ell \doteq 2.3521432 \text{ mpg.}$$

We discuss problems having to do with conversions in more detail in Section 10.6. As long as we have the need to deal with both types of rates, we need to know how to go back and forth from one system to the other.

•1. A car's driving range is the number of miles it can be driven on a tank of gas.

   a. A particular model of sports car is rated at 15.6 mpg. If its tank can hold 18 gallons, what is its driving range?

   b. A compact car with a 21 gallon tank has a driving range of 613.4 miles. What is the mpg performance on this car? (Round off to nearest hundredth mpg.)

   c. A small sedan has a fuel tank that would hold 54.89 litres. If it can get 9.35 km/ℓ, what is the driving range in kilometres?

2. A high performance sports car was listed as getting 2.17 km/ℓ on a run.

   a. How many mpg was this sports car getting?

   b. At 75¢ per gallon for premium gas, how much would it cost to drive this high-performance car 100 miles?

   c. Explain, if you can, whether this car can drive from San Francisco to Los Angeles without stopping for gas. The distance from San Francisco to Los Angeles is about 400 miles. If you can't, do you need more information?

•3. Refer to the chart in Example 10.9. On the first day, the odometer reading went from 26,539.8 miles to 26,989.6 miles.

   a. What was the distance Kevin covered on July 1?

   b. How much gas did he use driving the distance in Part a?

   c. Find the mpg for each day on this trip. An efficient way to do this problem would be to complete a table such as the one below.

Date	Odometer	Miles	Gallons	MPG
July 1	26539.8	——	18.8	——
July 1	26989.6	449.8	17.1	26.30
July 2	27521.3	?	?	? and so on.

10.4 The Old "70.1-Year-Life" Routine

A resident of the United States is expected to live 70.1 years.

Tom Phulerie is an A— student.

Mr. U.S. Citizen wears a size 9 shoe.

What do these statements have in common? The first states a longevity rate; the second might be considered an achievement performance rate; but it is hard to think of the third example as a rate.

What *can* be said is that each of these statements uses a single number (or measurement) to represent a set of numbers (or measurements), and thus each expresses an *average*. We might restate each in the following manner:

> The average lifespan of a U.S. resident is 70.1 years.
>
> Tom Phulerie's average grade is A−.
>
> The average shoe size for a U.S. citizen is size 9.

We shall see that different kinds of averages are given in these examples. The most widely used is also one of the most easy to compute. This is the *arithmetic average* (or the *arithmetic mean*).

Example 10.11: Maria's scores on five tests were

$$73, 62, 69, 91, 85$$

What is her average score?

Solution: The (arithmetic) average score is obtained by adding up the individual scores and then dividing by the number of scores.

$$(73 \boxplus 62 \boxplus 69 \boxplus 91 \boxplus 85) \div 5 \rightarrow 380 \boxdiv 5 \boxminus 76.$$

Thus, the average is 76.

In the above example, 76 is the representative score. It is "the rate at which she has been scoring on the tests." Indeed, if Maria had gotten 76 points on each of the tests, her total points would be

$$76 \boxplus 76 \boxplus 76 \boxplus 76 \boxplus 76 \boxminus 380.$$

or

$$5 \boxtimes 76 \boxminus 380.$$

Hence this idea: if each of Maria's scores were to be replaced by the same number so that the sum of the scores would be the same (380), what would this number be? (*Answer:* 76.)

Why are we interested in averages? What are they good for? Here is one example.

> If it is known that an average avocado weighs 1.3 pounds, how much will 2,000 avocados weigh? About
>
> $$2000 \boxtimes 1.3 \boxminus 2600. \text{ (pounds)}.$$

To ship 2,000 avocados, we need a truck capable of hauling 2,600 pounds.

Another example of an average is a "unit price," which is the average price paid for each unit (ounce, pound, biscuit, or exposure) purchased.

Example 10.12: A box of shredded wheat contains 18 biscuits and weighs about 425 grams. How much does each biscuit weigh on the average? If the box of cereal sells for 65¢, what does the ratio $\frac{425}{65}$ represent?

Solution: 425 grams for 18 biscuits:

$$425 \boxplus 18 \boxminus 23.611111 \doteq 23.6 \text{ grams/biscuit.}$$

The unit price is

$$65 \text{ (cents)} \div 425 \text{ (grams), or } \frac{65}{425} ¢/\text{gm.}$$

The ratio $\frac{425}{65}$ is the reciprocal of the unit price. It is the average weight (unit weight) one gets for each cent spent on the cereal:

$$\frac{425}{65} = 425 \boxplus 65 \boxminus 6.5384615, \text{ or about 6.5 gm/cent.}$$

This ratio can be used equally well for price comparison.

Averages are used as a basis for comparison in general.

Example 10.13: The same test was given to students in two algebra classes in the same school. The scores in each class were as follows.

Class A: 43, 55, 68, 98, 39, 76, 67, 89, 45
Class B: 88, 86, 72, 32, 59, 84, 56, 76, 68, 75, 97

Which class did better?

Solution: We shall use "class average" as a means of comparison.

Class A: $(43 \boxplus 55 \boxplus 68 \boxplus ... \boxplus 45) \boxdiv 9 \boxminus 64.444444$
Class B: $(88 \boxplus 86 \boxplus 72 \boxplus ... \boxplus 97) \boxdiv 11 \boxminus 72.090909$

The two averages indicate that Class B did better on this test than Class A. Note that Class A had two students who did better than anyone in Class B! In addition, the lowest scorer (32 points) was in Class B. We can be fooled just by glancing at the scores alone.

Sometimes the data permit us to shorten some of the arithmetic.

Example 10.14: A class of 24 students made these scores on a particular test:

$$72, 64, 85, 91, 72, 51, 33, 64, 72, 72, 48, 85,$$
$$51, 64, 48, 85, 72, 91, 51, 72, 48, 72, 64, 91$$

Find the class average for this test.

Solution: We can, of course, add the numbers as they appear. But notice a few numbers appearing repeatedly. We can make use of this fact. First make a frequency table.

Score	Tally	Number of Occurrences
33	|	1
48	|||	3
51	|||	3
64	||||	4
72	⦀| ||	7
85	|||	3
91	|||	3
		24 total

Now the sum may be written

$$(1 \times 33) + (3 \times 48) + (3 \times 51) + (4 \times 64) +$$
$$(7 \times 72) + (3 \times 85) + (3 \times 91)$$
$$\rightarrow 33 + 144 + 153 + 256 + 504 + 255 + 273 = 1618.$$

Because there are 24 scores, the average is

$$1618 \div 24 = 67.416666 \doteq 67.42$$

In working this way, there are fewer numbers to keep track of, and so there is less chance for error. (If your calculator has a memory register, it will be easy to carry out this computation of the average score in a single sequence of operations. You simply carry out each multiplication and add the result to the total in the memory; then bring the final total out of the memory and divide by 24 to get the average. See the instruction manual for your machine to find out exactly how to use the memory register if you have one.)

In this simple example, the use of the frequency table may not seem much of a simplification. However, as we shall see, in situations involving a great many individual "scores" or other numbers, this approach can be a very significant simplification of the computations.

$$(1 \times 33) + (3 \times 48) + (3 \times 51) + \cdots$$

shows that the number 48 must be given 3 times the weight of the single 33, the 51 must be weighted 3, the 64 weighted 4, and so on. Hence, we may think of 67.416666 as the *weighted average* of the numbers 33, 48, 51, 64, 72, 85, and 91.

Notice that the weights 1, 3, 3, 4, 7, 3, and 3 add up to 24, the total number of scores.

Here is another example of a "weighted average."

Example 10.15: A preference poll shows the percentages of those favoring presidential candidate "C" in various geographic regions:

Northeast	32% for Candidate "C"
Southeast	94% for Candidate "C"
Midwest	40% for Candidate "C"
West	38% for Candidate "C"

What percent of the whole population favor Candidate "C"?

Solution: Can we get a simple arithmetic average of these percents,

$$(32 \boxplus 94 \boxplus 40 \boxplus 38) \boxdiv 4 \boxminus 51.$$

and claim that 51% of the people surveyed favored this candidate? No! We need more information.

Suppose 10 million individuals were interviewed: 4 million in the Northeast, 1 million in the Southeast, 3 million in the Midwest, and 2 million in the West.

This means, for example, that in the Northeast,

$$32\% \times 4 \text{ million} = 1.28 \text{ million}$$

favored Candidate "C".

Each section must be assigned appropriate weights. These weights (in millions) are 4, 1, 3, and 2, for a total of 10 million persons.

$$\text{True average} = \frac{(4 \times 0.32) + (1 \times 0.94) + (3 \times 0.40) + (2 \times 0.38)}{4 + 1 + 3 + 2}$$

$$= (1.28 \boxplus 0.94 \boxplus 1.2 \boxplus 0.76) \boxdiv 10 \boxminus 0.418 = 41.8\%$$

This solution turns out to be more than 9% off the 51% we got using the simplistic method.

Another application comes from the study of levers. You know about levers. A teetertotter is a good illustration of one. Here two forces act, one on each end of a rigid plank supported somewhere along the plank (see Figure 10.2).

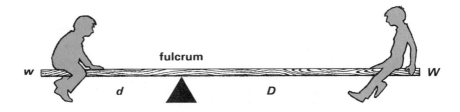

Figure 10.2

The support is called the *fulcrum*. The law governing levers states that, if the lever is to balance,

one weight (w) times its distance (d) from the fulcrum must be equal to the other weight (W) times its distance (D) from the fulcrum.

In Figure 10.2, we have

$$w \times d = W \times D.$$

You can try the following experiment to make a model of a teeter-totter.

Example 10.16: Balance a 12-inch ruler on a pencil as shown in Figure 10.3. The pencil acts as a fulcrum.

Figure 10.3
A ruler as a teetertotter.

Stack four pennies 3 inches to the left of the fulcrum; stack two pennies 6 inches to the right of the fulcrum. If done accurately, the ruler should balance. Notice that

$$4 \times 3 = 2 \times 6$$

Each side of this equation expresses

the product of the number of pennies and their distances from the fulcrum.

Hence, the numbers of pennies assign weights to the distances.

Here's a variation of the experiment.

Example 10.17: Stack pennies from the fulcrum as follows.

Left Side	Right Side
1 penny, 2 inches away	5 pennies, 3 inches away
2 pennies, 3 inches away	1 penny, 5 inches away
3 pennies, 4 inches away	

If done accurately, again the ruler should balance.

Taking a cue about the number of pennies assigning weights, on the left side:

$$(1 \times 2) + (2 \times 3) + (3 \times 4) = 2 + 6 + 12 = 20;$$

on the right side:

$$(5 \times 3) + (1 \times 5) = 15 + 5 = 20$$

Example 10.17 indicates how weights come into the picture. We still have not accounted for a weighted average. The weighted average on the left is

$$\frac{(1 \times 2) + (2 \times 3) + (3 \times 4)}{1 + 2 + 3} = 20 \boxed{\div} 6 \boxed{=} 3.3333333$$

Recall from Chapter 8 that $0.3333333 = \frac{1}{3}$ (check this on your calculator!). The weighted average is $3\frac{1}{3}$.

Now stack all 6 pennies on the left side $3\frac{1}{3}$ inches from the fulcrum. Don't change the right side. The ruler should still balance. Try it!

Another good example of a weighted average is the grade-point average (gpa). In averaging grades, an A earned in a 4-unit course should carry more weight than an A in a 2-unit course. In calculating grade points, one common system assigns points to letter grades by this scale:

A 4 points	B$-$ $2\frac{2}{3}$ points	D$+$ $1\frac{1}{3}$ points	F 0 points
A$-$ $3\frac{2}{3}$ points	C$+$ $2\frac{1}{3}$ points	D 1 point	
B$+$ $3\frac{1}{3}$ points	C 2 points	D$-$ $\frac{2}{3}$ point	
B 3 points	C$-$ $1\frac{2}{3}$ points	F$+$ $\frac{1}{3}$ point	

Example 10.18: Tom Phulerie received the following grades during the semester.

Course	Number of Units	Grade
Physiology	4	A
Bacteriology	3	A
Mathematics	3	A−
Physical Education	1	C

What was his gpa for the semester?

Solution: The number of units gives the weight for each course. So his gpa can be calculated:

$$\frac{(4 \times 4) + (3 \times 4) + \left(3 \times 3\frac{2}{3}\right) + (1 \times 2)}{4 + 3 + 3 + 1} = \frac{41}{11};$$

$$\frac{41}{11} = 4 \boxed{1} \boxed{\div} \boxed{1} \boxed{1} \boxed{=} 3.72\,72\,72\,7$$

Because A− is $3\frac{2}{3}$ or 3.6666666 on the scale, Tom's gpa is a bit better than A−.

Exercises 10.4

•1. A city has six high schools with enrollments of 387, 646, 591, 872, 426, and 519, respectively.

a. What is their average enrollment?

b. Explain the decimal in your answer.

•2. Use your calculator to find the arithmetic mean for each set of numbers.

a. 8463, 9271, 5847, 2983, 6541, 8767, 2329

b. 43.7, 41.6, 53.1, 37.5, 49.2

c. 4.1261, 1.2950, 0.2173, 0.1414, 1.3027, 12.0561

3. Over a seven month period, Mr. Capp Gaines bought one share of Immoble Oil each month. The costs in dollars were 23, 37, 51, 55, 41, 20, and 2, respectively.

a. What was the average cost per share?

b. Interpret what the average cost means.

c. Ignoring any commission, at what average price per share must he sell all seven shares to break even?

d. Ignoring commission, at what price per share must he sell his shares to make a total profit of 70 dollars?

•4. Thus far this year, the baseball player Tad Pole Grabber has played in 18 games. Here is his record as a batter.

Game	1	2	3	4	5	6	7	8	9	10	11	12	13	14	15	16	17	18
Times at Bat	3	1	4	2	3	5	3	4	2	3	4	4	5	3	4	5	1	4
Hits	1	0	2	1	2	1	0	1	0	0	0	2	0	0	1	3	0	2

a. What is Tad's batting average up to now?

b. What percent of the time does he get a hit?

c. If Tad were at bat 100 times during the season, how many hits would you expect him to get?

d. At this rate, how many games would he have to play to reach 300 hits?

•5. For each set of numbers listed at the left, their corresponding weights are listed at the right. Calculate the weighted average for each set.

Numbers	Weights
a. 87, 53, 62	2, 3, 5
b. 21.4, 27.3, 18.1	2.1, 3.7, 4.9
c. 347, 216, 541, 728	$\frac{1}{5}, \frac{2}{7}, \frac{5}{9}, \frac{1}{3}$
d. 18.23, 15.59, 21.03, 25.97	$\frac{1}{7}, \frac{2}{7}, \frac{3}{7}; \frac{1}{7}$

6. Tom Phulerie applied to be a lab assistant and was told that, if he could get a gpa of B— or better this semester, he would be accepted. Use the scale in this section to determine whether he got the job if he received these grades.

Course	Units	Grade
Chemistry	5	B+
Mathematics	3	C+
English	3	C
Political Science	3	B

10.5 Some Exotic Averages

With each collection of numbers we examined in Section 10.4, we looked for a number that might be considered typical of the collection. In Example 10.11, we considered "76" to be the typical or "representative" score for the five test scores 73, 62, 69, 91, and 85. As we noted, one advantage of the replacement in this example is that, instead of the repeated additions, the total score could have been found by the simple multiplication:

$$5 \times 76 = 380$$

The arithmetic average is typical from yet another viewpoint. Consider the errors made by each replacement in Example 10.11. Find the "total error" introduced when each number is replaced by the average (76).

Number	Error	
73	76 − 73 =	3
62	76 − 62 =	14
69	76 − 69 =	7
91	76 − 91 =	−15
85	76 − 85 =	− 9

The total error is

$$3 \boxed{+} 14 \boxed{+} 7 \boxed{+} (-15) \boxed{+} (-9) \boxed{=} 0.$$

Thus, the arithmetic mean is a very natural and practical average to use. A rather exotic kind of average emerges from the natural problem of averaging speeds.

Example 10.19: Mr. Zipalong Casualty takes a 2-hour trip, driving 1 hours at 45 miles per hour and 1 hour at 55 miles per hour. What is his average speed on this trip?

Solution: Driving 1 hour at 45 miles per hour, Zipalong covered 45 miles during that hour. Driving 1 hour at 55 mph, he covered 55 miles on this stretch. So, altogether, he traveled

$$45 \boxed{+} 55 \text{ (miles)} \quad \text{in 2 hours.}$$

Hence, his average was

$$(45 \boxed{+} 55) \boxed{\div} 2 \boxed{=} 100 \boxed{\div} 2 \boxed{=} 50. \text{ (mph).}$$

Example 10.20: Mr. Zipalong Casualty takes a trip in which the speed limit is 45 mph along a 100-mile stretch of the road and 55 mph along the next 100 miles. If he drives at the speed limit during the entire time, what will be his average speed?

Solution: The first impulse might be to average thus:

$$\frac{45 + 55}{2} = \frac{100}{2} = 50$$

This is the same procedure used in answering the question in Example 10.19. However, in this case, it is not appropriate.

First, recall that the rate of speed is related to the distance and time in this manner:

$$\text{rate} = \frac{\text{distance}}{\text{time}} \quad \text{or} \quad r = \frac{d}{t}.$$

(Remember, to find miles per hour, divide the number of miles by the number of hours.)

We know the distance driven is 200 miles. We need to find the amount of time it takes Mr. Casualty.

The first part of the journey:

$$45 \text{ mph} = \frac{100 \text{ miles}}{? \text{ hours}}$$

or, $$45 \times t = 100$$

Now, $$45 \times t = 100 \quad \text{and} \quad t = 100 \div 45$$

state the same fact. So,

$$t = \frac{100}{45}$$

We can likewise find the time for the second portion of the journey:

$$55 \times t = 100$$

So $$t = \frac{100}{55}$$

The total time would be the sum of these:

$$\frac{100}{45} + \frac{100}{55} = (100 \boxed{\div} 45) + (100 \boxed{\div} 55)$$

$$\rightarrow 2.2222222 \boxed{+} 1.8181818 \boxed{=} 4.040404$$

Then, to find the average rate, we must divide this into the total distance, 200.

$$\text{rate} \doteq 200 \boxed{\div} 4.04 \boxed{=} 49.5$$

So the average speed is *not* the same as the arithmetic average of 45 and 55.

In concentrating on getting a numerical solution, some of the general pattern was lost. Let's retrace these steps to bring the pattern into sharper focus. Throughout we have used the basic distance–rate–time relationship.

$$d = r \times t.$$

or one of the equivalent forms.

$$r = \frac{d}{t} \quad \text{or} \quad t = \frac{d}{r}.$$

Now, let's follow through the above process again with some slight changes.

We found that the total time Mr. Casualty spent was

$$t = \frac{100}{45} + \frac{100}{55}$$

We can rewrite this as

$$t = \left(100 \times \frac{1}{45}\right) + \left(100 \times \frac{1}{55}\right)$$

$$= 100 \times \left(\frac{1}{45} + \frac{1}{55}\right).$$

Because he covered a total distance of 200 miles in this time, we find the rate by dividing $\left(r = \frac{d}{t}\right).$

So
$$\text{rate} = \frac{200}{100 \times \left(\frac{1}{45} + \frac{1}{55}\right)} = \frac{100 \times 2}{100 \times \left(\frac{1}{45} + \frac{1}{55}\right)}$$

$$= \frac{2}{\frac{1}{45} + \frac{1}{55}}$$

Although the fraction looks more complicated now, in this form, we can see more clearly where each number came from. The "45" and "55" came from the number of miles per hour for each portion of the trip. This is easily seen.

A hint to where the "2" came from can be seen from the step above where the 200 canceled with the 100. In more detail, it is really the result of a weighted average: each 100 miles is given a weight of 1 and two of these account for a total weight of 2. This kind of weighted average involving reciprocals is called a *harmonic average*.

$$\text{rate} = \frac{(1 + 1)}{\left(1 \times \dfrac{1}{45}\right) + \left(1 \times \dfrac{1}{55}\right)}$$

The effect of "weighting" comes out more prominently in the next example.

Example 10.21: Mr. Casualty drove 200 miles at 45 miles per hour (mph) and 100 miles at 55 mph. What is his average speed on this trip?

Solution: Using the harmonic average given above,

$$r = \frac{(2 + 1)}{\left(2 \times \dfrac{1}{45}\right) + \left(1 \times \dfrac{1}{55}\right)} = \frac{3}{\dfrac{2}{45} + \dfrac{1}{55}}$$

$3 \div [(2 \boxplus 45) + (1 \boxplus 55)] \rightarrow 3 \div (0.0444444 \boxplus 0.0181818)$

$\rightarrow 3 \boxplus 0.0626262 \boxminus 47.9032 73$

Therefore, the average speed was about 47.9 mph.

If half of a trip were made at x mph and the remaining half at y mph, then we have the special case where the average speed is given by this harmonic average,

$$r = \frac{1 + 1}{\left(1 \times \dfrac{1}{x}\right) + \left(1 \times \dfrac{1}{y}\right)} = \frac{2}{\dfrac{1}{x} + \dfrac{1}{y}}$$

Harmonic averages also enter into "tank-filling" problems and electrical resistance problems. Here are some examples.

Example 10.22: Two hoses are used to fill an empty tank having 3,400 gallons capacity. A smaller hose, delivering, 300 gallons per hour, is left running for 8 hours, then turned off. Immediately, a second hose starts delivering 500 gallons per hour until the tank is full (taking another 2 hours).

Suppose both hoses are replaced by a single hose. At what rate should this single hose deliver so that it would also fill the tank in 10 hours?

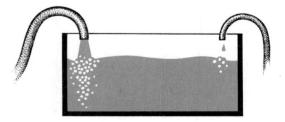

Figure 10.4
A tank filled by two different hoses.

Solution: The basic relationships we have to deal with here are similar to the distance–time–rate relationships. Here,

$$\text{rate} = \frac{\text{gallons}}{\text{hour}} \quad \text{or} \quad r = \frac{V}{t}$$

where V represents volume. The same fact can be stated as

$$V = r \times t \quad \text{or} \quad t = \frac{V}{r}.$$

It is required to fill 3,400 gallons in 10 hours, so the rate must be

$$\frac{3,400}{10} = 340 \text{ gallons per hour.}$$

Now, where does the harmonic average come in to the above example? To see this, consider the total time spent.

The small hose delivers 300 gal./hr. for 8 hours, so it delivers

$$300 \; \boxed{\times} \; 8 \; \boxed{=} \; 2400. \quad \text{(gallons).}$$

Similarly, the large hose delivers

$$500 \; \boxed{\times} \; 2 \; \boxed{=} \; 1000. \quad \text{(gallons).}$$

From these, we have

$$8 = \frac{2,400}{300} \quad \text{and} \quad 2 = \frac{1,000}{500}$$

The total time is

$$10 = 8 + 2 = \frac{2,400}{300} + \frac{1,000}{500}$$

$$= \left(2,400 \times \frac{1}{300}\right) + \left(1,000 \times \frac{1}{500}\right).$$

So the numbers $\frac{1}{300}$ and $\frac{1}{500}$ have weights 2,400 and 1,000, respectively. Hence the harmonic average,

$$\text{rate} = \frac{2,400 + 1,000}{\left(2,400 \times \frac{1}{300}\right) + \left(1,000 \times \frac{1}{500}\right)}$$

In general, if a hose delivers V gallons at x gallons per hour and V^* gallons at y gallons per hour, then the single hose replacement would deliver the $(V + V^*)$ gallons at the rate given by the harmonic average:

$$\text{rate} = \frac{V + V^*}{\dfrac{V}{x} + \dfrac{V^*}{y}}$$

So, weird as the harmonic average may seem at first, some applications call for its use. The harmonic average also comes into play with electric circuitry. In the example below, the mathematics can be followed without being an electrical wizard.

300 ohms

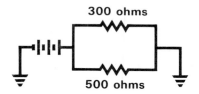

500 ohms

Figure 10.5
An electrical diagram of two resistors (〰), one with 300 ohm resistance and the other 500 ohms, connected in what is known as a parallel circuit.

We wish to replace these resistors by two other resistors having equal resistance (in ohms) and such that the current remains the same as before replacement.

Here, the basic relationship is Ohm's law:

$$\text{electromotive force} = \text{current} \times \text{resistance}$$

$$E \qquad = \qquad i \quad \times \quad r$$

where E is measured in volts;
$\qquad i$ is measured in amperes;
and r is measured in ohms.

Example 10.23: Suppose $E = 20$ in Figure 10.5. Replace the 300-ohm and 500-ohm resistors by resistors having equal ohms.

Solution: By Ohm's law, the current in the top half of the circuit is

$$20 \div 300 = \frac{20}{300} \text{ amps.}$$

Similarly, the current in the bottom half of the circuit is

$$20 \div 500 = \frac{20}{500} \text{ amps.}$$

So the total current is

$$\frac{20}{300} + \frac{20}{500} \quad \text{or} \quad 20 \times \left(\frac{1}{300} + \frac{1}{500}\right).$$

If each resistor is to be replaced by one having r ohms, the current would be

$$\frac{20}{r} + \frac{20}{r} \quad \text{or} \quad 20 \times \left(\frac{1}{r} + \frac{1}{r}\right) = 20 \times \frac{2}{r}$$

As the current is to remain the same, then we must have

$$20 \times \frac{2}{r} = 20 \times \left(\frac{1}{300} + \frac{1}{500}\right),$$

which means

$$\frac{2}{r} = \frac{1}{300} + \frac{1}{500}$$

It then follows that the reciprocals of both sides must be equal:

$$\frac{r}{2} = \frac{1}{\dfrac{1}{300} + \dfrac{1}{500}}$$

Now, multiplying by 2,

$$r = \frac{2}{\dfrac{1}{300} + \dfrac{1}{500}}$$

which we recognize as the harmonic average of 300 and 500.

$$r = 2 \div [(1 \div 300) + (1 \div 500)]$$

$$\rightarrow 2 \div [0.0033333 + 0.002]$$

$$\rightarrow 2 \div 0.0053333 = 375.00234$$

So, to 2-place decimals, $r = 375.00$ ohms.

In Example 10.23, we got 375.00234 because of the approximation 0.0033333 for $\frac{1}{300}$. We can show that the result should come out *exactly* 375. Watch closely!

$$r = \frac{2}{\dfrac{1}{300} + \dfrac{1}{500}} = \left(\frac{2}{\dfrac{1}{300} + \dfrac{1}{500}}\right)\left(\frac{300 \times 500}{300 \times 500}\right)$$

$$= \frac{2 \times 300 \times 500}{\left(\dfrac{1}{300} \times 300 \times 500\right) + \left(\dfrac{1}{500} \times 300 \times 500\right)}$$

$$= \frac{300,000}{500 + 300} = \frac{300,000}{800}$$

$$= 3000 \boxed{\div}\, 8 \boxed{=}\, 375.$$

Because $375 \times 8 = 3,000$, the exact quotient must be 375.

Exercises 10.5

•1. For each of these problems, decide whether to use the weighted average or the weighted harmonic average; then find the correct rate.

  a. On a trip, Mr. Zipalong Casualty drove 2 hours at 40 mph and 3 hours at 60 mph.

  b. On a trip, Zipalong drove 200 miles at 40 mph and 300 miles at 60 mph.

  c. Find the distances traveled at each rate in part a.

  d. Find the time traveled for each portion of the trip in part b.

  e. In which of the two situations (part a or b) is the ratio of the time spent driving 40 mph the greater?

2. The number of miles that Kevin traveled each day during his vacation (Example 10.11) is summarized below. Suppose his average speed each day is that appearing in column 3. Then, it is possible to find the number of hours driven per day. (Use your calculator.) From this information, find the speed he averaged over the entire trip.

Date	Miles	Speed
July 1	449.8	55.0
July 2	531.7	50.3
July 3	415.8	61.5
July 4	533.6	47.1
July 5	435.7	50.3
July 6	495.1	48.2
July 7	573.9	45.5

•3. Consider a parallel circuit (as in Figure 10.5) with one resistor giving 300 ohms and one giving 600 ohms resistance. Replace these with two resistors having the same number of ohms. How many ohms for each replacement? (Round off the seventh-place decimal in your intermediate calculations.)

4. Suppose that $E = 20$ in Figure 10.5. Then, by Example 10.23, the total current would be

$$\left(\frac{20}{300} + \frac{20}{500}\right) \text{ amperes.}$$

Using the calculator, show that replacing the 300-ohm and 500-ohm resistors by two 400-ohm resistors will not give the same current.

•5. Two hoses are used to fill an empty tank having 5,000 gallons capacity. The smaller hose, delivering 400 gallons per hour, is left running for 6 hours and then turned off. The second hose, delivering 650 gallons per hour, starts delivering 650 gallons per hour for 4 hours. If these hoses were replaced by a single hose so that it would fill the tank in the same time it takes the two hoses, at what rate should this single hose deliver?

6. Two hoses are used to fill an empty tank having 5,500 gallons capacity. The smaller hose delivers at the rate of 400 gallons per hour, and the larger delivers at the rate of 650 gallons per hour.

   a. If the smaller hose were left running 4 hours, then turned off, and the larger hose then starts delivering immediately, how long would it take to fill this tank?

   b. If the two hoses were replaced by a single hose so that it takes the same numbers of hours to fill the tank, at what rate should this single hose be delivering?

10.6   Units, Units, Units

In Europe, some speed-limit signs say "85." This means 85 km/hr. Autos made for European markets have speedometers calibrated in kilometres per hour. A driver of an American car in Europe needs to know that 85 km/hr is about 52.8 miles per hour. If he should go "85," he would greatly exceed the speed limit!

The system of measurement in the United States has units that are different from those used in other countries. Until the United States has completely converted to the metric system, we need to know how to change from one unit to another.

Even within the same system, we need to know how to convert from one unit to another.

Example 10.24(a): A city block is about $\frac{1}{10}$ of a mile. If the block is divided equally into 13 lots, how wide is each lot?

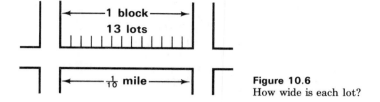

Figure 10.6
How wide is each lot?

Solution:

$$\frac{1}{10} \div 13 = ? \quad \text{or} \quad \boxed{0.1} \boxed{\div} \boxed{13} \boxed{=} \boxed{0.0076923} \text{ (miles).}$$

Each frontage, then, is 0.0076923 mile.

The answer, 0.0076923 mile, is not an easy length to visualize. It is more conventional to give lot sizes in feet. Before proceeding to convert the measurement to feet, let us briefly discuss the system of measures we use.

The U.S. system of weights and measures was modeled after the English. Historically, the units developed rather helter-skelter. Then a standardized system was decreed. As the stories went,

an inch was the length of the first joint of the thumb;

a foot was the length of the king's foot;

a yard was the length from the king's nose to the end of his outstretched fingers;

a rod was the length of 12 men, "*tried and true*," lined up in front of church on a Sunday morning.

A king's foot or a king's reach may change from time to time, and so such a system cannot continue for long. Eventually, definite relationships were established. A standard "yard" was adopted, and other units of length were described in terms of this standard.

We know many of these relationships. For example,

$$12 \text{ inches} = 1 \text{ foot,}$$

$$3 \text{ feet} = 1 \text{ yard,}$$

$$5{,}280 \text{ feet} = 1 \text{ mile.}$$

Using these few facts, let's return to example 10.24 once more.

Example 10.24(b): A city block is about $\frac{1}{10}$ of a mile. If the block were divided equally into 13 lots, how many feet wide is each lot?

Solution: 1 mile = 5,280 feet, so

$$0.0076923 \text{ miles} = \boxed{5280} \ \boxed{\times} \ \boxed{0.0076923} \ \boxed{=} \ \boxed{40.615344}$$

Each lot is about 40.6 feet in width.
   We can visualize 40.6 feet more easily than 0.0076923 miles. An average adult's step is about 3 feet in length. Because

$$\boxed{40.6} \ \boxed{\div} \ \boxed{3} \ \boxed{=} \ \boxed{13.533333} \doteq 13.5,$$

pace off about $13\frac{1}{2}$ steps to get an idea of the frontage for those houses.

Example 10.25: A quarter-mile relay was split into equal lengths for each of 4 runners. Does each runner cover more or less than 100 yards in his stretch? How much more or less?

Solution: 1 mile = 5,280 feet. Because 3 feet = 1 yard, to find the number of yards in 1 mile (= 5,280 feet), divide

$$\boxed{5280} \ \boxed{\div} \ \boxed{3} \ \boxed{=} \ \boxed{1760}. \quad \text{(yards in a mile).}$$

Hence for $\frac{1}{4}$ mile,

$$\boxed{1760} \ \boxed{\div} \ \boxed{4} \ \boxed{=} \ \boxed{440}. \quad \text{(yards).}$$

The 440 yards is again divided into 4 lengths:

$$\boxed{440} \ \boxed{\div} \ \boxed{4} \ \boxed{=} \ \boxed{110}.$$

So each runner covered 110 yards; 10 yards more than 100.

The common unit we use for auto speed is miles per hour (mph). But when it comes to stopping the car for emergency (breaking speed), a common unit is feet per second (fps).

Example 10.26: What is the equivalent in feet per second of 60 mps?

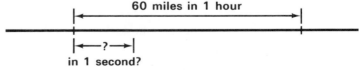

Figure 10.7
How many feet in 1 second?

Solution: 60 miles = 60 × 5,280 feet.

1 hour = 60 minutes = 60 × 60 seconds.

So $\dfrac{60 \text{ miles}}{1 \text{ hour}} = \dfrac{\cancel{60} \times 528\cancel{0} \text{ feet}}{\cancel{60} \times 6\cancel{0} \text{ seconds}}$

$\rightarrow$ 𝟝𝟚𝟠 ⊕ 𝟞 ⊜ 𝟠𝟠.   (fps).

Therefore, 60 mph = 88 fps.

The equivalent 60 mph = 88 fps, is a convenient one to remember. From this, we can get others directly without repeating the calculations of Example 10.26.

60 mph = 88 fps

30 mph = 44 fps   (half as much)

15 mph = 22 fps   (a quarter as much)

20 mph = 88 ÷ 3 = 29.333333

$55 \text{ mph} = 88 \times \left(\dfrac{55}{60}\right) \text{ fps} \doteq 80.666667 \text{ fps}$

Example 10.27:   The estimated time for reaction in emergency is about $\frac{3}{4}$ second (the *"thinking time"*). At 60 mph, how many feet does a car travel before the driver begins to brake?

Solution: Because 60 mph = 88 fps, the distance traveled at this speed in $\frac{3}{4}$ seconds is

$$88 \times \frac{3}{4} = 𝟠𝟠 \boxtimes 𝟘.𝟟𝟝 ⊜ 𝟞𝟞.   \text{(feet)}.$$

This means that 66 feet have been traveled before the brakes are applied. And then, of course, it takes quite some distance before the car can come to a complete stop. The estimate for the total, reaction and coming to a halt, for 60 mph is about 264 feet with "perfect 4-wheel brakes and ideal driving conditions." This is half a block in length. (See Example 10.24.)

In the metric system, there are

10 centimetres (cm) to 1 decimetre (dm),

10 decimetres to 1 metre (m).

So there are 10×10 (or 100) centimetres to a metre. The word *"centimetre"* means "1 one-hundredth of a metre." In this system, units are easily converted to one another. Conversion is simply through multiplying or dividing by powers of 10.

$$1 \text{ centimetre} = 10 \text{ millimetres,}$$

$$1 \text{ decimetre} = 10 \text{ centimetres} = 10^2 \text{ millimetres,}$$

$$1 \text{ metre} = 10 \text{ decimetres} = 10^2 \text{ centimetres}$$

$$= 10 \times 10^2 \text{ millimetres} = 10^3 \text{ millimetres,}$$

$$1 \text{ kilometre} = 1{,}000 \text{ metres} = 10^3 \text{ metres.}$$

One reason this system of units is preferred in many countries is this very ease of conversion. The United States has made plans to adopt this system in due course. The units in the metric system are called the S.I. units, for *"Système International."* We need to learn the relation between those units and the ones we use. Here is a brief summary.

For all practical purposes, we may define

$$1 \text{ inch} = 2.54 \text{ centimetres (abbreviated } cm).$$

$$1 \text{ foot} = 12 \times 2.54 \text{ centimetres} = 30.48 \text{ cm,}$$

So $\qquad$ $1 \text{ yard} = 3 \text{ feet} = 3 \times 30.48 = 91.44 \text{ cm.}$

Note that 1 metre = 100 centimetres, so 1 yard is less than a metre. In fact,

$$91.44 \div 100 = 0.9144,$$

so $\qquad\qquad$ $1 \text{ yard} = 0.9144 \text{ metre.}$

In turn, because

$$0.9144 \text{ m} = 1 \text{ yard,}$$

dividing by 0.9144,

$$1 \text{ metre} = \frac{1}{0.9144} \text{ yards}$$

$$= 1 \div 0.9144 = 1.0936132 \text{ (yards).}$$

Therefore,

$$1 \text{ metre} \doteq 1.0936132 \text{ yards.}$$

Similarly, 1 yard = 36 inches,

so $\qquad$ 1 metre = $/.0936/32$ ⊠ 36 ⊜ $39.37000 \, 75$

$$\doteq 39.37 \text{ inches.}$$

Example 10.28(a): How many metres are there in a 100-yard dash?

Solution: 1 yard = 0.9144 m, so

$$100 \text{ yards} = 91.44 \text{ metres.}$$

Example 10.28(b): How many yards are there in a 100-metre dash?

Solution: 1 metre = 1.0936132 yards, so

$$100 \text{ metres} = 109.36132 \text{ yards} \doteq 109.36 \text{ yards.}$$

Example 10.29: In Europe, a standard unit of distance is the kilometre (km). How does the kilometre compare with the mile?

Solution: 1 km = 1,000 m, and 1 m $\doteq$ 1.0936132 yards. So,

$$1{,}000 \text{ m} = 1.0963132 \times 1{,}000 = 1{,}093.6132 \text{ yards.}$$

From Example 10.25, we know

$$1 \text{ mile} = 1{,}760 \text{ yards.}$$

Hence, to find the number of miles per kilometre, divide:

$$/093.6/32 ⊞ /760 ⊜ 0.62/37//$$

Therefore, 1 kilometre is about 0.62 mile. Compare this figure with the number in Example 10.10.

Example 10.30: A long-distance runner ran 42 miles in 6 hours and 23 minutes. At what average rate in mph did he run?

Solution: Recall that

$$\text{rate} = \frac{\text{distance}}{\text{time}}.$$

Thus,   rate = (42 miles) ÷ (6 hours, 23 minutes).

To perform this division, we must convert time into a single unit—say, "hours." Because 1 minute = $\frac{1}{60}$ hour,

$$6 \text{ hours} + 23 \text{ minutes} = 6 + \frac{23}{60} \text{ hours}$$

$$= (23 \div 60) + 6 = 6.3833333 \text{ (hours)}.$$

Now $42 \div 6.3833333 = 6.579634$ (mph).

The racer averaged about 6.58 mph.

In addition to measures of length, we use measures of *capacity*. For liquid measures, the U.S. system has gills, cups, pints, quarts, and gallons (also hogsheads). The equivalents within the English system are

4 gills = 1 pint,

2 cups = 1 pint,

2 pints = 1 quart,

4 quarts = 1 gallon,

63 gallons = 1 hogshead.

Example 10.31(a): How many cups are in a quart of milk?

Solution: 1 quart = 2 pints, and 1 pint = 2 cups, so

$$1 \text{ quart} = 2 \times 2 = 4 \text{ (cups)}.$$

Example 10.31(b): How many cups are in a gallon?

Solution: 1 gallon = 4 quarts, and 1 quart = 4 cups, so

$$1 \text{ gallon} = 4 \times 4 = 16 \text{ (cups)}.$$

In the metric system, there are 10 decilitres in a litre (ℓ) and 1,000 millilitres (ml) in a litre. Therefore, there are

$$1,000 \div 10 = 100$$

millilitres in a decilitre. (A millilitre is also known as 1 cubic centimetre, or 1 cc, because a litre originally was defined simply as 1,000 cubic centimetres.)

How do we convert U.S. measures of capacity to metric ones? Some useful equivalents are

$$1 \text{ litre} = 1,000 \text{ millilitres},$$

$$1 \text{ quart} = 960 \text{ millilitres}$$

So 1 quart is slightly less than 1 litre. In fact,

$$1 \text{ quart} = 960 \boxplus 1000 \boxminus 0.96 \text{ (litre)}.$$

Some other useful equivalents are

$$1 \text{ gallon} = 4 \text{ quarts} = (4 \times 0.96) \text{ litres}$$

$$= 3.84 \text{ litres},$$

$$1 \text{ pint} = \frac{1}{2} \text{ quart} = 480 \text{ millilitres},$$

$$1 \text{ cup} = \frac{1}{2} \text{ pint} = 240 \text{ millilitres},$$

$$1 \text{ pint} = 16 \text{ fluid ounces}.$$

So
$$1 \text{ fluid ounce} = \frac{1}{16} \text{ pint} = 1 \text{ jigger}.$$

(A jigger is a common unit in mixing drinks.)

Although our system of weights and measures was modeled after the English system, slight variations began to show from the Colonial times. Each system has become nationally standardized by now, with the variations kept as they had been. One of the big differences between the English and U.S. systems is in liquid measure.

Example 10.32: The U.S. gallon contains 231 cubic inches. The English imperial gallon contains 277.418 cubic inches (cu. in.). We know that 1 in. = 2.54 cm, so

$$1 \text{ cu. in.} = 2.54 \times 2.54 \times 2.54 \text{ cubic centimetres}$$

$$= (2.54)^3 \text{ cc.}$$

Then,

$$1 \text{ U.S. gallon} = 231 \times (2.54)^3 \text{ cc}$$

$$\doteq 3785.4118 \text{ cc}$$

$$= 3785.4118 \text{ ml} \quad (\text{because } 1 \text{ cc} = 1 \text{ ml}).$$

Or, because 1,000 ml = 1 l,

$$1 \text{ U.S. gallon} = 3.7854118 \; l \doteq 3.785 \; l.$$

It is of interest to note that in Canada gas is sold by the imperial gallon. As in Example 10.32, we can convert the imperial gallon to litres.

$$1 \text{ imperial gallon} = 277.418 \text{ cu. in.}$$

$$= 277.418 \times (2.54)^3 \text{ cc}$$

$$= [277.418 \times (2.54)^3] \div 1,000 \; l$$

$$= [(2.54 \;\boxed{\times}\; 2.54 \;\boxed{\times}\; 2.54) \;\boxed{\times}\; 277.418] \;\boxed{\div}$$

$$1000 \;\boxed{=}\; 4.5460665$$

$$\doteq 4.546 \; l$$

Weights, of course, also occupy an important part of our living experiences. The weight units in the U.S. system are ounces, pounds, hundredweights, and tons. (Hundredweights are rarely used except on special jobs.)

$$1 \text{ pound (lb.)} = 16 \text{ ounces (oz.)}$$

$$1 \text{ ton (T.)} = 2,000 \text{ pounds}$$

In England there is also a *long ton* that is defined to weigh 2,240 pounds. It is sometimes used for shipping.

Metric weights are the gram and powers of tens of grams:

$$1,000,000 \text{ micrograms } (\mu g) = 1 \text{ gram (gm)},$$

$$1,000 \text{ milligrams (mg)} = 1 \text{ gram},$$

$$100 \text{ centigrams (cg)} = 1 \text{ gram},$$

$$10 \text{ decigrams (dg)} = 1 \text{ gram},$$

$$1 \text{ gram (the basic defined weight unit)},$$

$$10 \text{ grams} = 1 \text{ decagram (dkg)},$$

$$100 \text{ grams} = 1 \text{ hectogram (hg)},$$

$$1,000 \text{ grams} = 1 \text{ kilogram (kg)},$$

$$1,000,000 \text{ grams} = 1 \text{ metric ton (M.T.)} = 1,000 \text{ kg}.$$

Of these units, the decigram, decagram (dekagram), and the hecto-gram have been included to make the list a bit more complete. In practice, these units are seldom used. In most calculations powers of ten or scientific notation are used because they are so natural for this system.

Example 10.33: How many milligrams are there in a kilogram?

Solution:

$$1,000 \text{ milligrams} = 1 \text{ gram},$$

$$1,000 \text{ grams} = 1 \text{ kilogram}.$$

Thus, $1 \text{ kg} = 1,000 \times 1,000 = 10^3 \times 10^3$

$$= 10^6 \text{ (milligrams)}.$$

So there are 10^6 mg in 1 kg. In other words, there are 1 million milligrams in a kilogram.

As before, in working with metric units, conversion is simple, al-though we do have to know whether we want to multiply or to divide by appropriate powers of tens.

But we often have to cross over from one system to another. In dealing with weights, the magic passport is the number of pounds in a kilogram:

$$1 \text{ kilogram} \doteq 2.2046226 \text{ pounds}.$$

So $1 \div 2.2046226 = 0.45359124$ (kg) $= 1$ pound.

Example 10.34:   Convert 97 pounds to kilograms.

Solution: 1 lb. $\doteq$ 0.4535924 kg. To find the number of kilograms in 97 pounds,

$$9\ 7\ \boxed{\times}\ 0.4535924\ \boxed{=}\ 43.9984\,63\ \text{(kg)}.$$

So a child weighing 97 pounds also weighs 44 kilograms.

Usually, the estimates

$$2.2\ \text{lb./kg}\quad\text{and}\quad 0.45\ \text{kg/lb.}$$

are good enough for fast everyday mental or calculator arithmetic.

Exercises 10.6

•1. Convenient rapid estimates for converting between kilometres and miles have been adopted so that people can approximate kilometres quickly while driving in foreign countries.

    a. If a kilometre is approximated by $\frac{3}{5}$ = 0.6 mile, find the error from the number in Example 10.29 to 4 places.

    b. Using the approximation in Example 10.29, find the number of km per mile.

    c. If 1 km $\doteq$ $\frac{3}{5}$ mi., then 1 mi. $\doteq$ $\frac{5}{3}$ km. (Why?) Find the difference between $\frac{5}{3}$ and your answer in part b to 4 places.

2. An event in international swimming competition is the 100-metre race. How much more or less than 100 yards is this distance?

•3. The 1974 bicycle race, *Tours de France,* was 4,078.1 km long.

    a. Find the number of miles in the race.

    b. Bicyclist Eddie Merkx completed the race in 116 hours, 16 minutes, 58 seconds. Find his average speed in km/hr. (Remember 1 hr. = 60 min. = 3,600 sec.)

    c. Express Eddie Merkx's speed in mph.

    d. The distance was covered in about 24 days of riding. About how many miles per day did Eddie average?

•4. The "thinking time" is from the instant a driver sees a need to stop until the time he applies the brakes. This has been estimated to be about $\frac{3}{4}$ of a second. The distance covered by the car during the "thinking time" is called the "thinking distance."

    a. Find the thinking distance for a car going 65 mph.

    b. If a car requires another 231 feet before coming to a stop at the speed in part a, what is the total stopping distance?

5. Using the relation 60 mph = 88 fps, find the equivalent in fps when the speedometer reads 45 (mph).

6. Use the table on liquid measure to convert the following.

   a. A bottle of salad dressing gives its volume as 1 pint, 8 ounces. Find the number of quarts.

   b. A bottle of cold drink is labeled 32 oz. Find the number of qts.

   c. Convert the volume in part a to ml.

   d. Convert the volume in part b to cc.

•7. If gas sells for $1.06 (U.S. dollars) per (imperial) gallon in Canada, what is its equivalent price by the U.S. gallon?

•8. It is estimated that the total amount of blood in the body of a person is about 90 ml per kg of body weight.

   a. Find the amount of blood in a woman weighing 48 kg.

   b. If a man weighs 150 pounds, how many quarts, pints, and fractions of pints of blood does he have?

9. The engine of a *"1900"*-model Opel has four cylinders with a total volume of about 1,900 millilitres.

   a. If the volume for the 4 cylinders is actually 1,896 ml, what is the volume for each cylinder?

   b. Find the number of cubic inches in the 1,896-ml engine.

   c. The total volume of the cylinders of a particular "muscle car" is claimed to be 440 cu. in. What is this volume in ml?

10.7 Similarity and Ratios

In Chapter 5 we considered many geometric shapes in the plane and in space. We found formulas for the area and volume of some of these figures. In that discussion, we used the principle that figures having the same size and shape (congruent figures) have the same area and volume. In this section, we consider figures that have the same *shape,* but not necessarily the same size. Such figures are said to be *similar* to each other.

To keep things simple, we consider only figures in the plane. Even so, the applications to scale drawings and map reading are very important.

Figure 10.8
Similar figures.

Figure 10.8 shows two drawings of a horse. Although one drawing is larger than the other, both drawings have exactly the same shape. In the mathematical sense, we say that each drawing is similar to the other. (Note that our mathematical definition of the word "similar" is more restricted than the everyday usage of the word. In casual speech, we might say that any two drawings of running horses are "similar." In the mathematical sense of the word, the drawings are similar only if they have *exactly* the same shape.)

An important class of figures in mathematics is that of plane figures such as triangles, rectangles, and squares—figures that are composed of line segments joined end to end. Such figures are called *polygons* (from two Greek words meaning "many angles"). Let us examine examples of polygons having the same shape. The ones shown in Figure 10.9 surely fit this description.

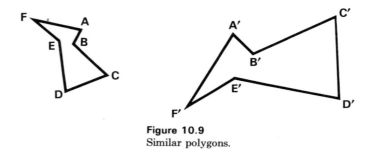

Figure 10.9
Similar polygons.

In Figure 10.10 we show four triangles. Three of them are similar.

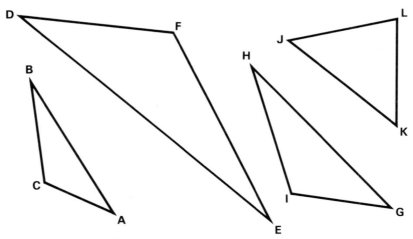

Figure 10.10
Triangles ABC, DEF, and GHI are similar to one another.
Triangle JKL is not similar to any of the other three.

Now, how can we check for similarity? One way is to measure angles that correspond. All corresponding angles must match in size. For this, we need a protractor. An easier way just makes use of an ordinary ruler and some calculations.

> If the figures are similar, the ratio of any two corresponding sides is constant.

Here are some measurements for triangles ABC and DEF in Figure 10.10.

$$\overline{AB} = 44 \text{ mm} \qquad \overline{DE} = 88 \text{ mm}$$

$$\overline{BC} = 29 \text{ mm} \qquad \overline{EF} = 58 \text{ mm}$$

$$\overline{AC} = 21 \text{ mm} \qquad \overline{DF} = 42 \text{ mm}$$

Now we set up ratios of corresponding sides:

$$\frac{AB}{DE} = \frac{44}{88} = \frac{1}{2}$$

$$\frac{BC}{EF} = \frac{29}{58} = \frac{1}{2}$$

$$\frac{AC}{DF} = \frac{21}{42} = \frac{1}{2}$$

In this case, each side of figure DEF is twice as long as the corresponding side of figure ABC. (We can, of course, say that each side of ABC is $\frac{1}{2}$ as long as the corresponding side of DEF.) The two figures, ABC and DEF, are similar because we can pair the corresponding sides in a way that gives a constant ratio. Try the same thing with figures DEF and JKL, and convince yourself that there is no way to pair corresponding sides to obtain a constant ratio. Figures DEF and JKL are *not* similar.

Example 10.35: Find the ratios of sides that seem to correspond in the triangles ABC and DEF shown in Figure 10.11.

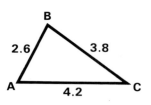

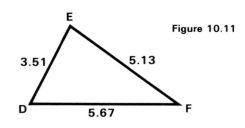

Figure 10.11

Solution: The ratios between triangles are the following.

$$\frac{AB}{DE} = \frac{2.6}{3.51} = 0.7407407$$

$$\frac{BC}{EF} = \frac{3.8}{5.13} = 0.7407407$$

$$\frac{AC}{DF} = \frac{4.2}{5.67} = 0.7407407$$

Therefore, triangle ABC is similar to triangle DEF. In symbols,

$$\triangle ABC \sim \triangle DEF.$$

In this case, the sides of ABC are about 0.74 of those of DEF. So ABC is a reduction of DEF, or DEF is an enlargement of ABC.

Example 10.36: A triangle with sides

$$AB = 2.6, \quad BC = 3.8, \quad AC = 4.2,$$

is enlarged by a ratio of 2.34 to 1. (See $\triangle ABC$ in Example 10.35.) Find the lengths of the sides of the enlarged triangle.

Solution: If $\triangle DEF \sim \triangle ABC$, then

$$DE = 2.34 \times 2.6 = 6.084$$

$$EF = 2.34 \times 3.8 = 8.892$$

$$DF = 2.34 \times 4.2 = 9.828$$

In Example 10.35, notice the following pairs of ratios within each triangle.

$$\frac{AB}{BC} = \frac{2.6}{3.8} = 0.6842105 \qquad \frac{DE}{EF} = \frac{3.51}{5.13} = 0.6842105$$

$$\frac{BC}{AC} = \frac{3.8}{4.2} = 0.9047619 \qquad \frac{EF}{DF} = \frac{5.13}{5.67} = 0.9047619$$

$$\frac{AC}{AB} = \frac{4.2}{2.6} = 1.6153846 \qquad \frac{DF}{DE} = \frac{5.67}{3.51} = 1.6153846$$

The ratios $\dfrac{AB}{BC}$, $\dfrac{BC}{AC}$, and $\dfrac{AC}{AB}$ are not constant. But they can be paired with the corresponding ratios $\dfrac{DE}{EF}$, $\dfrac{EF}{DF}$, and $\dfrac{DF}{DE}$ for the other triangle. In general, corresponding sides of similar polygons have the same ratio. In this example, we see that

$$\frac{AB}{BC} = \frac{DE}{EF}; \qquad \frac{BC}{AC} = \frac{EF}{DF}; \qquad \frac{AC}{AB} = \frac{DF}{DE}.$$

Example 10.37: Are the triangles shown in Figure 10.12 similar?

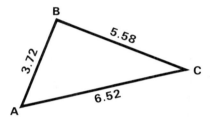

Figure 10.12

Solution: Note that, in the attempt to match, the shortest side is made to correspond with the shortest side, the medium side with the medium side, and so on. Now compute ratios.

$$\frac{AB}{DE} = \frac{3.72}{2.4} = 1.55$$

$$\frac{BC}{EF} = \frac{5.58}{3.6} = 1.55$$

$$\frac{AC}{DE} = \frac{6.52}{4.2} = 1.5523809$$

Although the triangles look somewhat similar here, by our calculations we can see that the ratios of corresponding sides are not constant. The last one is a bit off. $1.55 \neq 1.5523809$! So we conclude that the triangles are *not* similar.

We can also check by the ratios within each triangle.

$$\frac{AB}{BC} = \frac{3.72}{5.58} = 0.6666667 \qquad \frac{DE}{EF} = \frac{2.4}{3.6} = 0.6666667$$

$$\frac{BC}{AC} = \frac{5.58}{6.52} = 0.8558282 \qquad \frac{EF}{DF} = \frac{3.6}{4.2} = 0.8571429$$

$$\frac{AC}{AB} = \frac{6.52}{3.72} = 1.7526882 \qquad \frac{DF}{DE} = \frac{4.2}{2.4} = 1.75$$

Again, the test fails for these ratios. Only one pair of ratios matches between the two triangles. If the triangles were similar, all three pairs of ratios would match.

10 Rates, Ratios, and Similarity

Example 10.38: An enlargement is made from a snapshot portrait. The distance between the eyes in the print is 3.4 mm. In the enlargement this distance is 10.06 mm. If the distance from the top of the forehead to the tip of the nose is 7.24 mm in the print, how far would this be in the enlargement?

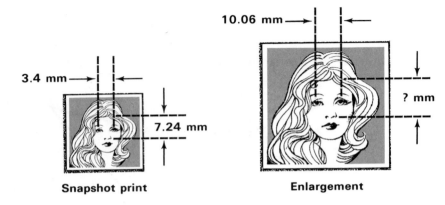

3.4 mm

7.24 mm

Snapshot print

10.06 mm

? mm

Enlargement

Figure 10.13
Similar figures.

Solution:

Let e be the distance between the eyes in the print;
let E be the distance between the eyes in the enlargement;
let f be the distance from the forehead in the print; and
let F be the distance from the forehead in the enlargement.

Then $\qquad \dfrac{e}{E} = \dfrac{f}{F};$ so $\dfrac{3.4}{10.06} = \dfrac{7.24}{F}$

Cross multiplying,

$$3.4 \times F = 7.24 \times 10.66$$

Dividing both sides by 3.4,

$$\frac{3.4 \times F}{3.4} = \frac{7.24 \times 10.06}{3.4}$$

$F = 7.24 \; \boxed{\times} \; 10.06 \; \boxed{\div} \; 3.4 \; \boxed{=} \; 21.421882$

So the forehead to nose distance in the enlargement is about 21.4 mm, or about 2.14 cm. Check it with your ruler.

In the above examples we always measured straight-line distances. However, we often have to work with geometric figures that are curved. Think again about a photo print and its enlargement. Suppose there is an image of a circle, like a ball, on the small print. What would it look like in the enlargement? Exactly; it would also be circular.

In other words,

Any two circles are similar.

Likewise,

Any two semicircles are similar;

Any two quarter-circles are similar.

Example 10.39: Are these two pie wedges similar?

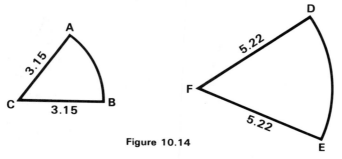

Figure 10.14

Solution: Lengths of curves are hard to measure. How can we determine whether these figures are similar? It turns out that we need only measure the angles of C and F. If these angles are the same size, then these wedges, are similar. (We assume here that the curved lines in both figures are parts of circles.)

Angles may be measured by an instrument called a protractor. Does the above test for similarity mean that we have to use a protractor? No. Even without protractors, we can check these wedges for similarity. Draw the segments (chords) AB and DE. Now, we just check the triangles to see if they are similar. (See Figure 10.15.)

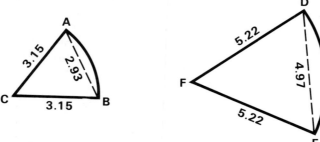

Figure 10.15
Pie wedges with chords drawn to determine similarity.

$$\frac{AC}{AB} = \frac{3.15}{2.93} = 1.0750853 \qquad \frac{DF}{DE} = \frac{5.22}{4.97} = 1.0503018$$

So, although the ratios are close, they agree only to one-place decimals and are not equal. We conclude that the pie wedges are not similar.

CAUTION: Sometimes, because of round-off errors or because we cannot measure very accurately, we may conclude by the results of our calculations that certain figures are not similar when they might actually be.

Example 10.40: Construction of similar wedges. We wish to construct wedges ABC and DEF so that they will be similar. What length should we make the chord DE?

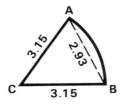
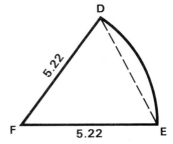

Solution: We require that $\dfrac{DE}{AB} = \dfrac{DF}{AC}$. Hence

$$\frac{DE}{2.93} = \frac{5.22}{3.15}$$

and

$$DE = 2.93 \times \frac{5.22}{3.15}$$

$$= 2.93 \;\boxed{\times}\; 5.22 \;\boxed{\div}\; 3.15$$

$$\boxed{=}\; 4.8554286$$

Now suppose we have constructed wedges (according to appropriate angles and so on) so that they would be similar. Suppose also that the measurements have been read to the nearest hundredths. Then, we may have gotten

$$AB = 2.93, \quad AC = 3.15, \quad DE = 4.86, \quad DF = 5.22$$

So, with these measurements, we "check" for similarity.

$$\frac{DE}{AB} = \frac{4.86}{2.93} = 1.6587031 \qquad \frac{DF}{AC} = \frac{5.22}{3.15} = 1.6571429$$

Hence, although these ratios agree only to two-place decimals, we must recognize that this came about due to rounding off 4.8554286 to 4.86, or because we were only able to read our ruler to hundredths.

Example 10.41: A man wants to know how tall a certain flagpole is. He stands close by and measures the length of his shadow. Then he measures the length of the shadow of the flagpole. If he is 5′ 9″ tall and casts a shadow that is 3′ 4″ while the shadow of the flagpole is 42′ 5″, what is the length of the flagpole?

Solution: The rays of the sun may be considered as parallel. Because the man stood close to the base of the flagpole, we may assume that the ground is flat. Then the man and his shadow and the pole and its shadow form similar triangles. Hence,

$$\frac{\text{height of the pole}}{\text{length of the pole's shadow}} = \frac{\text{height of man}}{\text{length of man's shadow}}.$$

Let h stand for the height of the pole. Then,

$$\frac{h}{42'\ 5''} = \frac{5'\ 9''}{3'\ 4''}$$

Converting all measurements to inches,

$$\frac{h}{509} = \frac{69}{40}$$

So $h = 509 \times \dfrac{69}{40} = $ 509 ⊠ 69 ⊞ 40 ⊟ 878.025 (inches).

To convert to feet,

878.025 ⊞ 12 ⊟ 73.16875 (feet).

Now 0.16875 feet is

0.16875 ⊠ 12 ⊟ 2.025 (inches).

So the flagpole is 73 feet, 2.025 inches, or a bit more than 73 feet tall.

Exercises 10.7

- 1. For each pair of figures below, measure to the nearest mm and use the calculator to determine whether the figures are similar.

a.

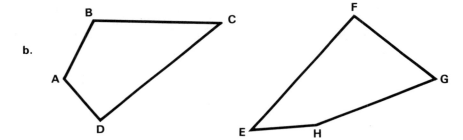

b.

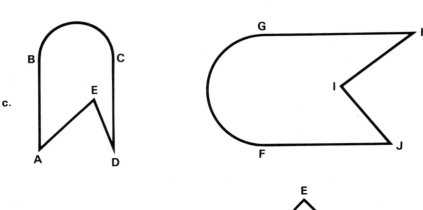

c.

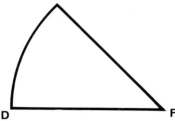

d.

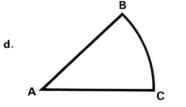

- 2. For each pair of similar figures in Exercise 1, match the corresponding corners.

3. If triangle ABC in Example 10.36 is reduced to three-fourths size, give the dimensions of the reduced triangle A′B′C′ (where the primed letters are matched with the ones unprimed).

- 4. ABCD and A′B′C′D′ are similar parallelograms, and $b' = 3 \times b$.

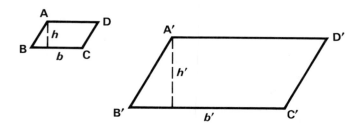

  a. How is the side A′B′ related to the side AB?

  b. How is the perimeter of A′B′C′D′ related to the perimeter of ABCD?

  c. How are the altitudes, h and h', related?

  d. How are the areas of the two parallelograms related?

5. Consider the triangles ABC and A′B′C′ in Exercise 4.

  a. How are the altitudes of these triangles related?

  b. How are the areas of these triangles related?

- 6. Consider the triangles ABC and DEF.

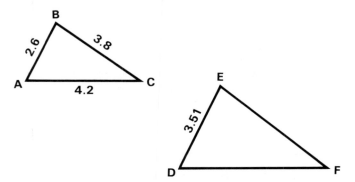

  a. Find the lengths of EF and DF to 2-place decimals so that △DEF ∼ △ABC.

  b. Use the ideas of Exercise 5 to state (to 4-place decimals) the relation of the area of triangle DEF to that of triangle ABC.

7. In geometry, there is a property that "the segment joining the midpoints of two sides of a triangle is half the length of the third side." That is, if D and E are midpoints, then

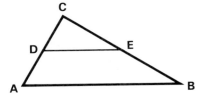

$$DE = \frac{1}{2}\,AB.$$

a. Explain why $\triangle DCE \sim \triangle ABC$, and name the points that match.

b. Locate the midpoint of AB in $\triangle ABC$ and label it F. Draw DF and EF, and identify four triangles that are similar to $\triangle ABC$.

c. For triangles ABC and DFE above, match the vertices.

8. Can you say that a triangle is similar to itself? That is, is it true that $\triangle ABC \sim \triangle ABC$?

a. Form ratios between triangles to check the above claim.

b. What is the ratio of the similarity in part a?

• 9. Find the length of the flagpole in Example 10.41 if all the measurements remain the same, except that the shadow of the flagpole measures 24′5″.

•10. Suppose a man, 5′ 9″ tall, casts a shadow 3′ 4″ long, How long would the shadow of a 60-foot pole be (provided the shadow is not interrupted).

10.8 Scaling

In discussing similarity, we considered a snapshot and an enlargement of this print. In map making, the map is a model of an actual section of geography, and so should be similar to it.

In a map, a scale is usually given. For instance,

$$1'' = 1 \text{ mile} \quad \text{or} \quad 1'' = 50 \text{ miles.}$$

Example 10.42: Suppose the number of air miles between San Francisco and Los Angeles is 340. ("Air miles" means straight "as the crow flies.") With a scale of $1'' = 50$ miles, how many inches would there be between the two cities on the map?

Solution:

$$\frac{\text{map distance in inches}}{\text{actual distance in miles}} = \frac{1 \text{ inch}}{50 \text{ mi.}}$$

So

$$\text{map distance} = \text{actual distance} \times \frac{1}{50}$$

$$= \text{actual distance} \div 50$$

$$= \boxed{340} \boxed{\div} \boxed{50} \boxed{=} \boxed{6.8}$$

The distance on the map should be 6.8 inches.

Example 10.43: A map is scaled 1 cm = 150 km. About how many kilometres are there between Los Angeles and New Orleans if it measures about 18.05 cm on the map?

Solution: Applying the scale factor 150,

$$\boxed{18.05} \boxed{\times} \boxed{150} \boxed{=} \boxed{2707.5} \text{ (km).}$$

So there are about 2707.5 km between the two cities.

Example 10.44: The distance from San Francisco to New York is 2,580 air miles. If this distance is 8.6 inches on the map, what is the scale of the map?

Solution: If 8.6 inches = 2,580 miles, then

$$1 \text{ inch} = \boxed{2580} \boxed{\div} \boxed{8.6} \boxed{=} \boxed{300} \text{ (mi.).}$$

So 1″ = 300 miles is the scale of this map.

The same idea is used in dictionary and encyclopedia drawings. The illustration is a model of the actual object, and so it should be similar to the real thing.

Example 10.45: An aoudad is pictured in a dictionary as 25 mm long. If the scale is given as $\frac{1}{36}$, how long is this animal?

Figure 10.16
Scale: $\frac{1}{36}$

Solution: The scale $\frac{1}{36}$ means that, for every millimetre, the animal in the drawing is $\frac{1}{36}$ of the real object. That is, the real object is 36 mm for every millimetre in the drawing:

$$\frac{\text{picture}}{\text{animal}} = \frac{1}{36}$$

or, the reciprocal,

$$\frac{\text{animal}}{\text{picture}} = \frac{36}{1}$$

So

$$\text{animal} = 36 \times \text{picture}$$

$$= 36 \boxed{\times} 25 \boxed{=} 900 \text{ (mm)}$$

Therefore the animal is 900 mm, or 90 cm, long. To find the number of inches it is long, divide 90 by 2.54 (because $1'' = 2.54$ cm).

$$90 \boxed{\div} 2.54 \boxed{=} 35.433071 \text{ (inches),}$$

or a bit shorter than a yard.

Exercises 10.8

•1. The scale in a map is $1'' = 4$ miles.

    a. If the widest part of Manhattan is 16.5 miles, how many inches would this width occupy in the map?

    b. If the greatest length (from northernmost Bronx to southernmost Staten Island) is 36 miles, why can't all of New York City fit into a map measuring $5''$ by $7''$ using the $1'' = 4$ miles scale?

2. A map is scaled 1 cm = 25 miles. If the distance from Atlanta to Philadelphia measures 267 mm on the map, how many air miles are there between the two cities?

•3. The distance from San Francisco to Honolulu is 2,400 air miles. If this distance measures 16 centimetres on a map, what is the scale?

4. Find the height of the aoudad in Example 10.45 (Figure 10.16) if the height in the drawing measures 22 mm.

•5. An illustration of a chamois measures 3 cm tall with a scale of $\frac{1}{45}$.

    a. Give the height of a typical chamois in centimetres.

    b. Convert the answer in part a to feet and inches.

6. A typical haddock is about 21 inches in length. If a space of $1\frac{1}{4}''$ is the most you can use for an illustration of this fish, what is the scale to use that would give the largest possible picture?

11 1

More About Integers

11.1 What Is a Prime?

Our discussion of numbers has stressed properties that helped us to operate with them. For example, we noticed that

$$a + b = b + a, \quad a \times b = b \times a,$$

and
$$a \times (b + c) = (a \times b) + (a \times c).$$

To learn still more about these numbers, we may ask how each is made up. We know that a rational number is the quotient of two integers. How are integers made up? For example, are there some numbers that make up the number 6?

The answer depends on the operations to be used. Under addition, we may say, 6 is

$$0 + 6, \quad 1 + 5, \quad 2 + 4, \quad \text{or} \quad 3 + 3$$

Another natural way to think about making up 6 is by way of multiplication. We may say 6 is

$$1 \times 6 \quad \text{or} \quad 2 \times 3$$

The first (1×6) is typical of *any* number:

Every number n may always be expressed as $1 \times n$.

The second way seems more interesting:

$$6 = 2 \times 3$$

Here, both 2 and 3 are less than 6, and we can think of 6 as being split up into smaller parts (2, 3). Many positive integers can be split into a product of two or more smaller positive integers. For example,

$$24 = 2 \times 12 \quad \text{or} \quad 24 = 3 \times 8 \quad \text{or} \quad 24 = 4 \times 6$$

Such integers are called *composite numbers* because they are composed from smaller integers. The smaller integers are called *factors;* the process is called *factoring.*

What about the number 5? Can it be factored?

Example 11.1: Factor 5 into integers.

Solution: We consider the possible breakdowns of 5 as products. (Use your calculator.)

$$? \times 1 = 5 \text{ means } ? = 5 \boxdot / \boxdot 5.$$

$$? \times 2 = 5 \text{ means } ? = 5 \boxdot 2 \boxdot 2.5 \qquad (*)$$

$$? \times 3 = 5 \text{ means } ? = 5 \boxdot 3 \boxdot 1.6666666 \quad (*)$$

$$? \times 4 = 5 \text{ means } ? = 5 \boxdot 4 \boxdot 1.25 \qquad (*)$$

$$? \times 5 = 5 \text{ means } ? = 5 \boxdot 5 \boxdot 1.$$

Because we are looking for integers to compose 5, we discard the lines indicated by an asterisk (*). From this list, we find that the only way to express 5 as a product of integers is

$$1 \times 5 \quad \text{or} \quad 5 \times 1$$

We naturally regard these two ways as equivalent.

There are many integers like 5 that cannot be split into two smaller positive integers (the "1" is smaller, but the "5" is not.) For example, 11, 13, 19, and so on. Such integers, if they are greater than 1, are called *prime numbers.*

The number 1 is neither a prime number nor a composite number. The number 1 is called a *unit.*

It is useful to be able to recognize small prime numbers on sight. To this end, we have listed a few primes in some boxes below. They are listed in increasing order from left to right. The blank space below the box for "2" is for the next prime after "11", and the next box to its right for the prime after this.

2	3	5	7	11
	17			
		41		
			67	
				97

Figure 11.1
Can you complete this table of primes less than 100?

We can easily test that 2 is a prime and 3 is a prime. Four is composite because $4 = 2 \times 2$. So is 6 composite. After 11, we can ask about 12, 13, and so on. But we can do better. Can we discard all the even numbers from here on? Yes; even numbers are divisible by two.

Exercises 11.1

•1. Are prime numbers necessarily odd numbers? Explain.

•2. Factor each of the following in all possible ways.

   a. 15   e. 23

   b. 17   f. 24

   c. 21   g. 25

   d. 22   h. 36

•3. Complete the entries in Figure 11.1. These are all the prime numbers less than 100.

4. Explain why 2 is the only even number in Figure 11.1.

•5. Suppose a number is the product of three distinct primes. (Thirty is one such number.) In how many ways can it be written as the product of two integers?

6. Suppose a number is the product of a prime squared and another prime. (Examples: $12 = 2^2 \times 3$ and $45 = 3^2 \times 5$.) In how many ways may such a number be written as the product of two numbers?

11.2 Prime Factorization

In Section 11.1 we discovered prime numbers as we searched for the "building blocks" of the positive integers. Prime numbers behave very much like the atoms of chemistry. Just as any chemical compound, however complicated, may be broken down into atoms, any natural number (greater than 1) may be factored into primes.

Every positive integer greater than 1 can be expressed as the product of primes and, except for the order in which the primes appear, this factorization can be done in only one way.

The decomposition of a number as a product of primes is called its *prime factorization*. The prime factorization of certain integers, even large ones, may be found quickly. For example,

$$21,000 = 21 \times 1000$$

$$= 21 \times 10 \times 10 \times 10$$

$$= (3 \times 7) \times (2 \times 5) \times (2 \times 5) \times (2 \times 5)$$

$$= 2^3 \times 3^1 \times 5^3 \times 7^1$$

Usually, the practice is to list the factors as powers of primes beginning with the least prime number. There was no need to resort to our calculator for the above factorization. Following is a more complicated example for which our calculator will prove quite useful. This example displays a methodical procedure for factoring any integer into its prime parts.

Example 11.2: Find the prime factorization of 23,373,441.

Solution: Let us follow the procedure in detail.

Try 2. Does 2 divide 23,373,441? No, 23,373,441 is an odd number.

Try 3. Does 3 divide 23,373,441?
$2337344I \div 3 = 779II47.$

Try 3 again. Does 3 divide 7,791,147?
$779II47 \div 3 = 2597049.$

Try 3 again. Does 3 divide 2,597,049?
$2597049 \div 3 = 865683.$

Try 3 again.
$865683 \div 3 = 28856I.$

Try 3 again.
$28856I \div 3 = 96I87.$

Try 3 once more.
$96I87 \div 3 = 32062.333$ (96,187 is not divisible by 3.)

Up until the last line, we have divided by 3 repeatedly five times. That is, we performed the division

$$23,373,441 \div 3^5$$

and got the result, 96,187. In other words, 3^5 is a factor of 23,373,441:

$$23,373,441 = 3^5 \times 96,187$$

We continue to determine other prime factors of 23,373,441. Now, no prime number other than 3 will divide into

$$3 \times 3 \times 3 \times 3 \times 3 = 3^5$$

So we need only to work with the factor 96,187.

The next prime number (after 3) is 5; how could we tell at once that 5 is not a factor of 96,187?

Next, try 7. Does 7 divide 96,187?

$$96187 \boxplus 7 \boxminus 13741.$$

Good; try 7 again.

$$13741 \boxplus 7 \boxminus 1963.$$

Try 7 again.

$$1963 \boxplus 7 \boxminus 280.42857 \quad \text{(Stop; 1963 is not divisible by 7.)}$$

So far, we have

$$23,373,441 = 3^5 \times 7^2 \times 1,963$$

Continue with 11 (why not 9?):

$$1963 \boxplus 11 \boxminus 178.45454 \quad \text{(Stop; 1,963 is not divisible by 11.)}$$

Next try 13. Does 13 divide 1,963?

$$1963 \boxplus 13 \boxminus 151.$$

Try 13 again.

$$151 \boxplus 13 \boxminus 11.615384 \quad \text{(Stop.)}$$

In Section 11.4 we shall show that 151 is a prime. Using this advance information, we can conclude that

$$23,373,441 = 3^5 \times 7^2 \times 13 \times 151$$

Exercises 11.2

1. Without using your calculator, find the prime factorization of the following integers.

   a. 100 c. 400,000 •e. 150,000 g. 10^{53}

   •b. 360 d. 16,000,000 •f. 250,000 •h. 125×10^{12}

2. Use the method of Example 11.2 to find the prime factorization of the following.

   •a. 377 d. 3,267 g. 55,010,340 •j. 44,789,760

   b. 720 •e. 264,627 •h. 60,415,875

   •c. 2,173 •f. 416,745 i. 20,711

11.3   The Sieve of Eratosthenes

One of the nicest ways to find primes was known to the ancient Greeks and is attributed to Eratosthenes (who lived between 276 and 194 B.C.). We illustrate the method by finding all the primes less than 51.

First, make a list of all integers from 2 to 51.

Table 11.1   The integers 2 through 51 ready to be sieved for primes

2	3	4	5	6	7	8	9	10	11
12	13	14	15	16	17	18	19	20	21
22	23	24	25	26	27	28	29	30	31
32	33	34	35	36	37	38	39	40	41
42	43	44	45	46	47	48	49	50	51

Now, we know that 2 is a prime. Moreover, any other multiple of 2 is not, and so may be crossed from the list of potential primes. Cross out (/) these multiples of 2 (see Table 11.2). As you can see, we cross out every other number after 2.

Table 11.2   The integers 2 through 51 with all multiples of 2 larger than 2 crossed out (/).

2	3	4̸	5	6̸	7	8̸	9	1̸0̸	11
1̸2̸	13	1̸4̸	15	1̸6̸	17	1̸8̸	19	2̸0̸	21
2̸2̸	23	2̸4̸	25	2̸6̸	27	2̸8̸	29	3̸0̸	31
3̸2̸	33	3̸4̸	35	3̸6̸	37	3̸8̸	39	4̸0̸	41
4̸2̸	43	4̸4̸	45	4̸6̸	47	4̸8̸	49	5̸0̸	51

The numbers that are not crossed out are now candidates for primes. In fact, the first number not crossed out, 3, must be a prime. Of course, we know it is a prime; but we also know from what we have just done. Because 3 was not crossed out as a multiple of 2, it is not divisible by 2. Furthermore, because 2 is the only number in the list less than 3, then 3 cannot be divided by any number less than itself, except 1.

Saving the 3, the next step is to cross out all other multiples of 3 because they are not primes. So, in Table 11.3, we cross out (\) every third number after 3. Note in Table 11.3 that some numbers have been crossed out twice—once as a multiple of 2, and again as a multiple of 3. These are, of course, the multiples of 6.

Table 11.3   The integers 2 through 51 with the multiples of 2 and 3 crossed out.

2	3	4̸	5	6̸	7	8̸	9̸	1̸0̸	11
1̸2̸	13	1̸4̸	1̸5̸	1̸6̸	17	1̸8̸	19	2̸0̸	2̸1̸
2̸2̸	23	2̸4̸	25	2̸6̸	2̸7̸	2̸8̸	29	3̸0̸	31
3̸2̸	3̸3̸	3̸4̸	35	3̸6̸	37	3̸8̸	3̸9̸	4̸0̸	41
4̸2̸	43	4̸4̸	4̸5̸	4̸6̸	47	4̸8̸	49	5̸0̸	5̸1̸

Now look for the next number after 3 that has not been crossed out. It is 5. Now, 5 must be a prime. Indeed, if it were not, then it would have been a multiple of a smaller prime; but these were 2 and 3, and we have just crossed out from the list all of their multiples. The number 5 remains so 5 must be a prime. Again, we may cross out (—) all multiples of 5; that is, every fifth number after 5 (Table 11.4).

Table 11.4 The integers 2 through 51 with the multiples of 2, 3, and 5 crossed out.

2	3	4	5	6	7	8	9	10	11
12	13	14	15	16	17	18	19	20	21
22	23	24	25	26	27	28	29	30	31
32	33	34	35	36	37	38	39	40	41
42	43	44	45	46	47	48	49	50	51

Now look for the next number after 5 that has not yet been crossed out. It is 7. By the reasoning we used in each of the preceding cases, 7 is a prime. Indeed, if 7 were not a prime, it would have been a multiple of one of the preceding primes, 2, 3, or 5. But we have crossed out all multiples of these numbers. So we now cross out all multiples of 7 from 7×2 on. These multiples that are within the scope of our table are

$$14, 21, 28, 35, 42, \text{ and } 49,$$

and should be crossed out from the table.

We may continue in this way. The numbers not crossed out are the primes less than 51. The final result is displayed in Table 11.5.

Table 11.5 The numbers not crossed out are the primes less than 51.

2	3	4	5	6	7	8	9	10	11
12	13	14	15	16	17	18	19	20	21
22	23	24	25	26	27	28	29	30	31
32	33	34	35	36	37	38	39	40	41
42	43	44	45	46	47	48	49	50	51

Observe that this scheme to find all primes less than 51 *can be ended after crossing out the multiples of 7!* Thus, for example, we do not have to cross out the multiples of 13 because, as we now show, any multiple of 13 less than 51 must have already been crossed out! Listen to our argument.

Reconsider the situation for 7. Notice that most of the multiples of 7 have already been crossed out. Indeed,

$$7 \times 2 = 14, \quad 7 \times 3 = 21, \quad 7 \times 4 = 7 \times 2 \times 2 = 28,$$

$$7 \times 5 = 35, \quad \text{and} \quad 7 \times 6 = 7 \times 2 \times 3 = 42$$

have already been crossed out. Only 7×7 has not been crossed out. Thus, the first multiple of 7 that would not have been crossed out is $7 \times 7 = 49$. In other words, we can *begin* the computation of those multiples of 7 to be crossed out with 7×7. Cross out 49. Now, the next multiple is $7 \times 8 = 56$, but $56 > 51$, and so exceeds the scope of Table 11.1. So only $7 \times 7 = 49$ is to be crossed out.

To give another example of this point, consider the next number after 7 that was not crossed out in Table 11.5. It is 11. All the multiples of 11 from 11×2 to 11×10 would already have been crossed out in any table (because the second factor is less than 11). The next multiple, $11 \times 11 = 121$, is a number beyond 51 and not in Table 11.5. (In fact, any multiple of 11 from 11×5 on is beyond the scope of these tables.)

Finally, we then make a list of all those numbers of Table 11.5 not crossed out. This will be a list of all the primes less than 51:

$$2, 3, 5, 7, 11, 13, 17, 19, 23, 29, 31, 37, 41, 43, 47$$

This method is perhaps not an efficient one if you wish to test whether a certain number is a prime or to find all the prime factors of a given number. Methods to do these are discussed in the next section. But the sieve is about as efficient a method as there is to find all primes less than a certain number.

Incidentally, you can see why this method is called a *sieve*. The composite numbers "drop through" as they are crossed out. The sieve catches the primes that remain not crossed out.

Now we shall indicate how we may extend this method to find all the primes less than 1,200 with the help of two shortcuts and a calculator to ease the burden of computation. Together, they will enable you to use Table 11.6 from Problem 2, Exercises 11.3 and the sieve method to find these primes.

The first shortcut is to omit from the list the multiples of 2, 3, and 5. Because $2 \times 3 \times 5 = 30$, this suggests that we should think, for a moment, of the table of integers from 2 to 1,200 as listed in rows 30 numbers long:

2	3	4	5	6	7	8	9	10	11	12	$\cdots$	29	30
31	32	33	34	35	36	37	38	39	40	41	42	$\cdots$	59	60
61	62	63	64	65	66	67	68	69	70	71	72	$\cdots$	89	90

$\cdots$

Now the numbers that are multiples of 2 occur in columns 2, 4, 6, 8, 10, 12, ... , 30. Those integers that are multiples of 3 occur in columns 3, 6, 9, 12, 15, ... , 30. Those integers that are multiples of 5 occur in columns 5, 10, 15, ... , and 30. Thus, after we have deleted the multiples of 2, 3, and 5, only numbers in the eight remaining columns headed by will remain. So we need list only integers that appear in these columns. This is our first shortcut, and you will see that in Table 11.6 this is what we have done.

Given a number N, how can we tell which column it is in? Well, perform a division by 30 with remainder. (Recall Section 4.12.)

$$N = (30 \times Q) + R, \quad \text{where } 0 \leq R < N.$$

Notice that, because 2, 3, or 5 divides 30, then 2, 3, or 5 will divide N if and only if 2, 3, or 5 divides R. So if R is divisible by 2, 3, or 5, then N is not a prime and will not even be listed in Table 11.6. If $R = 1, 7, 11, 13, 17, 19, 23,$ or 29, then N is in the column headed by that same number in Table 11.6.

From these considerations, it should be clear that, aside from the primes 2, 3, and 5, the only candidates for primes are the integers having one of the following forms:

$$(30 \times Q) + 1, \quad (30 \times Q) + 7, \quad (30 \times Q) + 11, \quad (30 \times Q) + 13,$$
$$(30 \times Q) + 17, \quad (30 \times Q) + 19, \quad (30 \times Q) + 23, \quad (30 \times Q) + 29$$

Table 11.6 lists all the numbers from $7 [= (30 \times 0) + 7]$ to 1199 $[= (30 \times 39) + 29]$ having one of these forms.

You can apply the sieve method to this table to find all the primes starting with 7 and less than or equal to 1,199. For a complete list, you must add 2, 3, and 5—which, of course, do not appear in Table 11.6.

To show you how you can work with Table 11.6 and to point out another shortcut, Figure 11.2 gives a sketch of Table 11.6. [For short, instead of writing out $(30 \times Q) + 1$, we shall write "$30Q + 1$".]

Q	$30Q + 1$	$30Q + 7$	$30Q + 11$	$30Q + 13$	$30Q + 17$	$30Q + 19$	$30Q + 23$	$30Q + 29$
0	*	7	11	13	17	19	23	29
1	31	37	41	43	47	49	53	59
2	61	67	71	73	77	79	83	89
3	91	97	101	103	107	109	113	119
4	121	127	131	133	137	139	143	149
5	151	157	161	163	167	169	173	179
⋮								
39	1171	1177	1181	1183	1187	1189	1193	1199

Figure 11.2
Sketch of Table 11.6 for the sieve method to find all primes less than 1200.

To get started, let us cross out all multiples of 7 that appear in this table. You can do this easily with a calculator that permits you to carry out constant multiplication, say with a constant (K) switch. Set 7 as the constant, and then simply enter the multiples you wish to compute. Then look up the product in Table 11.6 and cross out that number.

Which multiples will you have to compute? Here is another short-cut: you certainly don't have to compute multiples of 7 that are even! Thus, 7×126 is even because 126 is even, and so 7×126 ($= 882$) is not listed in Table 11.6.

The multiples of 7 (or any other number) that you must compute and cross out from Table 11.6 are just those that appear in Table 11.6 itself! And of course you can stop computing multiples when the product becomes larger than 1,200 and so exceeds the contents of Table 11.6.

Thus, to find those multiples of 7 that must be crossed out from Table 11.6, use your calculator to compute as follows:

	7×7 $= 49$	7×11 $= 77$	7×13 $= 91$	7×17 $= 119$	7×19 $= 133$	7×23 $= 161$	7×29 $= 203$
7×31 $= 217$	7×37 $= 259$	7×41 $= 287$	7×43 $= 301$	7×47 $= 329$	7×49 $= 343$	7×53 $= 371$	7×59 $= 413$
7×61 $= 427$	$\ldots$						
$\vdots$							
7×151 $= 1057$	$\ldots$						7×179 $= 1253$

We may stop at $7 \times 169 = 1,183$ because $7 \times 173 = 1,211$, a number greater than the last entry in Table 11.6.

In Table 11.6, we have crossed out these multiples of 7 to get you started. Although the work may seem a bit tedious, actually 7 is the worst case computationally, and the rest of the work goes quickly. If you like primes, have a go at it!

Exercises 11.3

1. Make a table like Table 11.1 going up to 101.

   a. Use the sieve method to find all primes less than 101.

   •b. Show that you need to carry the "crossing-out" process only through multiples of 7. (Your table will show that there are 25 primes less than or equal to 101.)

•2. Use Table 11.6 to find all primes less than 1,200.

Table 11.6 Integers of the Form $(30 \times Q) + R$

Q \ R	1	7	11	13	17	19	23	29
1	31	37	41	43	47	~~49~~	53	59
2	61	67	71	73	~~77~~	79	83	89
3	~~91~~	97	101	103	107	109	113	~~119~~
4	121	127	131	~~133~~	137	139	143	149
5	151	157	~~161~~	163	167	169	173	179
6	181	187	191	193	197	199	~~203~~	209
7	211	~~217~~	221	223	227	229	233	239
8	241	247	251	253	257	~~259~~	263	269
9	271	277	281	283	~~287~~	289	293	299
10	~~301~~	307	311	313	317	319	323	~~329~~
11	331	337	341	~~343~~	347	349	353	359
12	361	367	~~371~~	373	377	379	383	389
13	391	397	401	403	407	409	~~413~~	419
14	421	~~427~~	431	433	437	439	443	449
15	451	457	461	463	467	~~469~~	473	479
16	481	487	491	493	~~497~~	499	503	509
17	~~511~~	517	521	523	527	529	533	~~539~~
18	541	547	551	~~553~~	557	559	563	569
19	571	577	~~581~~	583	587	589	593	599
20	601	607	611	613	617	619	~~623~~	629
21	631	~~637~~	641	643	647	649	653	659
22	661	667	671	673	677	~~679~~	683	689
23	691	697	701	703	~~707~~	709	713	719
24	~~721~~	727	731	733	737	739	743	~~749~~
25	751	757	761	~~763~~	767	769	773	779
26	781	787	~~791~~	793	797	799	803	709
27	811	817	821	823	827	829	~~833~~	839
28	841	~~847~~	851	853	857	859	863	869
29	871	877	881	883	887	~~889~~	893	899
30	901	907	911	913	~~917~~	919	923	929
31	~~931~~	937	941	943	947	949	953	~~959~~
32	961	967	971	~~973~~	977	979	983	989
33	991	997	~~1001~~	1003	1007	1009	1013	1019
34	1021	1027	1031	1033	1037	1039	~~1043~~	1049
35	1051	~~1057~~	1061	1063	1067	~~1069~~	1073	1079
36	1081	1087	1091	1093	~~1097~~	1099	1103	1109
37	~~1111~~	1117	1121	1123	1127	1129	1133	1139
38	1141	1147	1151	~~1153~~	1157	1159	1163	~~1169~~
39	1171	1177	~~1181~~	1183	1187	1189	1193	1199

11.4 Another Method for Finding Primes

Our next example shows how to test a number to see if it is a prime.

Example 11.3: Show that 151 is a prime number.

Solution: We could follow the procedure of the sieve and try to divide 151 by each of the prime numbers less than 151. However, as we said before, this is not the most efficient way to proceed. Moreover, we have indicated in our shortcuts how we can do much better.

Let us split the primes less than 151 into two groups: those whose squares are less than 151, and those whose squares are greater.

Group I	Group II
2, 3, 4, 7, 11	13, 17, 19, ...
squares are less	squares are greater
than 151	than 151

Testing each of the five primes in Group I, we find that none of them divides into 151 exactly, so 151 cannot be written as a product involving one of these primes.

The information we got in our first series of testing rules out primes in the second group from being factors of 151. Why? Because we know that, if 151 is to be factored into a product of two or more primes less than 151, then these would have to come from Group II because 151 is divisible by none of the primes from Group I. But the product of two or more numbers in Group II is always greater than 151. Even the product of the smallest of these and itself is greater than 151. Hence, such a factorization is not possible.

As 151 cannot be factored into primes from either Group I or Group II, 151 is prime.

The point to notice in the above example is the following: To see that 151 is a prime number, we need only test that 151 is not divisible by primes whose squares are less than 151.

As a general rule,

An integer is prime whenever none of the primes whose squares are less than that integer divides it.

Following is a list of the first one hundred prime numbers. It will be useful in the exercises that follow.

2	3	5	7	11	13	17	19	23	29
31	37	41	43	47	53	59	61	67	71
73	79	83	89	97	101	103	107	109	113
127	131	137	139	149	151	157	163	167	173
179	181	191	193	197	199	211	223	227	229
233	239	241	251	257	263	269	271	277	281
283	293	307	311	313	317	331	337	347	349
353	359	367	373	379	383	389	397	401	409
419	421	431	433	439	443	449	457	461	463
467	479	487	491	499	503	509	521	523	541

Exercise 11.4

1. Determine which of the following numbers are prime.

- •a.   569
- •b.   593
- c.   1,189
- •d.   3,713
- e.   739
- •f.   953
- •g.   2,347

- h.   3,351
- •i.   9,379
- •j.   9,967
- k.   13,589
- •l.   38,293
- •m.   210,757
- n.   386,497

11.5   Greatest Common Divisors

Let us suppose that there are twelve small communities distributed more or less evenly around a large lake, and that a rapid transit system connects these communities (one station in each community) in the manner shown in Figure 11.3.

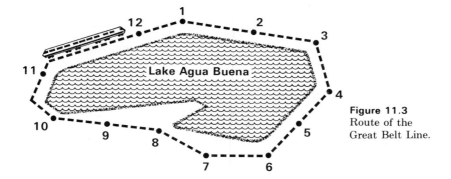

Figure 11.3
Route of the
Great Belt Line.

For convenience, we shall name these stations numerically as shown. Starting at Station 12, a train has several options for traveling clockwise around the lake. It could move forward one station at a time and thus stop at each station. It could move forward two stations at a time and stop only at Stations 2, 4, 6, 8, 10, and 12. Check this last statement by placing your finger at Station 12, and then following the path as you move forward two stations at a time.

The train could also move forward three stations each time before stopping. What stations would it stop at? (3, 6, 9, and 12.) As a matter of fact, it could move forward any number of stations, skipping the same number of stations between stops.

Now the question is: Given any one of these alternatives, at how many different stations will the train stop? Here is a table of the results.

Table 11.7

Number of Stations the Train Moves Forward Before Stopping	Number of Different Stops
1	12
2	6
3	4
4	3
5	12
6	2
7	12
8	3
9	4
10	6
11	12
12	1
13	12
14	6
15	4

You can verify that the numbers in the second column are correct. For example, for the 4-station moves, the train stops at the 3 stations numbered 4, 8, and 12.

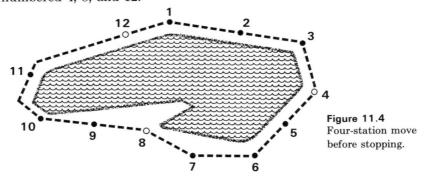

Figure 11.4
Four-station move before stopping.

For the first four rows in the table, things seem to be easy enough. We observe that the product of the pairs of numbers is always 12. So we simply have to divide 12 by the number appearing in the first column to obtain the number in the second. (Notice that 12 is the total number of stations in the problem.)

Thus,

$$\frac{12}{1} = 12, \quad \frac{12}{2} = 6, \quad \frac{12}{3} = 4, \quad \text{and} \quad \frac{12}{4} = 3$$

However, beginning with row 5, things seem to go haywire (row 6 and row 12 being exceptions).

The table is not an accidental distribution of numbers as one might think; there is actually a pattern to be uncovered. In order to make some sense out of the table, we now introduce the concept of *"greatest common divisor."*

> REMARK: In keeping with common practice, we shall use the phrases, "*a* is a divisor of *b*" and "*b* is divisible by *a*" to mean that *a* divides into *b* a whole number of times with remainder zero.

Thus

13 is a divisor of 26;

but

7 is not a divisor of 23 (there is a remainder of 2.)

Note that 6 has

1, 2, 3, and 6 as divisors,

while all of the divisors of 36 are

1, 2, 3, 4, 6, 9, 12, 18, and 36

Example 11.4: Given the integers, 24 and 36, let us examine the divisors of each of these numbers.

The divisors of 24 are: **1, 2, 3, 4, 6,** 8, **12,** and 24.

The divisors of 36 are: **1, 2, 3, 4, 6,** 9, **12,** 18, and 36.

The numbers appearing in boldface divide both 24 and 36, and are thus the *common* divisors. Hence the common divisors of 24 and 36 are 1, 2, 3, 4, 6, and 12.

As 12 is the greatest of these common divisors, it is the *greatest common divisor* of 24 and 36.

In general, we have the following:

Let a and b be positive integers. A positive integer d is said to be the *greatest common divisor* of a and b if d is the largest integer that is a divisor of both a and b.

We denote the greatest common divisor of a and b by the symbol GCD (a, b). We can equally well write the symbol as GCD (b, a); both versions mean the same. Thus,

$$\text{GCD } (24, 36) = 12 = \text{GCD } (36, 24).$$

Example 11.5: Find the greatest common divisor of 12 and each positive integer less than or equal to 12. In other words, find GCD $(n, 12)$ for each natural number n such that $1 \leq n \leq 12$.

Solution: Recall that the divisors of 12 are 1, 2, 3, 4, 6, and 12. Now let us try the various values of n from 1 through 12.

n	*Divisors of n*	*Common Divisors of n and 12*	*GCD(n, 12)*
1	1	1	GCD(1, 12) = 1
2	1, 2	1, 2	GCD(2, 12) = 2
3	1, 3	1, 3	GCD(3, 12) = 3
4	1, 2, 4	1, 2, 4	GCD(4, 12) = 4
5	1, 5	1	GCD(5, 12) = 1
6	1, 2, 3, 6	1, 2, 3, 6	GCD(6, 12) = 6
7	1, 7	1	GCD(7, 12) = 1
8	1, 2, 4, 8	1, 2, 4	GCD(8, 12) = 4
9	1, 3, 9	1, 3	GCD(9, 12) = 3
10	1, 2, 5, 10	1, 2	GCD(10, 12) = 2
11	1, 11	1	GCD(11, 12) = 1
12	1, 2, 3, 4, 6, 12	1, 2, 3, 4, 6, 12	GCD(12, 12) = 12

Now look again at Table 11.7. Let n be the number of stations moved (from the first column of the table) and s be the number of different stops (from the second column). Verify that, for each line of the table,

$$s \times \text{GCD}(n, 12) = 12$$

In other words, we can find the number of different stops (s) by dividing

$$s = \frac{12}{\text{GCD}(n, 12)}$$

For instance, for 5-station moves, we have

$$s = \frac{12}{\text{GCD }(12, 5)} \cdot \frac{12}{1} = 12$$

For 8-station moves,

$$s = \frac{12}{\text{GCD }(12, 8)} = \frac{12}{4} = 3$$

Similarly, all other entries (even for more than 12 stations) can be verified by this relationship.

In our work, the numbers involved were small, so it was rather easy to determine the greatest common divisor for the various pairs of numbers there. For larger numbers, however, finding the greatest common divisor can become quite a difficult chore. Here is where our knowledge of prime factorization comes in handy.

We have previously observed that GCD $(24, 36) = 12$. We can determine this result another way. Observe that the prime factorizations of 24 and 36 are

$$24 = 2^3 \times 3^1 \quad \text{and} \quad 36 = 2^2 \times 3^2$$

Note that 2^2 is the largest power of 2 that is a divisor of both numbers, and that 3^1 is the largest power of 3 that is a divisor of both numbers. Hence, $2^2 \times 3^1 = 12$ is the largest integer dividing 24 and 36. This means that 12 is the greatest common divisor of 24 and 36.

Let us try this method on still larger numbers.

Example 11.6: Find GCD $(3{,}920,\ 8{,}232)$.

Solution: Using our calculator, we find the prime factorizations of 3,920 and 8,232:

$$3{,}920 = 2^4 \times 5^1 \times 7^2;$$

$$8{,}232 = 2^3 \times 3^1 \times 7^3$$

Using the fact that $a^0 = 1$ for any number a, it will be more convenient to rewrite these factorizations as follows:

$$3{,}920 = 2^4 \times 3^0 \times 5^1 \times 7^2;$$

$$8{,}232 = 2^3 \times 3^1 \times 5^0 \times 7^3$$

Now determine the largest power of each of the primes (2, 3, 5, and 7) that is a divisor of both 3,920 and 8,232. They are, respectively,

$$2^3,\ 3^0,\ 5^0,\ \text{and } 7^2$$

The product of these numbers will be the largest divisor of 3,920 and 8,232. Thus,

$$\text{GCD }(3{,}920, 8{,}232) = 2^3 \times 3^0 \times 5^0 \times 7^2$$

$$= 8 \times 1 \times 1 \times 49$$

$$= 392$$

Over two thousand years ago, Euclid (well known to high school students from geometry) discovered a very efficient method for computing greatest common divisors that does not require prime factorization as used above.

To see how Euclid's algorithm for computing greatest common divisors works, we begin experimenting with some small numbers. (The ancient Greeks no doubt discovered this algorithm through such experimentation.)

Without any computation (or at least with very little), one can see that

$$GCD\ (20, 6) = 2$$

If we remove 6 from 20, we have

$$20 - 6 = 14$$

Note that

$$GCD\ (14, 6) = 2$$

Thus, removing 6 from 20 did not change the greatest common divisor: it remained 2. Was that a fluke or a happy accident? Well, let us see. Let us remove another 6 from 20. We have

$$20 - (2 \times 6) = 8$$

Again,

$$GCD\ (8, 6) = 2$$

How about once more?

$$20 - (3 \times 6) = 2$$

Still,

$$GCD\ (2, 6) = 2$$

We started out with GCD $(20, 6) = 2$. The GCD remained the same on repeated subtraction of 6 from 20. This works in general. In other words,

If $a > b$, then GCD $[(a - b), b] = $ GCD (a, b).

Let us look at another example. First, we shall find the value of GCD $(495, 140)$ by the method using prime factorization.

Example 11.7: Find GCD $(495, 140)$.

Solution: Using our calculator, we find the prime factorizations:

$$495 = 3^2 \times 5 \times 11 \quad \text{and} \quad 140 = 2^2 \times 5 \times 7$$

Picking the largest power of each of the primes that is a divisor of both 495 and 140,

$$GCD\ (495, 140) = 2^0 \times 3^0 \times 5^1 \times 7^0 \times 11^0$$

$$= 5$$

Thus, GCD (495, 140) = 5

Now, to say that GCD (495, 140) = 5 means that 5 is the greatest divisor shared by both 495 and 140. What is the effect of this claim in the subtraction,

$$495 - 140 = 355?$$

Because

$$495 = 5 \times 99 \quad \text{and} \quad 140 = 5 \times 28,$$

then

$$495 - 140 = (5 \times 99) - (5 \times 28).$$

By the distributive law,

$$495 - 140 = 5 \times (99 - 28).$$

The important thing to see is this:

the common divisor of two numbers is a divisor of their difference.

Now we have the following chain of reasoning.

GCD of 495 and 140 is the GCD of $(495 - 140)$ and 140; that is, of 355 and 140.

GCD of 355 and 140 is the GCD of $(355 - 140)$ and 140.

GCD of 215 and 140 is the GCD of $(215 - 140)$ and 140.

In short,

$$GCD\ (495, 140) = GCD\ (355, 140) = GCD\ (215, 140) = \cdots$$

The next subtraction gives

$$GCD\ (215, 140) = GCD\ (75, 140).$$

Now we can reverse the role of the 140 and subtract, instead, by 75 to get $140 - 75 = 65$. Thus,

$$GCD\ (75, 140) = GCD\ (75, 65).$$

Subtracting 65,

$$\text{GCD}\,(75, 65) = \text{GCD}\,(10, 65)$$

$$= \text{GCD}\,(10, 55) = \cdots$$

$$= \text{GCD}\,(10, 5) = \text{GCD}\,(5, 5)$$

$$= 5$$

If we had the insight, we could have shortened this chain by noticing at the second stage, where we had

$$\text{GCD}\,(355, 140),$$

that $355 = 5 \times 71$. Because both 5 and 71 are prime numbers (see Section 11.4) and

$$5 \text{ is a divisor of } 140$$

but $\qquad\qquad$ 71 is not a divisor of 140,

we could have concluded that

$$\text{GCD}\,(355, 140) = 5$$

We have observed an illustration of Euclid's algorithm in which we see that, for positive integers a and b with $a > b$, the GCD remains unchanged with subtraction of multiples of b from a. That is,

$$\text{GCD}(a, b) = \text{GCD}[(a - b), b] = \text{GCD}[(a - 2b), b]$$

$$= \text{GCD}[(a - 3b), b] = \cdots$$

(By the notation "$2b$," for example, we mean "$2 \times b$.")

Hereafter we will refer to this simply as the "GCD rule." In practice, we merely short-circuit all the repeated subtractions by performing a division. Now, let us try this rule on some fairly large numbers.

Example 11.8: Find GCD (5,371, 4,387) without using prime factorization.

Solution: By the method of Section 4.12, divide 5,371 by 4,387 to find the quotient and remainder:

$$5,371 = (1 \times 4,387) + 984$$

So $\qquad 5{,}371 - (1 \times 4{,}387) = 984$

Our rule says

$$GCD(5{,}371,\ 4{,}387) = GCD(984,\ 4{,}387).$$

Now using 984 as the divisor, we find that

$$4{,}387 = (4 \times 984) + 451,$$

so $\qquad GCD(984,\ 4{,}387) = GCD(984,\ 451).$

Continuing now with 451 as the divisor,

$$984 = (2 \times 451) + 82$$

Here we conclude that

$$GCD(984,\ 451) = GCD(82,\ 451).$$

Then $\qquad 451 = (5 \times 82) + 41,$

so $\qquad GCD(82,\ 451) = GCD(82,\ 41).$

From beginning to this point, the chain of equalities say that

$$GCD(5{,}371, 4{,}387) = GCD(82, 41).$$

Because 41 is a divisor of 82 ($82 \div 41 = 2$,) we have

$$GCD(82, 41) = 41$$

Thus,

$$GCD(5{,}371, 4{,}387) = 41$$

At this point, you might say, "Good grief, I would rather use prime factorization!" The above procedure looks long and messy because of all the discussion. Stripped of all details, we need only keep track of each of the quotients and remainders that were calculated. If we arrange all the divisions in a table, it will all become clear.

Dividend	Divisor	Quotient	Remainder
5,371	4,387	1	984
4,387	984	4	451
984	451	2	82
451	82	5	41
82	41	2	0

Our answer, 41, appears as the last non-zero remainder. It is simple as that!

We should certainly check that our answer, 41, is correct. If correct, 41 should be a divisor of both 5,371 and 4,387. Because

$$5,371 \div 41 = 131 \quad \text{and} \quad 4,387 \div 41 = 107,$$

41 is *a* common divisor.

From the division, we have

$$5,371 = 41 \times 13 \quad \text{and} \quad 4,387 = 41 \times 107$$

Referring to the list in Section 11.4, we see that both 107 and 131 are prime numbers. So 41 is indeed the GCD.

Another example will be helpful in showing how rapid the computations can really be, compared with prime factorization.

Example 11.9: Find the value of GCD(3,818, 18,011).

Solution:

Dividend	Divisor	Quotient	Remainder
18,011	3,818	4	2,739
3,818	2,739	1	1,079
2,739	1,079	2	581
1,079	581	1	498
581	498	1	83
498	83	6	0

Hence, GCD(3,818, 18,011) = 83. This certainly beats factorization!

Exercises 11.5

•1. After a few years, several of the cities around Lake Agua Buena (Figure 11.3) had to add some more station stops. If the total number of stations became 108, at how many different stations will the train stop if—

a. it moves forward 9 stations after each stop?

b. it moves forward 18 stations after each stop?

c. it moves forward 86 stations after each stop?

d. it moves forward 55 stations after each stop?

2. Use the method of Example 11.6 to find the greatest common divisor of the following pairs of numbers.

•a.	55	91	•f.	1,728	3,888
b.	308	819	g.	2,800	1,144
•c.	576	486	•h.	6,468	6,678
•d.	1,178	1,235	•i.	5,005	29,233
e.	1,225	1,715	j.	1,923,940	29,704,961

3. Using the Euclidean-algorithm method, find the greatest common divisor for the following pairs of numbers.

•a. 33,957   4,557   •d. 63,011,844   9,144,576

•b. 84,672   72,576   e. 261,140,697   7,628,403

 c. 284,029   249,061

4. Find the greatest common divisor by the Euclidean algorithm and compare the time involved with that used for parts h, i, and j of Exercise 2.

•a. 6,468   6,678

 b. 5,005   29,233

•c. 1,923,940 29,704,961

11.6 The Least Common Multiple

You may remember working with least common denominators when you have had to add or subtract fractions.

Example 11.10: Find the sum $\frac{1}{6} + \frac{3}{4}$.

Solution: First, convert each fraction to fractions having the least common denominator, 12:

$$\frac{1}{6} = \frac{2}{12} \quad \text{and} \quad \frac{3}{4} = \frac{9}{12}$$

So

$$\frac{1}{6} + \frac{3}{4} = \frac{2}{12} + \frac{9}{12} = \frac{11}{12}$$

A least common denominator is just an example of a *"least common multiple."* For now, let us concentrate on the concept of the least common multiple.

Consider the following two lists of numbers. The first consists of multiples of 3, and the second consists of multiples of 4:

$$3, 6, 9, \textbf{12}, 15, 18, 21, \textbf{24}, 27, 30, 33, \textbf{36}, \cdots$$
$$4, 8, \textbf{12}, 16, 20, \textbf{24}, 28, 32, \textbf{36}, \cdots$$

The numbers appearing in boldface are common to both lists. For this reason, they are called the *common multiples* of 3 and 4.

As usual, the three dots at the end of each list means that the list continues in this manner without end. So there is no greatest multiple for either number (3 or 4), and hence also no greatest multiple common to both lists. However, among the common multiples (12, 24, 36, 48, . . .) observe that 12 is the smallest. That is, 12 is the *least common multiple* of 3 and 4. (Notice, too, that every number in the list of common multiples is a multiple of 12.)

The least common multiple of two positive integers a and b may be denoted by $LCM(a, b)$. Thus

$$LCM(3, 4) = 12 = LCM(4, 3).$$

The above procedure can be used to produce the least common multiple of any pair of positive integers. For instance, for $LCM(8, 10)$, we begin with the lists,

$$8, 16, 24, 32, \mathbf{40}, 48, \ldots$$
$$10, 20, 30, \mathbf{40}, 50, \ldots$$

and see that 40 is the least common multiple of 8 and 10. Hence the common multiples of 8 and 10 are

$$40, 80, 120, 160, 200, \ldots$$

Similarly, if we start with 15 and 24, then the two lists of multiples become

$$15, 30, 45, 60, 75, 90, 105, \mathbf{120}, \ldots$$
$$24, 48, 72, 96, \mathbf{120}, \ldots$$

Here, we see that 120 is the least common multiple of 15 and 24.

Although the above procedure will always yield the least common multiple of any two integers, it is not very practical when the integers are large. For example, what is the least common multiple of 7,733 and 18,073? In writing each of the lists of multiples of these numbers, we would have to write several hundred terms before we could see what the least common multiple is. As with the GCD, clearly we need a better way.

Example 11.11: Find $LCM(3, 4)$, $LCM(8, 10)$, and $LCM(15, 24)$.

Solution: Begin with $LCM(3, 4)$. Note that $3 = 3^1$ and $4 = 2^2$. If the least common multiple is to be a multiple of 3, it must contain the prime number 3, at least to the first power. If the least common multiple is to be a multiple of 4, it must contain at least two factors of the prime number 2.

So
$$LCM(3,4) = 2^2 \times 3^1$$
$$= 4 \times 3 = 12$$

Now let's try $LCM(8, 10)$. Note that $8 = 2^3$ and $10 = 2^1 \times 5^1$. So $LCM(8, 10)$ must contain at least three factors of 2 and one factor of 5.

$$LCM(8, 10) = 2^3 \times 5^1 = 8 \times 5 = 40$$

Finally, we tackle $LCM(15, 24)$:

$$15 = 3^1 \times 5^1 \quad \text{and} \quad 24 = 2^3 \times 3^1,$$

or
$$15 = 2^0 \times 3^1 \times 5^1 \quad \text{and} \quad 24 = 2^3 \times 3^1 \times 5^0$$

Thus
$$\text{LCM}(15, 24) = 2^3 \times 3^1 \times 5^1$$
$$= 8 \times 3 \times 5 = 120$$

The pattern is clear. In general, we express each number by its prime decomposition. Then the least common multiple will consist of each factor to the higher of the two powers.

Example 11.12: Find LCM(3,920, 8,232).

Solution: From Example 11.6, we have the factorizations:

$$3,920 = 2^4 \times 3^0 \times 5^1 \times 7^2;$$
$$8,232 = 2^3 \times 3^1 \times 5^0 \times 7^3$$

Therefore, the LCM must have four factors of 2, one factor of 3, one factor of 5, and three factors of 7. That is,

$$\text{LCM}(3,920, 8,232) = 2^4 \times 3^1 \times 5^1 \times 7^3$$
$$= \boxed{16} \boxed{\times} \boxed{3} \boxed{\times} \boxed{5} \boxed{\times} \boxed{343} \boxed{=} \boxed{82320}.$$

Although the least common multiple of two numbers can be found by prime factorization, this is not an efficient method if the numbers are large. For the next improvement, let us examine the GCD together with the LCM for pairs of numbers that we have studied.

Example 11.13: Compare the GCD and the LCM of number pairs.

Number pair	GCD	LCM
(3, 4)	1	12
(8, 10)	2	40
(15, 24)	3	120
(3920, 8232)	392	82,320

Notice that for (3, 4) $\quad 3 \times 4 = 1 \times 12$
and for (8, 10) $\quad 8 \times 10 = 2 \times 40$

That is, the product of the number pair equals the product of the GCD and LCM. Test this relationship on the other pairs:

$$15 \times 24 = 360, \qquad 3 \times 120 = 360;$$
$$3920 \times 8232 = 32,269,440, \quad 392 \times 82,320 = 32,269,440$$

Here's the general rule:

If a and b are positive integers, then the product of their least common multiple and their greatest common divisor is equal to the product of the integers. That is,

$$\text{LCM}(a, b) \times \text{GCD}(a, b) = a \times b.$$

From this, we see that

$$\text{LCM}(a, b) = \frac{a \times b}{\text{GCD}(a, b)}$$

Now, we can find readily the least common multiple of such large numbers as 7,733 and 18,073. By the Euclidean algorithm, we find that GCD (7,733, 18,073) = 11. The rest is a simple job for the calculator.

Example 11.14: Find LCM(7,733, 18,073).

Solution: We know that GCD(7,733, 18,073) = 11.

$$\text{LCM}(7,733, 18,073) = \frac{7,733 \times 18,073}{\text{GCD}(7,733, 18,073)}$$

$$= \frac{7,733 \times 18,073}{11}$$

$$= 12,705,319$$

REMARK: In calculating the value of $\dfrac{7,733 \times 18,073}{11}$, what happens if you first multiply 7,733 by 18,073? What can you do to avoid this difficulty?

Let us now consider the problem of adding two fractions—say, $\frac{11}{273}$ and $\frac{19}{455}$. We need a common denominator. We can always try the product, 273×455.

Example 11.15: Find the sum $\frac{11}{273} + \frac{19}{455}$.

Solution: Express each fraction in terms of a common denominator—say, 273×455.

$$\frac{11}{273} = \frac{11}{273} \times \frac{455}{455} \quad \text{and} \quad \frac{19}{455} = \frac{19}{455} \times \frac{273}{273}$$

So
$$\frac{11}{273} + \frac{19}{455} = \left(\frac{11}{273} \times \frac{455}{455}\right) + \left(\frac{19}{455} \times \frac{273}{273}\right)$$

$$= \frac{10,192}{124,215}$$

To reduce this fraction to lowest terms, we find the GCD of the numerator and the denominator. Using the Euclidean algorithm, we have

$$GCD(10{,}192,\ 124{,}215) = 637$$

Because

$$10{,}192 \div 637 = 16 \quad \text{and} \quad 124{,}215 \div 637 = 195,$$

we conclude that

$$\frac{10{,}192}{124{,}215} = \frac{16 \times 637}{195 \times 637} = \frac{16}{195},$$

and we are sure that our answer cannot be reduced any further.

Now let us work the same problem again making use of the least common multiple of 273 and 455.

Example 11.16: Find the sum $\frac{11}{273} + \frac{19}{455}$.

Solution: Because $GCD(273, 455) = 91$ (verify), we have

$$LCM(273, 455) = \frac{273 \times 455}{91} = 1{,}365$$

To use 1,365 as a common denominator, first note that

$$1{,}365 \div 273 = 5 \quad \text{and} \quad 1{,}365 \div 455 = 3$$

Hence,

$$\frac{11}{273} + \frac{19}{455} = \left(\frac{11}{273} \times \frac{5}{5}\right) + \left(\frac{19}{455} \times \frac{3}{3}\right)$$

$$= \frac{55}{1{,}365} + \frac{57}{1{,}365} = \frac{112}{1{,}365}$$

Again, by the method of Section 11.6, we find

$$GCD(112,\ 1{,}365) = 7$$

So we express 112 and 1,365 with 7 as a factor:

$$112 \div 7 = 16 \quad \text{and} \quad 1{,}365 \div 7 = 195$$

So

$$\frac{112}{1{,}365} = \frac{16 \times 7}{195 \times 7} = \frac{16}{195}$$

From which,

$$\frac{11}{273} + \frac{19}{455} = \frac{16}{195}$$

The advantage in using the LCM is in working with smaller numbers. Then, for instance, finding the GCD in order to reduce the fraction to lowest terms, less work would be involved.

Let us use the calculator to approximate each fraction in Example 11.16 and thus verify our result.

$$\frac{11}{273}: \quad 11 \div 273 = 0.0402930$$

$$\frac{19}{455}: \quad 19 \div 455 = 0.0417582$$

$$\frac{11}{273} + \frac{19}{455}: \quad 0.0402930 + 0.0417582 = \boxed{0.0820512}$$

$$\frac{16}{195}: \quad 16 \div 195 = \boxed{0.0820513} \leftarrow$$

Exercises 11.6

1. For each fraction, determine whether it is reduced. If not, reduce it to lowest terms.

   •a. $\frac{91}{133}$ •e. $\frac{540}{1,008}$

   b. $\frac{107}{756}$ f. $\frac{2,079}{7,371}$

   •c. $\frac{403}{949}$ •g. $\frac{305,613}{1,440,747}$

   d. $\frac{151}{203}$ h. $\frac{40,259}{140,431}$

2. Calculate the least common multiple of each pair of numbers.

   a. 12, 18   •d. 51, 315

   •b. 65, 91   •e. 4,235, 13,475

   c. 48, 192   f. 30,397, 85,273

3. Find each sum and express the result as a reduced fraction.

   a. $\frac{3}{17} + \frac{5}{23}$ c. $\frac{7}{12} + \frac{35}{54}$

   •b. $\frac{5}{42} + \frac{7}{74}$ •d. $\frac{130}{1323} + \frac{589}{1701}$

4. Two dials, one with 4 and the other with 6 numerals, equally spaced, show 1 and 1 at the start as in the diagram below.

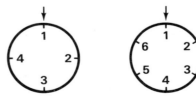

If each one advances one position per minute, when will they again show 1 and 1?

•5. A large shopping center operates two trams that start from the same point where riders may change trams.

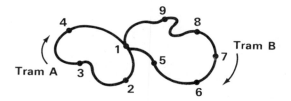

Tram A makes 4 stops; Tram B makes 6 stops. Each tram takes 3 minutes to go from one stop to the next. If both trams start their routes at noon, when will they both arrive at Station 1 at the same time again?

6. In Exercise 5, suppose Tram A makes one circuit in 15 minutes and Tram B makes one circuit in 20 minutes. When will both trams arrive at Station 1?

•7. A Christmas tree has three blinking lights. One blinks 20 times a minute; another blinks 15 times a minute; a third blinks 25 times a minute. How often will they blink together?

11.7 Fibonacci Numbers

An Italian mathematician, Leonardo Fibonacci (ca. 1175–1250), discovered a very interesting sequence of numbers, which has since been named after him. The first few terms of the *Fibonacci sequence* are

$$1, 1, 2, 3, 5, 8, 13, 21, 34, 55, \ldots$$

After the first two terms, each term is the sum of the two previous terms. Thus,

$$2 = 1 + 1, \quad 3 = 1 + 2, \quad 5 = 2 + 3, \quad \text{and so on.}$$

The next term is $34 + 55 = 89$.

Besides having many curious properties, the Fibonacci sequence has made unexpected appearances in several branches of science. In this section we will discuss only one of the many situations where this sequence makes its appearance.

In Section 11.2, a positive integer was factored as a product of primes. Now, we start off with two basic "addition building-blocks," 1 and 2. In how many ways can we express a positive integer as the sum of 1's and 2's?

Example 11.17: How many ways are there of expressing 3 as sums of 1's and 2's?

Solution: Note that

$$3 = 1 + 1 + 1$$
$$= 1 + 2$$
$$= 2 + 1$$

So altogether, we count 3 different ways of summing to 3 using only 1's and 2's. (Notice that we now make a distinction between $1 + 2$ and $2 + 1$.)

In a like manner, we can split other integers. The following table shows such splitting of the integers from 1 to 5, and counts the number of splittings for each integer.

Integer	Expressions	Number of Ways of Expressing as Sums of 1's and 2's
1	$1 = 1$	1
2	$2 = 1 + 1$ $= 2$	2
3	$3 = 1 + 1 + 1$ $= 1 + 2$ $= 2 + 1$	3
4	$4 = 1 + 1 + 1 + 1$ $= 1 + 1 + 2 = 1 + 2 + 1$ $= 2 + 1 + 1 = 2 + 2$	5
5	$5 = 1 + 1 + 1 + 1 + 1$ $= 1 + 1 + 1 + 2 = 1 + 1 + 2 + 1$ $= 1 + 2 + 1 + 1 = 2 + 1 + 1 + 1$ $= 1 + 2 + 2 = 2 + 1 + 2$ $= 2 + 2 + 1$	8

Continue this table for the integers, 6, 7, and 8. You will see that the sequence of numbers in the right-hand column is

$$1, 2, 3, 5, 8, 13, 21, \ldots$$

Compare this sequence with the Fibonacci sequence. Aside from the extra 1 at the beginning of the Fibonacci sequence, the rest of the sequences agree perfectly.

Exercises 11.7

•1. Use your calculator to find the first 24 terms of the Fibonacci sequence. Save your results for use in the exercises that follow.

2. How many ways are there of expressing the integer 15 as sums of 1's and 2's?

•3. Using the Fibonacci sequence, look at the following sums:

$$1 + 1 = 2$$
$$1 + 1 + 2 = 4$$
$$1 + 1 + 2 + 3 = 7$$
$$1 + 1 + 2 + 3 + 5 = 12$$

Can you guess how the next two sums are going to turn out? Make a conjecture, then test it.

•4. Starting with the first term of the Fibonacci sequence, begin adding every other term:

$$1 + 2 = 3$$
$$1 + 2 + 5 = 8$$
$$1 + 2 + 5 + 13 = 21$$

Make a conjecture about how the sums go, then test it on the next two terms.

5. Starting with the second term of the Fibonacci sequence, begin adding every other term:

$$1 + 3 = 4$$
$$1 + 3 + 8 = 12$$
$$1 + 3 + 8 + 21 = 33$$

Make a conjecture and test it.

•6. Here, we look at sums of squares of consecutive terms:

$$1^2 + 1^2 = 2 = 1 \times 2$$
$$1^2 + 1^2 + 2^2 = 6 = 2 \times 3$$
$$1^2 + 1^2 + 2^2 + 3^2 = 15 = 3 \times 5$$
$$1^2 + 1^2 + 2^2 + 3^2 + 5^2 = 40 = 5 \times 8$$

What is always going to happen?

7. Look at the sequence

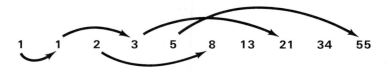

Each arrow indicates that the number it starts from divides the number the arrow points to. Can you see a pattern? What is the pattern?

11.8 Continued Fractions

In order to make a clock that indicates phases of the moon, certain gears must be brought together in precise ratios. The same is true when an astronomer wants to track a star with his telescope. Suppose the exact ratio of two gears is to be, say, $\frac{1,247}{859}$. Now, here is the problem. It is not practical to make gears with so many teeth. We seek a fraction $\frac{a}{b}$ that is nearly equal to $\frac{1,247}{859}$, but where a and b are relatively small.

This is where continued fractions enter the picture. According to this theory, we will find that we can replace $\frac{1,247}{859}$ by $\frac{16}{11}$ or by $\frac{45}{31}$. Your calculator gives

$$1247 \div 859 = 1.4516880$$

$$16 \div 11 = 1.4545454$$

$$45 \div 31 = 1.4516129$$

We see that $\frac{16}{11}$ is in error by about 0.003, while $\frac{45}{31}$ is off by only about 0.00007. Not bad!

Where did $\frac{16}{11}$ and $\frac{45}{31}$ come from? Well, that is one application of continued fractions.

When we see the fraction $\frac{19}{7}$, we can say

$$\frac{19}{7} = 2 + \frac{5}{7}$$

or

$$\frac{19}{7} = 2.7142857$$

The first representation yields more information. Let us experiment with this fraction. We begin with the equation,

$$\frac{19}{7} = 2 + \frac{5}{7} \tag{1}$$

Now

$$\frac{5}{7} = \frac{1}{\frac{7}{5}}$$

By expressing $\frac{5}{7}$ in this form, we can continue with another division; namely,

$$\frac{7}{5} = 1 + \frac{2}{5}$$

This means we can rewrite equation (1) as

$$\frac{19}{7} = 2 + \frac{1}{1 + \frac{2}{5}} \qquad (2)$$

Can we go on with yet another division? Yes!

$$\frac{2}{5} = \frac{1}{\frac{5}{2}} = \frac{1}{2 + \frac{1}{2}}$$

so equation (2) can be written as

$$\frac{19}{7} = 2 + \frac{1}{1 + \frac{1}{2 + \frac{1}{2}}} \qquad (3)$$

Now, however, if we try to repeat the same stunt with $\frac{1}{2}$, we find that

$$\frac{1}{2} = \frac{1}{\frac{2}{1}} = \frac{1}{2},$$

so at this point the process terminates.

The right-hand side of equation (3) is called the *continued-fraction* representation of $\frac{19}{7}$.

Example 11.18: Find the continued-fraction representation of $\frac{69}{13}$.

Solution:

$$\frac{69}{13} = 5 + \frac{4}{13} = 5 + \frac{1}{\frac{13}{4}} = 5 + \frac{1}{3 + \frac{1}{4}}$$

The process stops. So the desired representation is

$$\frac{69}{13} = 5 + \frac{1}{3 + \frac{1}{4}}$$

The continued-fraction representation for most fractions is a typesetter's nightmare. To simplify the notation, we will write [5, 3, 4] in place of

$$5 + \frac{1}{3 + \frac{1}{4}}$$

The numbers 5, 3, and 4 in [5, 3, 4] are called the *partial quotients* of the continued fraction.

With this notation, we write

$$\frac{69}{13} = [5, 3, 4] \quad \text{and} \quad \frac{91}{7} = [2, 1, 2, 2].$$

So far, we have not used our calculator in connection with continued fractions. Let us bring it into action by returning to the already familiar fraction $\frac{19}{7} = [2, 1, 2, 2]$. The routine uses the division with remainder process at each step.

Example 11.19: Use the calculator to find the continued fraction for $\frac{19}{7}$.

Solution:

Step	Find the Quotient and Remainder for	Quotient	Remainder
1	19 divided by 7	2	5
2	7 divided by 5	1	2
3	5 divided by 2	2	1
4	2 divided by 1	2	0 STOP

The numbers appearing in the quotient column of this table give the continued fraction representation of $\frac{19}{7}$; that is,

$$\frac{19}{7} = [2, 1, 2, 2].$$

Previous calculations have shown that $\frac{69}{13} = [5, 3, 4]$. We will now apply the above method to $\frac{69}{13}$.

Example 11.20: $\frac{69}{13}$ as a continued fraction by the calculator.

Solution:

Step	Find the Quotient and Remainder for	Quotient	Remainder
1	69 divided by 13	5	4
2	13 divided by 4	3	1
3	4 divided by 1	4	0 STOP

Again, the numbers appearing in the quotient column yield the correct representation for $\frac{69}{13}$. You will get a chance to try the above procedure in the problems that follow.

Whenever a new process or procedure is discovered in mathematics that turns out to be useful or interesting, mathematicians make it a common practice to investigate and determine if the process can be reversed. Moreover, if the process can be reversed, the mathematician will then look for algorithms that will make such a reversal as simple as possible.

We now know a process for converting a positive fraction into a continued fraction and, in fact, this process can be reversed. Suppose we want to determine the fraction represented by the continued fraction [2, 3, 4, 5]. Let us resort to brute force for the moment in the hope that we might uncover a more elegant plan of attack. We use the numbers 2, 3, 4, and 5 on purpose in order to be able to trace these partial quotients in our computations.

First, observe that the notation [2, 3, 4, 5] represents the continued fraction

$$[2, 3, 4, 5] = 2 + \cfrac{1}{3 + \cfrac{1}{4 + \frac{1}{5}}}$$

As $4 + \frac{1}{5}$ is approximately equal to 4, we see that

$$2 + \cfrac{1}{3 + \frac{1}{4}} \doteq 2 + \cfrac{1}{3 + \cfrac{1}{4 + \frac{1}{5}}}$$

In like manner, we might think of [2, 3, 4, 5] as being approximated (roughly) by each of these:

$$[2, 3, 4] = 2 + \cfrac{1}{3 + \frac{1}{4}}; \quad [2, 3] = 2 + \frac{1}{3}; \quad [2] = 2$$

Each of these is called a *convergent* of the continued fraction:

[2] is the first convergent;
[2, 3] is the second convergent;
[2, 3, 4] is the third convergent.

Of course, in this case, the fourth convergent is equal to the continued fraction itself.

With these ideas in mind, we can go on with the (brute force) process of determining the fraction represented by [2, 3, 4, 5].

Example 11.21: Find the common fraction for [2, 3, 4, 5].

Solution: The first convergent is

$$2 = \frac{2}{1}$$

The second convergent is

$$2 + \frac{1}{3} = \frac{(3 \times 2) + 1}{3}$$

Notice that the numerator is the product of the second partial quotient and the previous numerator, plus 1; the denominator is the product of the second partial quotient and the previous denominator.

The third convergent is

$$2 + \cfrac{1}{3 + \cfrac{1}{4}} = 2 + \cfrac{1}{\cfrac{(4 \times 3) + 1}{4}}$$

$$= 2 + \cfrac{4}{(4 \times 3) + 1}$$

$$= \frac{2[(4 \times 3) + 1] + 4}{(4 \times 3) + 1}$$

Notice that the denominator is the product of the third partial quotient and the denominator of the second convergent, plus the denominator of the first convergent.

To see the pattern for the numerator, we do a bit of juggling. Keep your eye on the third partial quotient, 4.

$$2[(4 \times 3) + 1] + 4 = (2 \times 4 \times 3) + 2 + 4$$

$$= 4 \times (3 \times 2) + 4 + 2$$

$$= 4[(3 \times 2) + 1] + 2$$

| 3rd partial quotient | numerator of 2nd convergent | numerator of 1st convergent |

We can go on to the fourth convergent. Through the maze of numbers, we would find that the pattern shown in the third convergent persists:

the numerator is the product of the fourth partial quotient and the numerator of the third convergent, plus the numerator of the second convergent.

In Example 11.21,

the fourth partial quotient is **5;**
the numerator of the third convergent is

$$4[(3 \times 2) + 1] + 2 = 30;$$

the numerator of the second convergent is

$$(3 \times 2) + 1 = 7$$

So the numerator of the fourth convergent is:

$$(5 \times 30) + 7 = 157$$

For the denominator,

the fourth partial quotient is **5;**
the denominator of the third convergent is

$$(4 \times 3) + 1 = 13;$$

the denominator of the second convergent is **3.**

So the denominator of the fourth convergent is

$$(5 \times 13) + 3 = 68$$

Hence, the fourth convergent is $\frac{157}{68}$.

Exercises 11.8

1. Without using your calculator, find the continued-fraction representation for the following fractions, and use the simplified notation to express your answer.

    •a. $\frac{39}{5}$ •d. $\frac{12}{5}$ •g. $\frac{131}{110}$

    b. $\frac{15}{2}$ e. $\frac{109}{33}$ •h. $\frac{1}{5}$

    •c. $\frac{24}{19}$ •f. $\frac{57}{13}$ i. $\frac{8}{1}$

2. Express the following continued fractions in long form and, by hand calculation, determine the fractions that they represent.

    a. $[4, 1, 3]$ •e. $[2, 3, 3, 2, 2]$

    •b. $[1, 3, 3, 2]$ f. $[0, 1, 2, 1, 2]$

    •c. $[7, 5, 2, 2]$ g. $[2, 2, 2, 2, 2]$

    d. $[2, 1, 2, 1, 2]$ •h. $[0, 1, 1, 1, 1, 3]$

Actually, the wording for the procedure in the preceding section is more complicated than the arithmetic. A scheme has been devised to make the arithmetic painless. First, start out with a two-line form with the partial quotients heading the columns. Thus for the continued fraction, [2, 3, 4, 5]:

Partial Quotient	2	3	4	5
Numerator				
Denominator				

The numbers in the boxes will be the numerators and denominators of the particular convergents. To follow the pattern, let us use letters, $a, b, c, \ldots$, to represent numbers in these boxes:

2	3	4	5
a	b	c	d
e	f	g	h

The third partial quotient is 4. The numerator of the third convergent is then

$$(4 \times b) + a.$$

Schematically,

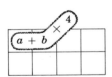

Similarly for the denominator:

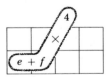

To start things going, we annex four boxes at the left to give the correct numerator and denominator of the first convergent by this pattern:

			2	3	4	5
0	1	a	b			
1	0					

So,

$$a = (2 \times 1) + 0 \qquad b = (3 \times a) + 1 = (3 \times 2) + 1$$
$$= 2 \qquad\qquad\qquad = 7$$
$$e = (2 \times 0) + 1 \qquad f = (3 \times e) + 0 = (3 \times 1) + 0$$
$$= 1 \qquad\qquad\qquad = 3$$

Now, we can complete the chart to get the following:

		2	3	4	5
0	1	2	7	30	157
1	0	1	3	13	68

From this, we can read each convergent immediately:

$$\text{1st convergent:} \quad \frac{2}{1} = 2.0000000$$

$$\text{2nd convergent:} \quad \frac{7}{3} = 2.3333333$$

$$\text{3rd convergent:} \quad \frac{30}{13} = 2.3076923$$

$$\text{4th convergent:} \quad \frac{157}{68} = 2.3088235$$

As a check, we can use our previous method to expand $\frac{157}{68}$ as a continued fraction and, indeed, it will be found to be exactly [2, 3, 4, 5]! The advantage in this scheme surely can be appreciated just *thinking* about what has to be done for the two different procedures in finding the fraction represented by, say,

$$[5, 3, 4, 2, 4, 4, 7, 6, 8, 4, 5].$$

By changing the chart from a horizontal to a vertical display, we can show the details of each step in equation form more conveniently and dispense with drawing the chart.

Example 11.22: Convert [5, 3, 7, 2, 2, 6] into a common fraction.

Solution:

	0		1
	1		0

$(5 \times 1) + 0 =$	5	$(5 \times 0) + 1 =$	1
$(3 \times 5) + 1 =$	16	$(3 \times 1) + 0 =$	3
$(7 \times 16) + 5 =$	117	$(7 \times 3) + 1 =$	22
$(2 \times 117) + 16 =$	250	$(2 \times 22) + 3 =$	47
$(2 \times 250) + 117 =$	617	$(2 \times 47) + 22 =$	116
$(6 \times 617) + 250 = 3{,}952$		$(6 \times 116) + 47 = 743$	

Hence, $[5, 3, 7, 2, 2, 6] = \dfrac{3{,}952}{743}$.

Check:

Step	Find the Quotient and Remainder for	Quotient	Remainder
1	3,952 divided by 743	5	237
2	743 divided by 237	3	32
3	237 divided by 32	7	13
4	32 divided by 13	2	6
5	13 divided by 6	2	1
6	6 divided by 1	6	0 STOP

Thus the continued fraction expansion of $\dfrac{3{,}952}{743}$ is [5, 3, 7, 2, 2, 6], which confirms our answer.

Earlier we mentioned that continued fractions were both interesting and useful. Up to this point, we have not seen much of either of these attributes. We are now ready to "put up or shut up."

Example 11.23: (Something of interest.) Convert [1, 2, 1, 2, 1, 2] into a fraction.

Solution: This is so simple that we do not need our calculator.

		1	2	1	2	1	2
0	1	1	3	4	11	15	41
1	0	1	2	3	8	11	30

Here is what is interesting (it also turns out to be useful). Pick any consecutive pair of columns from the table and mentally label it as follows.

Top Left	Top Right
Bottom Left	Bottom Right

The following is true for no matter which pair of columns you happen to pick:

(Top left × Bottom right) − (Bottom left × Top right) = ±1

We shall refer to the above property as the *double-column property* of continued fractions. For example, for

3	4
2	3

we have

$$(3 \times 3) - (2 \times 4) = 9 - 8 = 1$$

Try

15	41
11	30

which gives

$$(15 \times 30) - (11 \times 41) = 450 - 451 = -1$$

Using

4	11
3	8

we obtain −1.

It can be proved that, no matter which double column you pick, this property is there. Also, *any* continued fraction has this double-column property.

Now, how can continued fractions be useful? For one thing, they can be used to answer questions of the following nature: What whole-number multiples of 5, and 7, when added, equal 2?

In symbols, we are looking for whole numbers a and b such that $(a \times 5) + (b \times 7) = 2$. Such solutions, for example, may be required in equations for numbers of cattle, or people, or things that make sense only when counted in whole numbers (such as, how many teeth in a gear?).

Here is another example.

Example 11.24: Consider this puzzle about the number line.

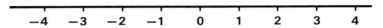

Suppose a coin is moved along this line with these restrictions:

1. any movement to the right must be made in steps of 5 units;

2. any movement to the left must be made in steps of 7 units.

Question: Is it possible to start at 0 and eventually land on 2?

Solution: After a bit of experimentation, we see that, starting at 0, if we make 6 moves to the right and 4 moves to the left, the final position will be 2, and we have our answer.

Certainly we may want to remove the guesswork and experimentation from the problem above. Especially, if the numbers are larger and more difficult to juggle around mentally. Here is where continued fractions can be useful.

Recall that we are required to make steps of 5 and 7 units. Now,

$$\frac{7}{5} = 1 + \frac{2}{5} = 1 + \frac{1}{2 + \frac{1}{2}}$$

Thus $\frac{7}{5} = [1, 2, 2]$, and we have the following table of its convergents.

		1	2	2
0	1	1	3	7
1	0	1	2	5

Applying the double-column property to the last two columns,

3	7
2	5

we obtain,

$$(3 \times 5) - (2 \times 7) = 1$$

Multiplying both sides of this equation by 2,

$$(6 \times 5) - (4 \times 7) = 2,$$

which can be written,

$$(6 \times 5) + [(-4) \times 7] = 2$$

We can interpret this last equation to mean:

"Take 6 moves to the right and 4 moves to the left to get to 2."

Of course, this was precisely the same solution we got by guesswork before. In our next example, we will not be able to guess the answer as readily.

Example 11.25: Here we repeat the puzzle of the previous example except that we are now restricted to moving to the right in steps of 122 units, and moving to the left in steps of 53 units. Question: Starting at zero, how do we arrive at 17?

Solution: The numbers 122 and 53 are probably awkward enough to discourage guessing. So let's start right in with the continued-fraction method of solution.

$$\frac{122}{53} = [2, 3, 3, 2].$$

So the convergents are

		2	3	3	5
0	1	2	7	23	122
1	0	1	3	10	53

Applying the double-column property to the last two columns, we have

$$(23 \times 53) - (10 \times 122) = -1$$

Multiplying both sides by -17 produces

$$-(391 \times 53) + (170 \times 122) = 17$$

This means that, if we start at zero, make 170 moves to the right, and then make 391 moves to the left, the final position will be at 17. Considering the size of the numbers, 391 and 170, it is little wonder that guessing or experimentation would not likely produce these results. However, there are other smaller solutions as we shall see.

We can apply a bit of sleight-of-hand to the solution in Example 11.25 for other results. Notice that, from

$$- (391 \times 53) + (170 \times 122) = 17,$$

if we *add* and then *subtract* the same thing in the left-hand member, we still have 17 as the result. Let us add and subtract

$$53 \times 122 \times t, \quad \text{for any integer } t.$$

Then we have

$$- (391 \times 53) + (170 \times 122)$$
$$= -(391 \times 53) - (53 \times 122 \times t) + (53 \times 122 \times t) + (170 \times 122)$$

$$= -(53) \times [391 + (122 \times t)] + (122) \times [170 + (53 \times t)]$$

$$= 17$$

This says

if we add a multiple of 122 to 391, and the same multiple of 53 to 170, we would get another solution!

Because subtraction is the same as addition of the negatives, we can equally well subtract multiples of 53 from 170 and multiples of 122 from 391 for other solutions. So,

$$170 - 53 = 117 \quad \text{and} \quad 391 - 122 = 269;$$

in other words, 117 moves to the right and 269 moves to the left will also land us on 17:

$$(117 \times 122) - (269 \times 53) = 14{,}274 - 14{,}257 = 17$$

Similarly,

$$117 - 53 = 64 \quad \text{and} \quad 269 - 122 = 147$$

gives another solution. Finally,

$$64 - 53 = 11 \quad \text{and} \quad 147 - 122 = 25$$

is also a solution. (We say "finally" because any further subtraction would give us negative solutions.)

Hence, the smallest number of moves landing us on 17 would be 11 moves to the right and 25 moves to the left.

$$(11 \times 122) - (25 \times 53) = 1{,}342 - 1{,}325 = 17$$

The procedures used for answering the questions in these examples can be used to answer (if there is any) the general question:

Given three whole numbers, A, B, C, what multiples of A added to what multiples of B equals C?

In symbols, find whole numbers x and y such that

$$Ax + By = C.$$

Note that, if $A = 22$, $B = 10$, and $C = 15$, then there is no solution to the equation,

$$(22 \times x) + (10 \times y) = 15$$

Why? (Note that 22 and 10 are even and 15 is odd.)

Here is another puzzler that can be solved through continued fractions.

You are given two containers, one measuring exactly 8 gallons and the other exactly 5 gallons. The containers have no other markings giving the number of gallons between full and empty. You can fill and pour from one container to another, and you can dump out as many times as you wish. How do you measure exactly four gallons of water from the river using these containers?

This is really the same problem as the game problem in Example 11.25. We can interpret "filling the 8-gallon container" as "taking 8 steps (say, to the right)." We interpret "dumping out the 5-gallon container" as "taking 5 steps (to the left)."

Then, we have

$$(8 \times a) - (5 \times b) = 4$$

Actually, we first examine

$$(8 \times a) - (5 \times b) = 1$$

by looking at the double-column property, and then adjust accordingly.

Using the fraction $\frac{8}{5}$ we have the continued fraction

$$\frac{8}{5} = [1, 1, 1, 2].$$

Its convergents are

		1	1	1	2
0	1	1	2	3	8
1	0	1	1	2	5

So $$(8 \times 2) - (5 \times 3) = 1$$

Multiplying by 4 on both sides,

$$(8 \times 8) - (5 \times 12) = 4$$

Using the same idea for reducing these numbers as before, the solution $(8, 12)$ can be trimmed down to

$$8 - 5 = 3 \quad \text{and} \quad 12 - 8 = 4;$$

which means filling the 8-gallon container 3 times and dumping out the 5-gallon container 4 times.

To check this, let us display what is in each container at each stage.

	Contents	
Action	*8-gal.*	*5-gal.*
Fill 8-gallon container.	→ 8	0
Pour from 8-gallon to 5-gallon jug.	3	5
Dump out from 5-gallon container.	3	0 →
Pour from 8 jug to 5 jug.	0	3
Fill 8 jug.	→ 8	3
Pour from 8 jug to 5 jug.	6	5
Dump out from 5 jug.	6	0 →
Pour from 8 jug to 5 jug.	1	5
Dump out 5 jug.	1	0 →
Pour from 8 jug to 5 jug.	0	1
Fill 8 jug.	→ 8	1
Pour from 8 jug to 5 jug.	4	5
Dump out from 5 jug.	4	0 →

Note the arrows showing the times you fill and dump out from the containers.

•1. Using the continued-fraction algorithm, rework the problems in Exercise 1 of Exercises 11.8.

•2. Check the double-column property in the second and third columns for parts b, c, and f in Exercise 2 of Exercises 11.8.

3. Find each of the convergents for part 2c of Exercises 11.8, and tell how much each convergent differs from the actual fraction.

•4. Suppose you have unmarked 8- and 5-gallon containers.

    a. Describe how you can measure out exactly 7 gallons using these containers.

    b. Give two additional pairs of solutions (if possible) for this problem.

5. Use the reverse algorithm for continued fractions to convert the following continued fractions into common fractions. Run a check on each of your answers.

    •a. [2, 1, 2, 3]   •f. [0, 1, 1, 2, 2, 3, 3]

    b. [1, 2, 1, 2, 1, 2]   •g. [0, 9, 8, 7, 6, 5]

    •c. [3, 4, 4, 3]   •h. [0, 1, 1, 1, 1, 2]

    •d. [2, 3, 4, 5, 6]   i. [1, 2, 2, 2, 2, 2, 3]

    e. [2, 3, 4, 3, 2]   j. [3, 3, 3, 3, 4, 4, 4, 4]

•6. Convert each of the following into a common fraction and compare the results with the result for part a of Exercise 5. What is the effect of adding each 0 to the left of the partial quotients?

    a. [0, 2, 1, 2, 3]

    b. [0, 0, 2, 1, 2, 3]

•7. According to the results of Exercise 6, guess the common fraction for [0, 0, 0, 0, 0, 2, 1, 2, 3], and check by obtaining the ninth convergent.

8. Use the algorithmic process to work the following problems.

    a. What multiple of 14 added to what multiple of 19 equals 7?

    •b. What multiple of 73 added to what multiple of 31 equals 25?

    c. Find whole numbers x and y such that

$$(23 \times x) + (55 \times y) = 9.$$

    d. Find whole numbers x and y such that

$$(47 \times x) + (114 \times y) = -5.$$

•9. At the beginning of Section 11.8, we mentioned that $\frac{1,247}{859}$ could be represented very closely by $\frac{16}{11}$ or $\frac{45}{31}$. Show how these approximations were obtained.

11.10   Magic Squares

In the prelude, we displayed two magic squares, each consisting of 3 rows and 3 columns. The "magic" lies in the fact that each row gives the same sum (15). The sum of each column is also this same row sum. So is the sum of each diagonal, from corner to corner.

2	7	6
9	5	1
4	3	8

Figure 11.5
A 3 × 3 magic square.

Such squares were known in ancient China over 4,000 years ago. The fact that numbers could be brought together in such a great harmony promoted a mystical attitude toward magic squares. In fact, during the Middle Ages, they appeared on medallions and were worn to ward off evil spirits or diseases.

Magic squares come in many sizes. The one in Figure 11.5 is a 3 × 3 (three-by-three) square and consists of nine *cells*. The entries here are the nine consecutive integers 1, 2, 3, 4, 5, 6, 7, 8, 9. Must magic squares be constructed using only consecutive integers? Let us examine a few examples, first making new magic squares from old ones. In these examples we will begin with the magic square in Figure 11.5.

Example 11.26:   Double each entry in Figure 11.5.

Solution:

4	14	12
18	10	2
8	6	16

Figure 11.6
Doubling each entry.

Partial check:

$$4 \boxplus 14 \boxplus 12 \boxminus 30.$$
$$18 \boxplus 10 \boxplus 2 \boxminus 30.$$

You can verify that Figure 11.6 is a magic square. In general, multiplying each entry by the same number (a constant multiplier) throughout will give a magic square. Notice what happens on multiplying. Consider the sum of any row (or column, or diagonal):

$$2 + 7 + 6 = 15 \quad \text{(top row of Fig. 11.5)}.$$

By the distributive law,

$$(2 \times 2) + (2 \times 7) + (2 \times 6) = 2 \times (2 + 7 + 6).$$

Because this law applies to any row of the new square, each sum is doubled (or multiplied by the chosen constant). This is surely a property of *any* magic square, regardless of its size. What about division?

Example 11.27: Add a constant (say, 4) to each entry in Fig. 11.5.

Solution:

6	11	10
13	9	5
8	7	12

Figure 11.7
Adding a constant to each entry.

Partial check:

6 ⊞ 11 ⊞ 10 ⊟ 27.

6 ⊞ 13 ⊞ 8 ⊟ 27.

6 ⊞ 9 ⊞ 12 ⊟ ?

As each entry is increased by a constant (4), the sum of the three is increased by 3 times this constant ($3 \times 4 = 12$).

$$2 + 7 + 6 = 15;$$

$$(2 + 4) + (7 + 4) + (6 + 4) = (2 + 7 + 6) + (4 + 4 + 4).$$

What about subtraction?

One way to construct a magic square with an odd number of cells can be found by following the path of the numbers, 1, 2, . . . , 9, in Figure 11.5. Notice that "1" is in the middle cell along an edge, and notice the direction of the sequence 4, 5, 6 (from southwest to northeast).

If we follow the same (southwest to northeast) direction of the square.

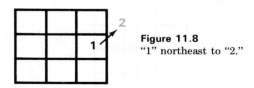

Figure 11.8
"1" northeast to "2."

When the path leads outside a row (as in Figure 11.8), the number is located at the opposite end of the row. (Similarly, for overshooting a column, as from "2" to "3.")

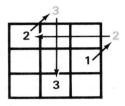

Figure 11.9
Relocating to the end of the line.

Now what happens if we try to proceed from "3"?

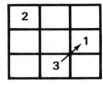

Figure 11.10
"4" is blocked.

The cell is already occupied. When this happens, the number (in this case, 4) is located to the left of the previous number (3). (Notice that *left* is opposite from the location of the first number, 1, which is the *right* edge, center.) The numbers 4, 5, and 6, are located according to the basic pattern.

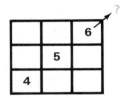

Figure 11.11
"7" overshoots row and column.

The next number, 7, overshoots both row and column, so should be located along the opposite end of the row and column. This cell is already occupied (by which number?). Therefore, the 7 goes to the left of the previous number (6). The rest of the numbers easily fall in place.

Example 11.28: Construct a 5 × 5 magic square according to the pattern just discussed and verify the sums.

Partial solution:

	3	25		
2				
				1
		4		

Figure 11.12

Sample check of diagonals:

$$9 + 21 + 13 + 5 + 17 = 65.$$

$$11 + 12 + 13 + 14 + 15 = 65.$$

The directions, ↗, ↘, ↙, ↖, are selected so that the first entry is always in the middle cell along an edge and such that the second entry leads out of the boundary.

Clearly, the directions for an even-celled magic square cannot be exactly as an odd-celled square. There is no middle cell along an edge. Here is a 4 × 4 magic square.

Example 11.29: A 4 × 4 magic square (Figure 11.13).

16	2	3	13
5	11	10	8
9	7	6	12
4	14	15	1

Figure 11.13
A 4 × 4 magic square

Compare the square in Figure 11.13 with one in which the cells are numbered in increasing order from left to right, row by row (Figure 11.14).

1	2	3	4
5	6	7	8
9	10	11	12
13	14	15	16

Figure 11.14
An orderly square.

Notice the interchanges of the positions for

1 and 16;   4 and 13;   6 and 11;   7 and 10

How are these positions related?

Here is a magic square consisting of only prime numbers. Verify the sums with your calculator.

839	557	761
641	719	797
677	881	599

Figure 11.15
A prime-number magic square.

Exercises 11.10

In Exercises 1 to 4, you are asked to combine some of the ideas in this section in making new squares out of old. Construct each square and verify that each is a magic square.

•1. Start with Figure 11.5; double each entry, then subtract 1.

    a. List the entries of the new squares in increasing order.

    b. What kind of numbers are in the new square?

2. Multiply each entry in Figure 11.5 by 5.

    a. Describe the new entries.

    b. What do you expect the sum of each row (or column) to be? Verify your guess.

•3. The first nine counting numbers are 1, 2, 3, . . . , 9. The first nine positive odd numbers are: 1, 3, 5, . . . , 17. Let the counting numbers be paired with the odd numbers as follows.

$$
\begin{array}{ccccccccc}
1 & 2 & 3 & 4 & 5 & 6 & 7 & 8 & 9 \\
\downarrow & \downarrow & \downarrow & \downarrow & \downarrow & \downarrow & \downarrow & \downarrow & \downarrow \\
1 & 3 & 5 & 7 & 9 & 11 & 13 & 15 & 17
\end{array}
$$

    a. Replace each entry of Figure 11.5 by the odd number paired with it. Is the result a magic square?

    b. If the answer to part a is yes, identify it with one that you might have constructed previously.

4. Start with Figure 11.5.

    a. Multiply or divide each entry by an appropriate constant so that the sum of each row is 1.

    b. Add to or subtract from each entry an appropriate constant so that the sum of each row is 1.

    c. Add to or subtract from each entry an appropriate constant so that the sum of each row is 0.

5. Start with 1 in a 3×3 square on the bottom row, middle cell. Follow a route going northeast and complete this square as you had completed other odd-celled squares. Show by calculations whether this is a magic square.

•6. Construct a magic square using the following numbers.

    a. 10, 11, 12, 13, 14, 15, 16, 17, 18.

    b. $\frac{1}{6}, \frac{1}{3}, \frac{1}{2}, \frac{2}{3}, \frac{5}{6}, 1, \frac{7}{6}, \frac{4}{3}, \frac{3}{2}$.

7. Here is a magic square with diagonal lines drawn in to give a hint to how it was made.

64	2	3	61	60	6	7	57
9	55	54	12	13	51	50	16
17	47	46	20	21	43	42	24
40	26	27	37	36	30	31	33
32	34	35	29	28	38	39	25
41	23	22	44	45	19	18	48
49	15	14	52	53	11	10	56
8	58	59	5	4	62	63	1

a. How many interesting properties can you discover about this magic square?

b. Use your calculator to verify the sums in this square.

c. Can you explain how this square might have been made? (See Example 11.29.)

•8. Here is another type of magic square.

512	4	128
16	64	256
32	1024	8

a. Verify that row sums are not all the same.

b. Try another operation (say, multiplication, subtraction, or division) and verify that this square is "magic" from that point of view. (Use your calculator.)

c. Build another square from this in which the "4" is replaced by "1," using the same operation.

*d. Replace each entry in Figure 11.5 by powers of 3:

$$
\begin{array}{ccccc}
1 & 2 & 3 & 4 & 5 \quad \cdots \\
\downarrow & \downarrow & \downarrow & \downarrow & \downarrow \\
1 & 3 & 9 & 27 & 81 \quad \cdots
\end{array}
$$

Is this a magic square by taking either sums or products?

e. Square each entry of the construction in part d, and test to see whether this new square is "magic" by either sums or products.

Roots

12.1 What Is a Root?

In Chapter 7 we studied exponents, the effect of raising a number to an integral power. For example, we found that

$$7^2 = 7 \times 7 = 49 \qquad 2^3 = 2 \times 2 \times 2 = 8$$

and

$$7^{-2} = \frac{1}{7^2} = \frac{1}{49} \qquad 2^{-3} = \frac{1}{2^3} = \frac{1}{8}$$

In this chapter, we shall reverse the process.

Example 12.1: Marian Green-Thumb wants to lay out her garden in the shape of a square with an area of 169 square feet. How long should her garden be on each side?

Pictorially: Given a square with area 169 square units, how long is each side?

L

L **169 sq. ft** L

L

Solution: Whatever number L is, we know that the area of the square can be computed as L^2 square feet. Thus, L is the number such that

$$L^2 = 169 \qquad (1)$$

What is L? A little trial-and-error experimentation, or a good memory, or a peek at the table of squares compiled in Section 7.2 will show that

$$169 = 13^2 = 13 \times 13$$

So we see that 13 is a solution of equation (1):

$$13^2 = 169$$

Example 12.2: Skip Marsman wants to build his spaceship in the shape of a cube with a volume of 343 cubic metres. How long should each side of the cube measure?

Pictorially: Given a cube with volume 343 cubic units, how long is each side?

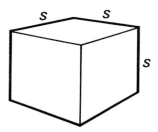

Solution: Whatever number S is, we know that the volume of the cube can be computed as S^3 cubic units. Thus, S is the number such that

$$S^3 = 343$$

What is S? A little trial-and-error experimentation, a good memory, or a peek at the table of cubes compiled in Section 7.2 will show that

$$7^3 = 343$$

In these problems, we have been finding roots.

Because $13^2 = 169$,   13 is called a *square root* of 169.
Because $7^3 = 343$,   7 is called a *cube root* of 343.
Because $2^5 = 32$,   2 is called a *fifth root* of 32.

In general, if n is an integer, and R and A are numbers such that

$$R^n = A, \quad \text{then } R \text{ is called an } nth \text{ root of } A.$$

Taking a root of a number is the reverse of raising to a power.

There is an important point to be remembered about roots, especially square roots. A square root is not uniquely determined! Because of the rule for multiplying signed numbers, we see, for example, that

$$5^2 = 5 \times 5 = 25 \quad \text{and} \quad (-5)^2 = (-5) \times (-5) = 25$$

so that both 5 and -5 are square roots of 25. The number 25 has two square roots.

In general,

$$\text{if } A^2 = B, \quad \text{then} \quad (-A)^2 = B \quad \text{also.} \tag{2}$$

Thus, either the sign of the square root of a number must be specified by outside considerations, or both alternatives must be considered. In Example 12.1, the answer had to be a positive number since it referred to the length of a garden plot—an "outside consideration."

By the rule for multiplying signed numbers, the *square* of any number is positive. This means that only positive numbers (aside from zero) have square roots belonging to the number system considered here. For example, (-1) has no square root. (To be complete, we should mention that, since the invention of *complex numbers* in the 16th century, every nonzero number has square roots in that number system. The complex-number system has many applications in mathematics, science, and engineering. This topic, however, must be left to more advanced books on mathematics.)

The same situation holds for *fourth* roots for similar reasons. For example,

$$(3)^4 = 81 = (-3)^4;$$

so that 3 and -3 are both fourth roots of 81. You can see that a similar situation holds for all even roots. Any positive number has two even-order roots. A negative number has no even-order root in our real-number system.

On the other hand, because the cube of a negative number is negative, a negative number does have a cube root. For example,

$$(-2)^3 = (-2) \times (-2) \times (-2) = -8$$

So (-2) is a cube root of (-8).

In general, the cube root of a negative number is negative, and the cube root of a positive number is positive.

The same is true for all roots of an odd order. For example,

$$(2)^5 = 32 \quad \text{and} \quad (-2)^5 = -32;$$

so that 2 is a fifth root of 32, and -2 is a fifth root of -32. Each number has exactly one root of odd order.

The notation for roots is very suggestive. It is an elongated letter "r." Imagine the evolution:

$$\text{from “r” to “} r \text{” to “} \sqrt{} \text{”}$$

The *positive* square root of a positive number A is denoted by

$$\sqrt{A}.$$

The negative square root of A is denoted by

$$-\sqrt{A}.$$

We define

$$\sqrt{0} = 0$$

Remember: $\sqrt{A} = B$ means $B^2 = A$ and B is positive or zero. In particular,

$$(\sqrt{A})^2 = A = \sqrt{A} \times \sqrt{A} = (-\sqrt{A}) \times (-\sqrt{A}). \qquad (3)$$

Example: 12.3

$$\sqrt{4} = 2 \qquad \text{because} \quad 2^2 = 4;$$
$$\sqrt{9} = 3 \qquad \text{because} \quad 3^2 = 9;$$
$$\sqrt{100} = 10 \quad \text{because} \quad 10^2 = 100;$$
$$\sqrt{-100} \boxed{=} \textbf{ERROR} \quad \text{no square root of } -100;$$
$$\sqrt{169} = 13 \quad \text{because} \quad 13^2 = 169;$$
$$\sqrt{2.25} = 1.5 \quad \text{because} \quad (1.5)^2 = 2.25;$$
$$\sqrt{0.01} = 0.1 \quad \text{because} \quad (0.1)^2 = 0.01$$

The cube root of a number A is denoted by

$$\sqrt[3]{A}.$$

(Note the position of the numeral "3" to show that it is the *cube* root that is being denoted.)

$\sqrt[3]{A}$ has the same sign (positive or negative) as A has.

Example 12.4:

$$\sqrt[3]{8} \quad = \quad 2 \quad \text{because} \quad 2^3 \quad = 8;$$
$$\sqrt[3]{-8} \quad = -2 \quad \text{because} \quad (-2)^3 = -8;$$
$$\sqrt[3]{27} \quad = \quad 3 \quad \text{because} \quad 3^3 \quad = 27;$$
$$\sqrt[3]{-343} = -7 \quad \text{because} \quad (-7)^3 = -343;$$
$$\sqrt[3]{1,000} = \quad 10 \quad \text{because} \quad 10^3 \quad = 1,000;$$
$$\sqrt[3]{0.001} = \quad 0.1 \quad \text{because} \quad (0.1)^3 = 0.001$$

Higher order roots are denoted by the same scheme. Pay attention to the numeral n in the symbol for the root:

$$\sqrt[n]{}.$$

Example 12.5:

$\sqrt[5]{32} = 2$.   The fifth root of 32 is 2 because $2^5 = 32$.
$\sqrt[6]{1,000,000} = 10$.   The positive sixth root of a million (1,000,000) is 10 because $10^6 = 1,000,000$.

REMEMBER: $\sqrt[n]{B} = A$ means $A^n = B$.
In particular, if n is a positive integer,

$$(\sqrt[n]{B})^n = \underbrace{\sqrt[n]{B} \times \sqrt[n]{B} \times \cdots \times \sqrt[n]{B}}_{n \text{ factors}} = B.$$

If n is even, $\sqrt[n]{B}$ denotes the positive root.

The rules for working with numbers and powers have extensions for roots.

Example 12.6:   Recall the rule $(2 \times 3)^2 = 2^2 \times 3^2 = 4 \times 9$. The analog for roots is

$$\sqrt{4 \times 9} = \sqrt{4} \times \sqrt{9} = 2 \times 3 = 6$$

Thus $\sqrt{36} = 6$

This, then, is the rule for factoring:

If A and B are positive numbers (or zero),

$$\sqrt{A \times B} = \sqrt{A} \times \sqrt{B}. \tag{4}$$

The rule extends to any number of factors, and it applies to roots of any order.

Example 12.7:

$\sqrt{12} \;\; = \sqrt{4 \times 3} \;\; = \sqrt{4} \times \sqrt{3} = 2 \times \sqrt{3};$
$\sqrt{225} = \sqrt{25 \times 9} = \sqrt{25} \times \sqrt{9} = 5 \times 3 = 15;$
$\sqrt{196} = \sqrt{49 \times 4} = \sqrt{49} \times \sqrt{4} = 7 \times 2 = 14;$
$\sqrt[3]{216} = \sqrt[3]{27 \times 8} = \sqrt[3]{27} \times \sqrt[3]{8} = 3 \times 2 = 6$

The same rule holds for determining the roots of fractions.

Example 12.8:

$$\sqrt{\frac{9}{4}} = \frac{\sqrt{9}}{\sqrt{4}} = \frac{3}{2}. \quad \text{In decimals: } \sqrt{2.25} = 1.5$$

$$\sqrt{\frac{16}{25}} = \frac{\sqrt{16}}{\sqrt{25}} = \frac{4}{5}. \quad \text{In decimals: } \sqrt{0.64} = 0.8$$

$$\sqrt[3]{\frac{729}{1,331}} = \frac{\sqrt[3]{729}}{\sqrt[3]{1,331}} = \frac{9}{11}. \quad \text{In decimals: } \sqrt[3]{0.54770848} = 0.8181818$$

Powers and roots appear in many situations in our natural world besides those of areas and volume. Here is a classical problem from physics.

Example 12.9: The distance (denoted by the letter s) an object falls increases with the length of the time (denoted by the letter t) it has been falling. For objects falling freely under the force of gravity near the earth's surface, the following simple formula relates s to t, provided that air resistance is neglected:

$$s = 16 \times t^2. \tag{5}$$

If Daree Debil jumps from a 25-foot diving platform, how long does it take her to hit the water?

Solution: We know the distance she falls: 25 feet. Thus, from formula (5), we have that

$$25 = 16 \times t^2,$$

or

$$t^2 = \frac{25}{16},$$

and so

$$t = \sqrt{\frac{25}{16}} = \frac{\sqrt{25}}{\sqrt{16}} = \frac{5}{4} = 1.25 \text{ seconds.}$$

Here again, outside considerations determine that we take the positive square root.

It is always easy to decide (or check) whether a particular number is a root.

Example 12.10(a): Is 4.5 the square root of 20? That is, is it true that

$$4.5 = \sqrt{20}?$$

Solution: If 4.5 is a square root of 20, then

$$(4.5)^2 = 20$$

But $(4.5)^2 = 20.25$ (an easy computation on your calculator). So,

$$4.5 \neq \sqrt{20}$$

Example 12.10(b): Is 12 the sixth root of 2,985,884? That is, is it true that

$$12 = \sqrt[6]{2,985,984}?$$

Solution: If so, then

$$12^6 = 2,985,984$$

A quick check on your calculator will confirm this.

Exercises 12.1

1. Decide whether the following indicated roots are correct. (Use your calculator.)
   - •a. $\sqrt{64} = 8$
   - b. $\sqrt{120} = 11$
   - •c. $\sqrt{20.25} = 4.5$
   - d. $\sqrt{0.16} = 0.4$
   - e. $\sqrt{292.41} = 17.1$
   - f. $\sqrt[3]{125} = 5$
   - •g. $\sqrt[3]{1,728} = 12$
   - •h. $\sqrt[3]{36} = 3.1$
   - •i. $\sqrt[3]{-64} = -4$
   - j. $\sqrt[4]{243} = 3$

2. By using the facts that $\sqrt{4} = 2$, $\sqrt{9} = 3$, $\sqrt{25} = 5$, and $\sqrt{49} = 7$, and the product rule of equation (4), find these square roots.
   - a. $\sqrt{16} =$
   - •b. $\sqrt{625} =$
   - •c. $\sqrt{196} =$
   - d. $\sqrt{441} =$
   - •e. $\sqrt{\dfrac{25}{9}} =$
   - f. $\sqrt{900} =$
   - •g. $\sqrt{1,225} =$
   - h. $\sqrt{144} =$
   - •i. $\sqrt{1,296} =$
   - •j. $\sqrt{1,024} =$

3. Using the information in Exercise 2 together with the approximations $\sqrt{2} \doteq 1.414$, $\sqrt{3} \doteq 1.732$, and $\sqrt{5} \doteq 2.236$, find these square roots. Round off your answer to the nearest two-place decimals. For each answer you obtain, compute its square to see how close the square of your estimate is to the correct number.

•a. $\sqrt{8} =$ f. $\sqrt{12} =$

•b. $\sqrt{27} =$ •g. $\sqrt{588} =$

c. $\sqrt{6} =$ •h. $\sqrt{882} =$

d. $\sqrt{10} =$ i. $\sqrt{1{,}000} =$

•e. $\sqrt{15} =$ •j. $\sqrt{0.9} = \sqrt{\dfrac{9}{10}} =$

•4. The formula for the period (P) of a pendulum is

$$P = 2\pi \sqrt{\frac{\ell}{32}}.$$

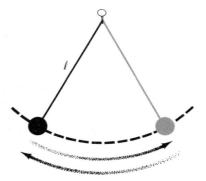

Here, ℓ is the length of the pendulum measured in feet, P is the time in seconds it takes the pendulum to make one complete swing from left to right and return, and π is the geometrical constant, approximately equal to 3.1416.

a. How long is the period of a pendulum 2 feet long?

b. Assuming one "tick" per complete swing, how many ticks would such a clock have in a minute?

c. How long should a pendulum be if its period is to be one second?

5. Let A be a positive integer. Show that, if $\sqrt[n]{A}$ is a rational number, then $\sqrt[n]{A}$ is an integer.

12.2   The Calculation of Square Roots

Although it is easy to compute the square of a number, it is a lot harder to compute its square root. In the preceding section, we selected easy examples for illustration. The square roots were easy to determine by squaring natural guesses. The situation is quite different if, for example, we want to find $\sqrt{3}$ to several decimal places.

If your calculator has a ☑ key, finding square roots is not difficult!

417

Example 12.11: Find $\sqrt{3}$ on calculators with a ☑ key.

Solution: *3 ☑ 1.7320508*

To check: **[K ON]** ☒ ⊟ *2.9999999*

or ☒ *1.7320508* ⊟ *2.9999999*

This shows that the calculator gives only an approximation: to 7-place decimals, $\sqrt{3}$ is about 1.7320508.

The check in Example 12.11 shows that the estimate is low, because its square is less than 3. If you try 1.7320509, you will find that

$$(1.7320509)^2 = 3.0000003$$

on your calculator. So 1.7320509 would be too large an estimate. Therefore, $\sqrt{3}$ is between these two numbers:

$$1.7320508 < \sqrt{3} < 1.7320509$$

In fact, because

$$(1.7320509)^2 - 3 = 0.0000003$$

and $$3 - (1.7320508)^2 = 0.0000001,$$

we conclude that 1.7320508 is closer to $\sqrt{3}$ than is 1.7320509.

If your calculator does not have a square root (☑) key, it is important to have an efficient algorithm to give an easy way for calculating square roots. Furthermore, it is useful to be able to calculate nth order roots. Advanced calculators, often called "scientist" or "slide-rule" models, have a key called (x^y). With these calculators, you can easily compute yth roots—but that's another story.

One method that always works is that of trial and error. We can use the natural order of numbers to reduce our trials. Suppose we want to estimate $\sqrt{3}$. Let us try 2. Well, $2^2 = 4$. We know that $4 > 3$, and so we reason that $\sqrt{3}$ must be less than 2. So for a better guess, we might try 1.5. Test it on your calculator:

$$(1.5)^2 = 2.25$$

We may therefore reason that 1.5 is less than $\sqrt{3}$.

It is suggestive to illustrate our attempts on the number line.

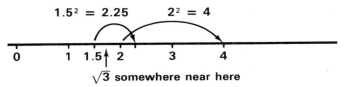

Figure 12.1
The location of some guesses for $\sqrt{3}$ and their squares.
Because $2.25 = (1.5)^2 < 3 < 2^2 = 4$, we conclude $1.5 < \sqrt{3} < 2$.

So we should try a number between 1.5 and 2; say, 1.75. Test it:

$$(1.75)^2 = 3.0625$$

Thus, 1.75 is still too large, but not by much. We can either make a guess like 1.7, slightly less than 1.7, or because we have

$$1.5 < \sqrt{3} < 1.75,$$

we could try the average of these two bracketing estimates:

$$\frac{1.5 + 1.75}{2} = 1.625$$

This procedure suggests a way to obtain better estimates. To continue:

$$(1.625)^2 = 2.640625$$

Too small. So we know that

$$1.625 < \sqrt{3} < 1.75$$

For the next estimate, compute

$$(1.625 + 1.75) \div 2 = 1.6875$$

Continuing in this manner, we could get as good an approximation as we please. This method is tedious. Fortunately, there is a very efficient algorithm for obtaining better estimates, partly using this idea of averaging. The one we now describe is efficient and may be the one built into those calculators with the ☑ key. It is commonly called the *divide-and-average* method.

To find improved estimates for $\sqrt{N}$, where $N > 0$.

Step 1   Make any initial guess, G. (You may start with $G = \dfrac{1 + N}{2}$ if you have no better information.)

Step 2   Improve your guess by computing (with your calculator)

$$G_1 = \frac{(N \div G) + G}{2}.$$

(Thus, the improved guess G_1 is the average of the old guess G and the quotient $N \div G$.)

Step 3   Improve this estimate by using G_1 as G in Step 2. That is, compute

$$G_2 = \frac{(N \div G_1) + G_1}{2}.$$

You may take as many steps as you wish, repeating Step 3. On your calculator, there is no point continuing on beyond the point where $G = N \div G$ according to your calculator.

Example 12.12:   Estimate $\sqrt{3}$ using the divide-and-average method.

Solution:

Step 1.   Make an initial guess—say, $G = 2$.

Step 2.   Compute

$$G_1 = \frac{(3 \div 2) + 2}{2} = 1.75$$

$$= ((3 \boxplus 2) \boxplus 2) \boxplus 2 \boxminus 1.75$$

Step 3.   Compute

$$G_2 = \frac{(3 \div 1.75) + 1.75}{2}$$

$$= ((3 \boxplus 1.75) \boxplus 1.75) \boxplus 2 \boxminus 1.7231428$$

Step 4.   Compute

$$G_3 = \frac{(3 \div 1.7231428) + 1.7231428}{2}$$

$$= ((3 \boxplus 1.7231428) \boxplus 1.7231428) \boxplus 2$$
$$\boxminus 1.7320508$$

Thus, in the above example, in three calculation steps (iterations) we have obtained the same accuracy for $\sqrt{3}$ as given by the ☑ key on an 8-digit calculator.

Example 12.13: Estimate $\sqrt{0.334}$.

Solution:

Step 1 Make an initial guess—say,

$$G = \frac{1 + 0.334}{2} = \frac{1.334}{2} = 0.667$$

Step 2 Compute

$$G_1 = \frac{(0.334 \div 0.667) + 0.667}{2} = 0.5838748$$

Step 3 Compute

$$G_2 = \frac{(0.334 \div 0.5838748) + 0.5838748}{2}$$

$$= 0.5779576$$

Step 4 Compute

$$G_3 = \frac{(0.334 \div 0.5779576) + 0.5779576}{2}$$

$$= 0.5779273$$

Now $0.334 \boxplus 0.5779273 \boxminus 0.5779273$

on an 8-digit machine, so there is no point in going any further.

The method deserves some consideration. Why does it work? How does it depend on the first guess? (You might enjoy seeing whether or not you can fool the method by trying some ridiculous first guesses.) What is the trend of the estimates obtained in this method? Some of the answers, you will discover empirically yourself. Consider the following example.

Example 12.14: Estimate $\sqrt{10}$.

Solution:

Step 1 Make a guess—say, $G = 0.1$. (This is a ridiculous guess; why?)

Step 2 Compute

$$G_1 = \frac{(10 \div 0.1) + 0.1}{2} = 50.05$$

(Note that G_1 is still a poor estimate!)

Step 3 Compute

$$G_2 = \frac{(10 \div 50.05) + 50.05}{2} = 25.1249$$

Step 4 Compute

$$G_3 = \frac{(10 \div 25.1249) + 25.1249}{2}$$

$$= 12.761455$$

Step 5 Compute

$$G_4 \quad \frac{(10 \div 12.761455) + 12.761455}{2}$$

$$= 6.772532$$

Step 6 Compute

$$G_5 = \frac{(10 \div 6.772532) + 6.772532}{2}$$

$$= 4.1245423$$

Step 7 Compute

$$G_6 = \frac{(10 \div 4.1245423 + 4.1245423}{2}$$

$$= 3.2745268$$

Step 8 Compute

$$G_7 = \frac{(10 \div 3.2745268) + 3.2745268}{2}$$

$$= 3.1642015$$

Step 9 Compute

$$G_8 = \frac{(10 \div 3.1642015) + 3.1642015}{2}$$

$$= 3.1622782$$

Step 10 Compute

$$G_9 = \frac{(10 \div 3.1622782) + 3.1622782}{2}$$

$$= 3.1622776$$

Finally, because

$$10 \div 3.1622776 = 3.1622776$$

on the calculator, the capacity of the calculator has been reached.

The slowness of the process here is due directly to the extremely poor initial guess. Note that a better guess would have been 3, because $3^2 = 9$ is almost 10. Indeed, once we were close to 3 (at Step 7), it took just three more steps to complete the process.

Another point to remember is that, if you should make an error in one of the steps, simply go on. The process is self-correcting! In particular, if you feel that your initial guess was bad when you view the results of the later steps, you may want to accelerate the process at any time by giving a better guess. That is, it is not necessary to use exactly the result of the preceding step in the next step! It is a good and powerful algorithm!

This method is a special case of a more general method for determining roots of even more complicated equations. The method is attributed to the great British genius of physics and mathematics, Isaac Newton. Newton, in the late seventeenth century, made profound discoveries in the theories of light and gravitational force and formulated many basic rules of physics. He invented, along with the German philosopher, Gottfried Leibnitz, the basic notions of the calculus. One of the by-products of the calculus is this algorithm.

Exercises 12.2

•1. Complete this table of square roots. Watch the growth of the square roots and make as good a first guess as you can. Keep track of the number of steps. Remember, the better the first guess, the fewer the steps.

A	$\sqrt{A}$	Number of Steps	A	$\sqrt{A}$	Number of Steps
0	0		2.8		
0.2			3.0		
0.4			3.5		
0.6			4.0		
0.8			4.5		
0.9			5.0		
1.0			5.5		
1.2			6.0		
1.4			6.5		
1.6			7.0		
1.8			7.5		
1.9			8.0		
2.0			8.5		
2.2			9.0		
2.4			9.5		
2.6			10.0		

2. Find the square roots of 350 and 0.035. Compare these with $\sqrt{3.5}$. Explain!

•3. If $\sqrt{A} < A$, is A greater or less than 1? (Consult your table. Prove your answer if you can!)

4. For what value(s) of A is $\sqrt{A} = A$?

5. What is $\sqrt{A} \times \sqrt{A} \times \sqrt{A} \times \sqrt{A}$?

6. How long will it take a bomb to fall 30,000 feet? (Use formula 5 in Section 12.1. Because bombs are "streamlined," the assumptions for the use of this formula are reasonable.)

7. A common barroom trick is to challenge a person to catch a dollar bill between his thumb and forefinger held an inch below the bill. A dollar bill is $7\frac{1}{8}$ inches long. How quick must the person's reaction time be to catch the bill?

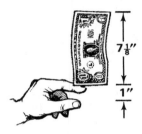

8. A clock makes 75 ticks per minute. How long is its pendulum? (Use the formula in problem 4 of Exercises 12.1.)

12.3 Graphical Methods: The Square-Root Function

An important and often helpful technique is to display the square-root function as a graph. This will give a way of making good estimates for square roots, and will give a good picture of how $\sqrt{A}$ grows as A grows.

We begin with the table compiled in Exercises 12.2. Now we locate some of these points on a graph. The construction may be made easier with the help of cross-hatched graph paper.

First, lay out an axis (number line) on your paper in a horizontal position. We shall locate the numbers A along this line. If you are using graph paper, select one of the horizontal lines near the bottom of the paper as the A axis.

Second, lay out an axis in the vertical position at right angles to the horizontal axis so that it crosses the horizontal axis at the 0 point. We shall locate the numbers $\sqrt{A}$ along this line. If you are using graph paper, select the vertical line through your 0 point.

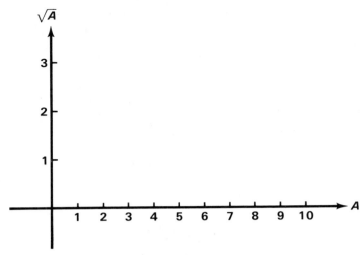

Figure 12.2
Horizontal and vertical axes for constructing a graph.

Now, corresponding to any row $(A, \sqrt{A})$ from the table in the Exercises, we shall record a point in the plane by marking a dot. We illustrate this with the pair $(0.4, 0.6324555)$. Because we cannot expect to plot more accurately than two decimal places, we first round off: $(0.4, 0.63)$. This point is located as follows. Above the number 0.4 on the horizontal axis (estimate the distance between 0 and 1), make a dot at the point corresponding to the number 0.63 on the vertical scale.

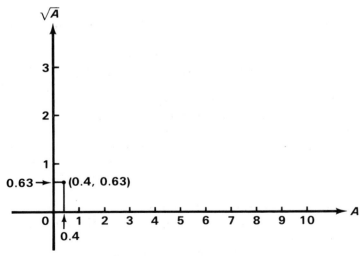

Figure 12.3
Plotting the point (0.4, 0.63).

If this is done for several of the entries of the table, the following picture is obtained. Actually, we have done it only for about a third of the points. The region from 0 to 1 is especially critical.

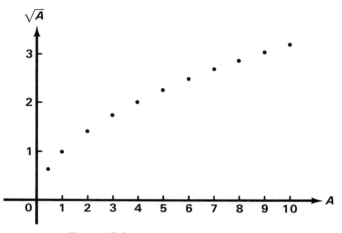

Figure 12.4
Selected points $(A, \sqrt{A})$ plotted.

Finally, draw a smooth curve through these points.

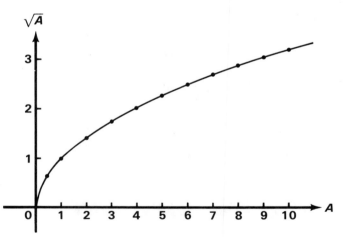

Figure 12.5
Graph of the square-root function drawn through plotted points.

This curve is the graph of the square-root function from 0 to 10. Now, we can obtain good estimates for other square roots.

Example 12.15: What is $\sqrt{2.25}$? (We know it is 1.5.)

Solution: Locate 2.25 on the horizontal axis. With either a ruler or your eye, locate the point above it on the graph. Then, using a ruler or your eye, "draw" a line parallel to the horizontal axis. This point is $\sqrt{2.25}$, and should be 1.5. (We got 1.53 when we did our own!)

There are several drawbacks to this procedure. One is that a graph is at best an approximation to within a couple of decimal places—even if the graph is drawn very carefully. Still, such an approximation makes an excellent tool for a first guess in the square-root algorithm.

Another fault is that the graph appears to be limited to the square roots of the numbers between 0 and 10. Of course, we could make a longer graph. However, this fault is reparable through the use of scientific notation and our knowledge of the square roots of the powers of 10.

Example 12.16: Find $\sqrt{300}$ and $\sqrt{22}$.

Solution: $\sqrt{300} = \sqrt{3 \times 100} = \sqrt{3} \times \sqrt{100} = \sqrt{3} \times 10 \doteq 1.732 \times 10$
$= 17.32$

$\sqrt{22} = \sqrt{2.2 \times 10} = \sqrt{2.2} \times \sqrt{10} \doteq 1.4832396 \times 3.1622776$
$\doteq 4.6904153$

It is interesting to note that, computing $\sqrt{22}$ in this way, errors incurred by the calculator due to rounding-off give an answer that is incorrect in the last decimal place.

Thus, it is useful to learn to compute square roots of powers of 10. The situation is quite easy for *even* powers of 10:

$$\sqrt{100} = \sqrt{10^2} = \sqrt{10 \times 10} = \sqrt{10} \times \sqrt{10} = 10;$$

$$\sqrt{10000} = \sqrt{10^4} = \sqrt{10^2 \times 10^2} = \sqrt{10^2} \times \sqrt{10^2} = 10^2;$$

$$\sqrt{1000000} = \sqrt{10^6} = \sqrt{10^3 \times 10^3} = \sqrt{10^3} \times \sqrt{10^3} = 10^3;$$

$$\sqrt{1} = \sqrt{10^0} = \sqrt{10^0 \times 10^0} = \sqrt{10^0} \times \sqrt{10^0} = 10^0 = 1;$$

$$\sqrt{0.01} = \sqrt{10^{-2}} = \sqrt{10^{-1} \times 10^{-1}} = \sqrt{10^{-1}} \times \sqrt{10^{-1}}$$
$$= 10^{-1} = \frac{1}{10} = 0.1;$$

$$\sqrt{0.0001} = \sqrt{10^{-4}} = \sqrt{10^{-2} \times 10^{-2}} = \sqrt{10^{-2}} \times \sqrt{10^{-2}}$$
$$= 10^{-2} = \frac{1}{100} = 0.01$$

Thus, for even powers, where $n = 2m$,

$$\sqrt{10^n} = 10^m. \tag{6}$$

The situation for *odd* powers differs only by a factor of 10. But unfortunately, $\sqrt{10}$ is approximately 3.1622776, which is not a convenient number to remember or to use in computation. However, here is the situation.

$$\sqrt{10} = \sqrt{10^1} \doteq 3.1622776 \doteq \frac{721}{228};$$

$$\sqrt{1,000} = \sqrt{10^3} = \sqrt{10 \times 10^2} = \sqrt{10} \times 10^1$$

$$\doteq 31.622776;$$

$$\sqrt{100,000} = \sqrt{10^5} = \sqrt{10 \times 10^4} = \sqrt{10} \times \sqrt{10^4}$$

$$= \sqrt{10} \times 10^2 \doteq 316.22776;$$

$$\sqrt{0.1} = \sqrt{10^{-1}} = \sqrt{10 \times 10^{-2}} = \sqrt{10} \times \sqrt{10^{-2}}$$

$$= \sqrt{10} \times 10^{-1} = \sqrt{10} \times 0.1 \doteq 0.31622776;$$

$$\sqrt{0.001} = \sqrt{10^{-3}} = \sqrt{10 \times 10^{-4}} = \sqrt{10} \times \sqrt{10^{-4}}$$

$$= \sqrt{10} \times 10^{-2} = \sqrt{10} \times 0.01 \doteq 0.031622776$$

Thus for odd powers, where $n = 2m + 1$ for some whole number m,

$$\sqrt{10^n} = \sqrt{10^{2m+1}} = \sqrt{10^{2m}} \times \sqrt{10^1}$$

$$\sqrt{10^n} = \sqrt{10} \times 10^m. \tag{7}$$

In this way, our graph of our table of square roots may be extended to both larger and smaller numbers than those that appear in the table.

Example 12.17: Find $\sqrt{221,367}$.

Solution: First estimate: $221,367 \doteq 2.2 \times 10^5$,

so $\qquad \sqrt{2212367} \doteq \sqrt{2.2} \times \sqrt{10^5}$

$$\doteq 1.48 \,\boxed{\times}\, 316.2 \,\boxed{=}\, 467.976 \doteq 467.98$$

This is a good first guess for our *divide-and-average* algorithm. Let us continue:

$$G_1 = \frac{(221367 \div 467.98) + 467.98}{2} \doteq 470.5;$$

$$G_2 = \frac{(221367 \div 470.5) + 470.5}{2} \doteq 470.49654,$$

which is the square root up to the limits of our calculator. Why?

Exercises 12.3

•1. The graph of $\sqrt{A}$ is very critical for values of A close to 0.

   a. Compute $\sqrt{A}$ for $A = 0.01, 0.06, 0.1, 0.15, 0.2, 0.25, 0.3, 0.35$, and so on, ending with $A = 1$.

   b. Use a large scale for the horizontal and vertical axes—say, 20 mm as the unit distance—and make a graph of the table in part a.

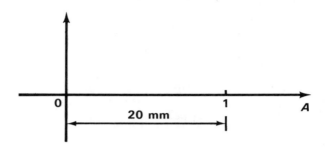

•2. Compute a table of square roots of A from $A = 10$ to $A = 50$ in intervals of 5.

 3. Plot the table in Exercise 2 as a graph, and check your work by using it to determine $\sqrt{36}$ and $\sqrt{49}$.

12.4   The Theorem of Pythagoras

One of the most useful applications of square roots occurs in geometric problems involving right triangles. The early Greek mathematician and philosopher Pythagoras (who lived during the sixth century B.C.) is credited with the discovery of the relationship between the squares of the sides of a right triangle. Here is a sample problem that shows how we can use this basic relationship.

Example 12.18:   Handi Mann wants to paint her house. She needs a ladder long enough to reach the eaves, which are 15 feet above the ground. The ladder will be braced against a low wall 7 feet from the house. How long a ladder must Handi rent?

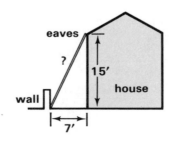

Solution: The geometry of the problem gives us a right triangle, with the two sides that meet at right angles having lengths of 15 and 7 units. We are to find the length (H) of the third side, which is called the *hypotenuse* of the triangle.

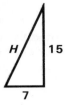

The theorem of Pythagoras tells us that the lengths of the sides are related as follows:

$H^2 = 7^2 + 15^2 = 49 + 225 = 274.$

So $\quad H = \sqrt{274} \doteq 16.552945$

As $0.552945 \times 12 \doteq 6.63534$, the distance H is thus less than 16 feet, 7 inches.

The theorem of Pythagoras is a theorem about right triangles. A right triangle is one in which one angle is a right angle—that is, measures $90°$. Equivalently, a right triangle is one in which two sides of the triangle are perpendicular to each other. (In Figure 12.6, each right angle is identified by the customary mark, ⌐, showing the angle "squared-off.")

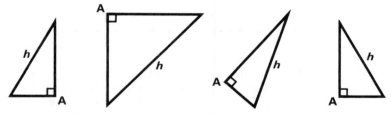

Figure 12.6
Some right triangles. The right angle is at A. The hypotenuse is h.

In each of the triangles in Figure 12.6, the right angle is shown at the point labeled "A." It is traditional to call the side opposite the right angle the *hypotenuse* of the triangle. We have labeled it as h in each case.

The theorem of Pythagoras states that

The *square* of the (length of the) hypotenuse equals the sum of the *squares* of the (lengths of the) other two sides. In symbols,

$$h^2 = r^2 + s^2.$$

This theorem can be verified in particular examples by measuring the sides and making the indicated computations. (We ask you to do this for the triangles of Figure 12.6 in the exercises.) Here's another example of how the theorem can be used.

Example 12.19: Roger and Dodger are flying jets that fly at speeds of 300 and 400 miles per hour, respectively. They rendezvous at point A. Then Roger flies due east; Dodger flies due south. How far apart are they after one hour?

Solution: Their positions relative to point A after one hour are shown as points R and D in the adjacent figure.

The courses are perpendicular, so the theorem of Pythagoras tells us

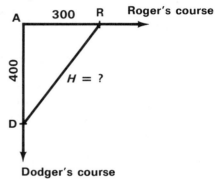

$$H^2 = 300^2 + 400^2$$

$$= 90,000 + 160,000$$

$$= 250,000.$$

So

$$H = \sqrt{250,000} = \sqrt{25 \times 10^4}$$

$$= 5 \times 10^2 = 500 \text{ miles.}$$

In the past, hundreds of proofs have been produced to demonstrate the truth of the Pythagorean theorem. Much of this had been inspired as mental exercises—meeting a sort of challenge to create a *different* proof. In the following, we show a proof that is reasonably understandable in case you wish to pursue the argument. If you prefer, of course, you can dive right into the problems in the Exercises instead.

A Proof of the Pythagorean Theorem

First, let us recall some of the elementary properties of triangles. What is a triangle? Simply, three points and three line segments joining these points. The sides that meet at a point form an angle. There are three of them: hence, the name *tri-angle*.

The most basic fact about a triangle is that the angles of a triangle sum to 180°, the number of degrees in an angle whose two sides lie on a straight line.

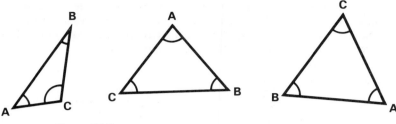

Figure 12.7
In every triangle, angle A + angle B + angle C = 180°.

You can demonstrate this fact easily with a paper model of a triangle. Cut a triangle of whatever size you please; better yet, try several!

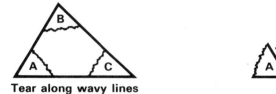

Tear along wavy lines

Figure 12.8
How to demonstrate that the sum of the angles in a triangle is 180°.

Now tear away the three angles, and then line up the three angles next to one another as indicated in Figure 12.8. You will see that the extreme edges lie on a line!

Now, a right triangle is one in which one angle measures 90°. Because all three angles sum to 180°, this means that, in a right triangle, the sum of the two angles, that are not right angles is also 90°. This is an important fact.

Models of right triangles can easily be formed by cutting a corner from a rectangle—for example, from a 3 × 5 card or from a sheet of typing paper. These cards and sheets have been carefully milled to have right-angled corners.

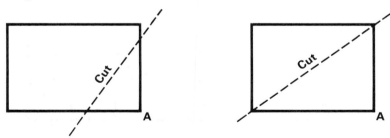

Figure 12.9
Making models of right triangles. On the right, two identical right triangles are formed by cutting along the diagonal.

If you take a rectangle, say a 3×5 card, and cut it along a diagonal as indicated on the right in Figure 12.9, you will form two identical right triangles. Conversely, if you take two identical right triangles and place them opposite from one another along the hypotenuse, a rectangle will be formed (Figure 12.10).

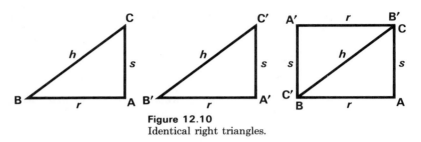

Figure 12.10
Identical right triangles.

Why is this so? First, the angles B and C add to 90°, a right angle. Then because angles B = B′ and C = C′, we have

$$B + C' = B' + C = 90°.$$

Thus a rectangle is formed.

An important consequence of this fact is that the area of a *right* triangle is one-half of the product of the lengths of the two sides meeting at the right angle.

Area of a right triangle $= \dfrac{1}{2} \times r \times s = \dfrac{r \times s}{2}$, where r and s are the lengths of the two sides meeting at the right angle.

This is so because the area of the rectangle formed by two of these triangles is, by definition, the product of the length and the width (see Chapter 5)—which are the lengths of the sides of the right triangle. Because two identical right triangles make up the rectangle, the area of one of them is one-half that of the rectangle.

These facts lead in turn to a proof of Pythagoras' theorem. Take any right triangle and make four identical copies. (This can be done by cutting four identically shaped corners from a 3×5 card. See Figure 12.11.)

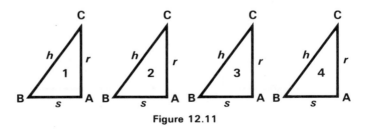

Figure 12.11

Next, we show how to arrange these identically shaped triangles on a square of side $(r + s)$.

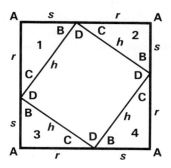

Figure 12.12
Proof of the theorem of Pythagoras. Four identical right triangles arranged on a square of side $(r + s)$. In the center is a square of side h. The area of the big square is $(r + s)^2$, and it is also the sum of the areas of the four triangles and the inside square. From $(r + s)^2 = (4 \times \frac{1}{2} \times r \times s) + h^2$, it follows that $r^2 + s^2 = h^2$.

First, we must show that the inside of the big square is itself a square. Certainly, each side has the same length, h. It remains to prove that each interior angle, labeled D in Figure 12.12, is a right angle.

To see that this is so, begin by recalling that, in each of the small triangles, the sum of angles A, B, and C is 180°. Because they are right triangles, A = 90°. Thus, the sum of angles B and C is also 90°. But now, observe from Figure 12.12 that angles B, D, and C fill out a line. So angles B, D, and C sum to 180° also. Finally, because the sum of B and C is 90°, we may conclude that D is 90° as well.

To complete our argument, we calculate the area of the big square in two ways.

(*i*)   Area = $(r + s)^2$

(*ii*)   Area = the sum of the areas of the 4 small triangles plus the area of the inside square

$$= \left({}^{2}\cancel{4} \times \frac{r \times s}{\cancel{2}} \right) + h^2 = (2 \times r \times s) + h^2$$

Hence

$$(r + s)^2 = (2 \times r \times s) + h^2 \qquad (8)$$

Now expand the left-hand side of (8):

$$(r + s)^2 = (r + s) \times (r + s)$$

$$= (r + s) \times r + (r + s) \times s \quad \text{(distributive law)}$$

$$= (r \times r) + (s \times r) + (r \times s) + (s \times s)$$

$$= r^2 + (2 \times r \times s) + s^2.$$

Now we can rewrite (8) as

$$r^2 + (2 \times r \times s) + s^2 = (2 \times r \times s) + h^2.$$

Subtracting $(2 \times r \times s)$ from both sides, we have

$$r^2 + s^2 = h^2,$$

which was to be proved—or *"Quod erat demonstrandum,"* as the geometers of the Middle Ages proclaimed in Latin! In short, Q.E.D.!

12 Roots

1. Measure the sides (in millimetres) of the right triangles in Figure 12.6 and verify the Pythagorean theorem:

$$(\text{hypotenuse})^2 = (\text{short side})^2 + (\text{middle side})^2.$$

2. Consider a right triangle as shown. In each problem, two of the three sides are given. Find the third side:

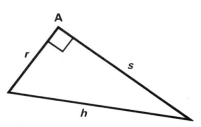

   • a. $r = 17$, $s = 33$; find h.

   b. $r = 8.7$, $s = 13.2$, find h.

   • c. $r = 51$, $h = 105$; find s.

   d. $s = 193$, $h = 230$, find r.

• 3. A group of scouts wish to determine the length of a lake. From point A, they sight down the lake to point B, and then start pacing at a right angle to point C.

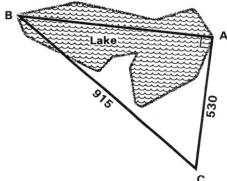

   Using a pedometer, they find that distance AC is 530 feet. Then they pace off the distance from C to B and find that it is 915 feet. Now they can find the length of the lake (without walking on water). What is the length to the nearest foot (pedometers aren't more accurate than this)?

4. Airport regulations require that small aircraft must keep 10 miles apart while in flight. On the radar scope, two small aircraft are seen to be 7.4 and 8.2 miles away from the control tower. At the same moment, the lines of sight to the planes are at a right angle. How far apart are the planes?

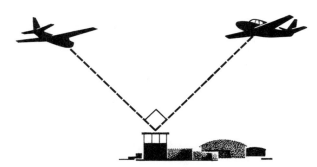

•5. Foggy Fumble measured the sides of several triangles. Sometimes he measured right triangles, sometimes not. Sometimes he made an error in measurement so bad that no triangle exists with those measurements! Here are his results. Which could come from a right triangle? Which could come from no triangle? (The sum of any two sides must be greater than the third side. Give Foggy the benefit of the doubt on small errors.)

a. Sides 3, 4, 5

b. Sides 3, 3, 5

c. Sides 3, 3, 7.9

d. Sides 7, 12, 13

e. Sides 1, 9, 13

f. Sides 2, 2, 2.82

g. Sides 1, 1, 2

h. Sides 1, 1.72, 2

i. Sides 6, 8, 1

j. Sides 5, 12, 13

6. A square has sides S units long.

a. In terms of S, how long is the diagonal?

b. If the diagonal is 5 units, how long is S?

7. An equilateral triangle is one with three equal sides. An equilateral triangle with sides 1 foot long is folded in half. How long is the crease? (The crease will be at a right angle with the side.)

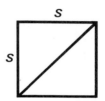

•8. Postal regulations require that the longest dimension, in any direction, of a parcel must not exceed 50 inches. (This means, of course, even along the diagonal DB in the figure.) Should a postal clerk accept a rectangular box that is 20 in. by 30 in. by 40 in.? (Make two applications of the theorem of Pythagoras. First, find the length of BC.)

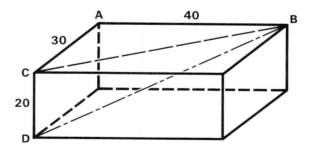

12.5 An Algorithm to Compute Cube Roots

A manufacturer wants to produce aluminum cubes having a volume of 10 cubic centimetres. What is the length of each side of the cube?

Volume = 10

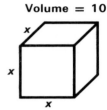

Figure 12.13
How long is x?

We know that the volume of the cube with side x is x^3. The manufacturer wants the volume to be 10 (cubic centimetres). So we want to determine an x such that

$$x^3 = 10$$

So we need to solve this equation.
We know that

$$\text{if } x^3 = 10, \quad \text{then } x = \sqrt[3]{10}$$

Now, what is a good estimate for this number? Let us see what else we know.

We know that $2^3 = 8$ and that $3^3 = 27$; also we know that 10 is between 8 and 27:

$$2^3 < 10 < 3^3$$

Therefore,

$$2 < \sqrt[3]{10} < 3$$

We can continue to experiment in this way. Your calculator makes it easy. For example, verify that

$$(2.1)^3 = 9.261 \quad \text{and} \quad (2.2)^3 = 10.648$$

Because

$$9.261 < 10 < 10.648,$$

$$(2.1)^3 < 10 < (2.2)^3$$

and so

$$2.1 < \sqrt[3]{10} < 2.2$$

Continued refinements of this type will give us a good approximation to $\sqrt[3]{10}$. However, we would like an algorithm that would give an efficient method for computing close approximations to $\sqrt[3]{10}$.

In Section 12.2, we developed this type of algorithm for finding square roots. Remember the general scheme! It consisted of a way to improve any estimate for the square root. In this section, we describe a similar algorithm for finding cube roots. Like the one for square roots, this algorithm is also "self-correcting." If you should make an error in any of the calculations, just continue on. The next stage will bring you closer to the correct answer!

Here is the algorithm to use on cube roots based on Newton's work. We display it here in a form that is convenient for use on a simple 8-place calculator.

Algorithm for computing cube roots.

Given an estimate G for $\sqrt[3]{A}$, calculate a better estimate G':

$$G' = ([(A \div 2 \div G \div G) + G] \times 2) \div 3$$

Let us use this method to obtain a good approximation to $\sqrt[3]{10}$.

Example 12.20: Find $\sqrt[3]{10}$.

Solution: Here, $A = 10$. We began with an initial guess—say, 2.1. After all, there is no point in ignoring our previous good work!

First improvement:

$$G' = ([(\boxed{10} \div \boxed{2} \div \boxed{2.1} \div \boxed{2.1}) + \boxed{2.1}] \times \boxed{2}) \div \boxed{3} = \boxed{2.1558578}$$

Thus $G' = 2.1558578$; its cube is 10.019828. For a better estimate, we repeat the algorithm, now with $G = 2.1558578$.

$$G' = ([(\boxed{10} \div \boxed{2} \div \boxed{2.1558578} \div \boxed{2.1558578})$$

$$+ \boxed{2.1558578}] \times \boxed{2}) \div \boxed{3} = \boxed{2.1544356}$$

Thus, our new estimate $G' = 2.1544356$. Checking, we have

$$(2.1544356)^3 = 10.000012$$

This is probably sufficient accuracy for our manufacturer friend. Indeed, it would take very sophisticated machinery to grind a cube accurate to 10^{-7} centimetres, which is the last place value in this estimate! Mathematics has again outstripped technology! (Or has it?)

Example 12.21: Find $\sqrt[3]{0.8}$.

Solution: Here, $A = 0.8$. Make an initial guess—say, 0.9. One reason for choosing this guess is that we know that the cube root of a number less than 1 is greater than the number. In symbols,

$$0 < c < 1 \quad \text{implies} \quad 0 < c^3 < c < 1$$

So if $c^3 = 0.8$, it follows that c must be larger than 0.8 (and it must be smaller than 1).

Now use the algorithm to compute a better estimate:

$$G' = ([(\boxed{0.8} \boxed{\div} \boxed{2} \boxed{\div} \boxed{0.9} \boxed{\div} \boxed{0.9}) \boxed{+} \boxed{0.9}] \boxed{\times} \boxed{2}) \boxed{\div} \boxed{3} \boxed{=} \boxed{0.9292181}$$

Thus $G' = 0.9292181$. We check: $(0.9292181)^3 = 0.8023298$. Not good enough? Repeat, using $G = 0.9292181$:

$$G' = ([(\boxed{0.8} \boxed{\div} \boxed{2} \boxed{\div} \boxed{0.9292181} \boxed{\div} \boxed{0.9292181})$$
$$\boxed{+} \boxed{0.9292181}] \boxed{\times} \boxed{2}) \boxed{\div} \boxed{3} \boxed{=} \boxed{0.9283186}$$

Thus the newest estimate for $\sqrt[3]{.8}$ is 0.9283186.

Check: $(0.9283186)^3 = 0.8000021$. Good enough!

The algorithm we have used for computing the improved estimate has been expressed in a notation most convenient for use with calculators. Notice that we did not have to pause to write down, or send to memory, any intermediate steps. (However, if your calculator has a memory, store the estimate G, and simply recall it for operations: $\boxed{\div} G \boxed{\div} G \boxed{+} G$.)

The repeated division by G can be performed by use of a constant key. CAUTION: Your calculator may vary on the precise stage at which you should turn off the K switch so that the next addition $(+G)$ will be performed correctly. It is best to consult your instruction manual.

Setting these considerations aside, we may express G' as follows:

$$G' = \frac{\dfrac{N}{G^2} + (2 \times G)}{3}.$$

In this form, it looks even more similar to the algorithm for computing square roots.

Exercises 12.5

1. An ancient problem calls for "doubling the cube." That is, given a cube with volume 1 (unit volume) and thus edge of 1, find the edge of the cube with volume 2.

2. If the volume of a cube is V, how long is the edge of a cube of volume, $2 \times V$? Express your answer in terms of the length of the cube of volume V.

3. Compute these cube roots.

•a. $\sqrt[3]{64}$ (The answer is an integer.)   d. $\sqrt[3]{0.5}$ (Try $G = 0.8$)

  b. $\sqrt[3]{12}$ (Try $G = 2.3$)   e. $\sqrt[3]{100}$

•c. $\sqrt[3]{36}$ Try $G = 3$)   •f. $\sqrt[3]{0.263}$

•4. Make a table of cube roots and plot its graph (for $A = 0$ to $A = 10$).

A	$\sqrt[3]{A}$	A	$\sqrt[3]{A}$
0	0	1.5	
0.2		2.0	
0.4		3.0	
0.6		4.0	
0.8		6.0	
1.0		8.0	
1.25		10.0	

HINT: As you work down the table, your experience should help you make excellent first guesses! Alternatively, do every fourth one; plot these points; draw a smooth curve through these points. Use this curve to make good guesses for the remaining values.

5. Determine rules for finding the cube roots of powers of 10. HINT: Determine the following.

$$\sqrt[3]{1,000} = \sqrt[3]{10^3} = ?$$
$$\sqrt[3]{100} = \sqrt[3]{10^2} = ?$$
$$\sqrt[3]{10,000} = \sqrt[3]{10^4} = ?$$
$$\sqrt[3]{0.1} = \sqrt[3]{10^{-1}} = ?$$
$$\sqrt[3]{0.01} = \sqrt[3]{10^{-2}} = ?$$

$$\sqrt[3]{0.001} = \sqrt[3]{10^{-3}} = ?$$
$$\sqrt[3]{1,000,000} = \sqrt[3]{10^6} = ?$$
$$\sqrt[3]{100,000} = \sqrt[3]{10^5} = ?$$
$$\sqrt[3]{10,000,000} = \sqrt[3]{10^7} = ?$$

12.6 Finding nth Roots

We can extend the methods we have used for square and cube roots to find the nth root for any integer n. That is, we can compute accurate approximations for $\sqrt[n]{A}$. (This procedure, again, is due to Isaac Newton.)

The strategy is the same as in the other cases. Given an estimate G for $\sqrt[n]{A}$, we shall compute a better estimate G'. The formula is

$$G' = \frac{\frac{A}{G^{(n-1)}} + (n-1) \times G}{n}$$

$$= \left(\frac{A}{(n-1) \times G^{(n-1)}} + G \right) \times (n-1) \div n.$$

Now we translate this expression into a form that is efficient as an operational sequence for our calculators:

$$G' = \{[A \boxdiv (n \boxminus 1) \underbrace{\boxdiv G \boxdiv G \boxdiv \cdots \boxdiv G}_{(n-1) \text{ times}}] \boxplus G\} \boxtimes (n \boxminus 1) \boxdiv n \boxeq$$

As you can see, it will be especially convenient to use the constant (K) switch for this repeated division. Here is an example showing the procedure on a machine with a K key.

Example 12.22: Find $\sqrt[5]{7}$.

Solution: Here, $A = 7$, $n = 5$; therefore, $n - 1 = 4$. Choose an initial guess—say, $G = 2$. Compute

$$G' = \{[(7 \div 4 \div 2 \div 2 \div 2 \div 2) + 2] \times 4\} \div 5$$

$$\underbrace{}_{4 \text{ times}}$$

7 ⊞ 4 ⊞ **[K ON]** 2 ⊟ ⊟ ⊟ ⊟ **[K OFF]** ⊞ 2 ⊟ ⊠ 4 ⊞ 5 ⊟ *1.6875*

The notation **[K ON]** means to turn on the K switch; **[K OFF]** means to turn off the K switch. BEWARE: On some calculators, the K switch must be turned off just *before* the fourth ⊟ operation. Consult your instruction manual.

We now have an estimate of 1.6875. Check:

$$(1.6875)^5 = 13.684183$$

This is not a very good estimate. Let's improve it by repeating these steps with $G = 1.6875$.

$$G' = 7 \boxplus 4 \boxplus \textbf{[K ON]} \; 1.6875 \; \boxeq \boxeq \boxeq \boxeq \; \textbf{[K OFF]} \; \boxplus \; 1.6875 \; \boxeq \boxtimes 4$$
$$\boxplus 5 \boxeq 1.522644$$

So our new estimate is 1.522644. Check:

$$(1.522644)^5 = 8.184505$$

This is still not sufficiently close to 7, though it is better than before. If we repeat these steps once more (details omitted), we obtain

$$\sqrt[5]{7} \doteq 1.4757836 \quad \text{and} \quad (1.4757836)^5 = 7.0002474$$

The preceding example illustrates how a very bad first guess will eventually "home" into the correct value. Here, we did that with just three applications of the algorithm. These algorithms have the happy property that they do converge quickly to the correct value. In the next example, we shall see what a good first guess can do for us!

Example 12.23: Find $\sqrt[10]{1,000}$.

Solution: Here, $A = 1,000$, $n = 10$; and so $n - 1 = 9$.
What shall we use as a first guess? We should recall from Chapter 7 on exponents, and from our work with the powers of 2, that $2^{10} = 1,024$, which is very close to 1,000. Thus, a first guess of 2 is not too bad for $\sqrt[10]{1000}$.

We want to compute

G'
$= \{[(1,000 \div 9 \div 2 \div 2 \div 2 \div 2 \div 2 \div 2 \div 2 \div 2 \div 2) + 2] \times 9\} \div 10.$

$G' = \boxed{1000} \div \boxed{9} \div [\textbf{K ON}]\ \boxed{2}\ \boxed{=}\boxed{=}\boxed{=}\boxed{=}\boxed{=}\boxed{=}\boxed{=}\boxed{=}\boxed{=}\ [\textbf{K OFF}]$
$\boxed{+}\ \boxed{2}\ \boxed{\times}\ \boxed{9}\ \boxed{\div}\ \boxed{10}\ \boxed{=}\ \boxed{1.9953124}$

Thus, the new estimate for $\sqrt[10]{1,000}$ is 1.9953124. Check:

$$(1.9953124)^{10} = 1,000.2509$$

Actually, our estimate is correct to three decimals on one run! Repeating this procedure using $G = 1.9953124$, we would get

$$\sqrt[10]{1,000} \doteq 1.9952623$$

Checking, $(1.9952623)^{10} = 999.99984$, which is very close to 1,000. Notice that this value (1.9952623) agrees with the first estimate (1.9953124) to three decimal places.

Exercises 12.6

1. Determine these roots.
   a. $\sqrt[4]{64}$ (Try 2.8)   •e. $\sqrt[5]{100}$
   •b. $\sqrt[5]{64}$ (Try 2.3)   •f. $\sqrt[10]{12,345,678}$ (Try 5)
   c. $\sqrt[6]{64}$ •g. $\sqrt[4]{28,561}$
   •d. $\sqrt[4]{123}$ (Try 3.3)   h. $\sqrt[5]{28,561}$ (Try 8)

2. •a. Find $\sqrt{64}$ and find *its* square root; that is, find $\sqrt{\sqrt{64}}$. Compare with $\sqrt[4]{64}$.

   •b. Find $\sqrt{100}$ and then find its square root; that is, find $\sqrt{\sqrt{100}}$. Compare with $\sqrt[4]{100}$.

   c. Prove in general, that $\sqrt{\sqrt{A}} = \sqrt[4]{A}$.

•3. a. Find $\sqrt{64}$ and then find its cube root; that is, find $\sqrt[3]{\sqrt{64}}$. Compare it with $\sqrt[6]{64}$.

   b. Find $\sqrt[3]{64}$ and then find its square root; that is, find $\sqrt{\sqrt[3]{64}}$. Compare it with $\sqrt[6]{64}$.

4. Examine the results from Exercises 2 and 3.

   a. What general laws are suggested by these problems?

   b. These laws are analogs of laws for exponents. Which ones?

12.7 Plotting Functions: Maxima and Minima

Closely allied to the graphical techniques described in Section 12.3 for finding square roots is the analysis of functions. This topic needs the tools of algebra and calculus for a satisfactory development. Yet, with the ease of calculation afforded us by our machines, we can satisfactorily approximate many of the exact techniques considered in these more advanced subjects.

Example 12.24: You are given a square piece of paper 12 inches on a side. Your task is to construct an open box from this sheet of paper by cutting out small squares from each corner and folding up the edges (Figure 12.14). Your challenge is to construct the box with the largest volume.

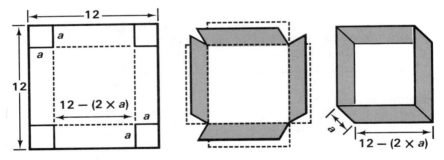

Figure 12.14

Solution: Let us suppose that, as suggested by the figure, you cut small squares a inches on each side from the sheet of paper. Then, you form a box. What, in terms of a, is the volume of this box? Then we shall have to see how the volume varies with different choices for a.

To find the volume of the box, we note that the bottom of the box is still in the shape of a square, of side $12 - (2 \times a)$ inches. This is because a inches have been cut from each end. The height of the box is a, because a side of width a has been turned up. Thus, the volume of the box is

$$V = a \times [12 - (2 \times a)]^2 = a \times [(2 \times 6) - (2 \times a)]^2$$
$$= a \times [2 \times (b - a)]^2 = 4 \times a \times (6 - a)^2.$$

Now we can make a table showing V for different values of a, the size of the cut-out square. Note that a can be at most 6 inches, in which case, the whole square paper has been cut away!

Here's the table, with calculations thanks to the trusty calculator.

a	$V = 4 \times a \times (6 - a)^2$
0	$0 = 6^2 \times 0 \times 4$
1	$100 = 5^2 \times 1 \times 4$
2	$128 = 4^2 \times 2 \times 4$
3	$108 = 3^2 \times 3 \times 4$
4	$64 = 2^2 \times 4 \times 4$
5	$20 = 1^2 \times 5 \times 4$
6	$0 = 0^2 \times 6 \times 4$

NOTE: In the display, we changed the last column to the equivalent form, $(6 - a)^2 \times a \times 4$. For machines with a squaring key (x^2), this form is more convenient.

As with the table of square roots, we can display the above information in graphical form (Figure 12.15).

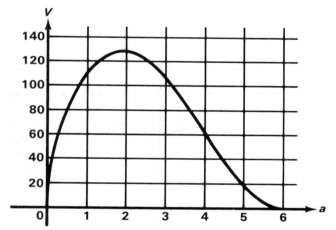

Figure 12.15
A graph of the volume of the box formed by cutting out corners from a square of side 12.

We should have been able to anticipate the general form of this graph. Certainly, if we don't cut out any corner, there is nothing to fold up, and so the volume is 0. This corresponds to the point $(0, 0)$.

Next, if we cut out corners of side 6, then there will be nothing left to fold up! Again, the volume will be 0. This is the point $(6, 0)$.

It is intuitive, then, that the volume of the box should increase from 0 to some value, and then decrease back to zero as corners of larger size are cut from the square.

Now, for what value(s) of a is the volume a maximum? As we can see from the table or from the graph, it appears that the volume is a maximum when $a = 2$, when the volume is 128 cubic inches.

This is, in fact, correct and we can test the reasonableness by including additional entries in our table: values of a just less than 2 and just more than 2.

a	V
1.9	127.756
2.1	127.764

We cannot prove with our techniques that the volume does what it appears to do; that is, that it increases to its maximum at $a = 2$, and then decreases. But that is what happens. Thus, in fact, 128 is the maximum volume possible.

The same technique for analysis will apply if the sheet of paper is rectangular instead of square. (In the Exercises, we ask you to try such a problem.)

Example 12.25: The old prospector, Des Aster, has just discovered Gwen Horne suffering from a snake bite miles from nowhere in the desert. Des locates the site of the ailing Gwen and radios to the nearest ranger aid station for help. (See Figure 12.16.)

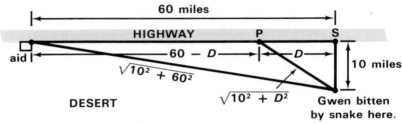

Figure 12.16

A jeep leaves the aid station for the accident site. The jeep can travel 60 miles per hour on the highway, but only 30 miles per hour over the desert. Where should the jeep leave the highway so as to reach Des and Gwen in the shortest possible time?

Solution: First, let's consider the two extreme possible routes. If the jeep comes by the shortest distance, all the driving would be in the desert. This distance, by ye old Pythagorean theorem, is

$$\sqrt{60^2 + 10^2} = \sqrt{3{,}700} = \sqrt{37} \times 10 \doteq 6.08 \times 10$$

$$= 60.8 \text{ miles, approximately.}$$

At 30 miles per hour, this would take a little longer than 2 hours:

$$t = 60.8 \div 30 = 2.0266666$$

If the jeep takes the longest reasonable route in terms of distance, by driving on the highway to point S (see map) and then the 10 miles to the accident site, the time would be 1 hour on the highway and $\frac{10}{30}$ ($= \frac{1}{3}$) hour on the desert. The total elapsed time would be $1\frac{1}{3}$ hour, or 1 hour and 20 minutes.

On the other hand, suppose the jeep drives on the highway to some point P (see map), and then heads directly to the accident site across the desert. How long would that take? Suppose that P is located D miles from S. Then, we have the following.

	Station to P + P to site
Distance (in miles)	$60 - D + \sqrt{10^2 + D^2}$
Time (in hours)	$\dfrac{60 - D}{60} + \dfrac{\sqrt{10^2 + D^2}}{30}$

Now let us compute a table of the times for various values of D. It is helpful to compute some intermediate values, and we show these in the table.

D	$\dfrac{60 - D}{60}$	$10^2 + D^2$ $= D^2 + 100$	$\sqrt{D^2 + 100}$	$\dfrac{\sqrt{D^2 + 100}}{30}$	Total time $=$ Col 2 + Col 5
1	$\frac{59}{60} = 0.983$	101	10.05	0.335	1.318
2	$\frac{58}{60} = 0.967$	104	10.20	0.340	1.307
3	$\frac{57}{60} = 0.950$	109	10.44	0.348	1.298
4	$\frac{56}{60} = 0.933$	116	10.77	0.359	1.292
5	$\frac{55}{60} = 0.917$	125	11.18	0.373	1.290
6	$\frac{54}{60} = 0.900$	136	11.66	0.389	1.289
7	$\frac{53}{60} = 0.883$	149	12.20	0.407	1.290
15	$\frac{45}{60} = 0.750$	325	18.03	0.601	1.351
30	$\frac{30}{60} = 0.500$	1000	31.62	1.054	1.554
60	$\frac{0}{60} = 0.000$	3700	60.83	2.028	2.027

Thus, from the table we can see that for $D = 6$ the minimum is achieved. Actually, the minimum occurs at $D = 5.77$, as determined by techniques from the calculus.

One of the interesting facts that come from the table is how little change is made by altering the distance D around the value 6. Thus, leaving the highway anywhere between 3 and 7 miles from point S (that is, anywhere from 53 to 57 miles from the aid station) will give a time close to the theoretical minimum possible—within 0.01 hour, or 0.6 minute.

Exercises 12.7

•1. What is the largest (in volume) open rectangular box that can be made from an $8\frac{1}{2} \times 11$ sheet of typing paper?

2. The map below shows the location of Dizzy World (DW), a lavish entertainment center to be built near Sumptamento. An expressway is to be built from the city to DW. The cost of building the expressway along an abandoned railway is only $250,000 per mile; otherwise, it is 1.5 million dollars per mile. What route should the expressway take?

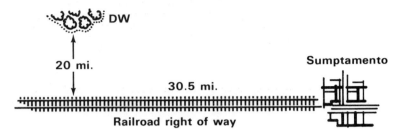

•3. What is the largest area of all rectangles that can be cut from a circular disk 11.5 inches in diameter?

HINT: Area $A = r \times s$.

Also, $(11.5)^2 = r^2 + s^2$

so $s = \sqrt{(11.5)^2 - r^2}$

and $A = r \times \sqrt{(11.5)^2 - r^2}$.

4. Two light bulbs are 50 feet apart. One bulb "puts out" 84 lumens, the other 260 lumens. (A "lumen" is a unit of intensity of light.) The intensity L of the light at any point D feet from the first bulb can be measured by the formula

$$L = \frac{84}{D^2} + \frac{260}{(50 - D)^2}$$

At what point is the intensity of the light the least?

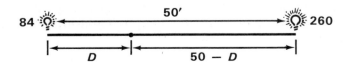

•5. A manufacturer wants to construct a cylindrical-shaped can (see figure) to contain $\frac{1}{2}$ litre ($= 500$ cubic centimetres) of soup.

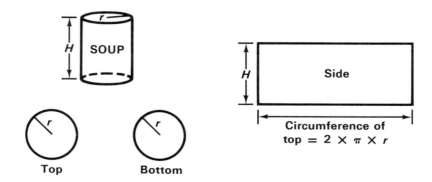

The can will be made from two circular pieces of alloy of radius r and a rectangular piece that will be rolled into a cylinder to fit the top and bottom pieces. What is the least amount of material that can be used to construct the can?

HINT: The volume of a cylinder is the height H times the area of the base, $\pi \times r^2$. Thus,

$$500 = \pi \times r^2 \times H. \qquad (1)$$

The area can be computed too:

$$\text{Area} = \text{Area of top} + \text{Area of bottom} + \text{Area of side}$$
$$= (\pi \times r^2) + (\pi \times r^2) + (2 \times \pi \times r \times H)$$
$$= (2 \times \pi \times r^2) + (2 \times \pi \times r \times H). \qquad (2)$$

From equation (1),

$$H = 500 \div (\pi \times r^2).$$

Using this in (2),

$$\text{Area} = (2 \times \pi \times r^2) + (2 \times \pi \times r) \times \left(\frac{500}{\pi \times r^2}\right)$$
$$= (2 \times \pi \times r^2) + \frac{1000}{r}$$

Plot the graph of the area of the can as a function of r. For what r will the area be a minimum? (ANS. $\doteq 4.3$ cm) What will the height of the can be then? (ANS. $\doteq 8.6$ cm) What is the minimum area? HINT: Choose these values for r: 1, 2, 3, 3.5, 4, 4.5, 5, 6, 8, 10. Then try $r = 4.1, 4.2, 4.3, 4.4.$

APPENDIX

A

"How Do You Think
of Numbers?"

Here are some answers provided by a class of elementary school teachers and teacher trainees.

1. The counting numbers look like a long line of soldiers marching over the hill in single file.

2. I need "things" to count on or with . . . like fingers, toes, candy bars, coins, etc., etc.

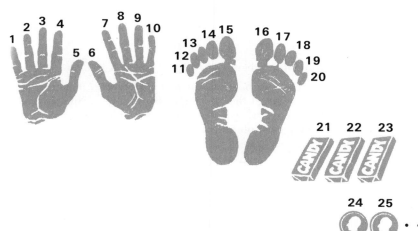

3. My visual image of the counting numbers is a picture of the symbols for the numbers (1   2   3   · · ·).

4. I remember when I was in elementary school, the long charts of the counting numbers stretched over the top of the blackboard. This has had a large effect on how I picture numbers. This is how I picture them:

5. To express my feelings on this problem, I would need to physically start writing the numbers consecutively starting with 1 *forever* since the set 1, 2, 3, 4, 5, 6, 7, 8, 9, 10, 11, 12, and so on is infinite.

6. When I think of the counting numbers, I picture the numbers in sequence written. I remember when I was in elementary school, the numbers on the book is [*sic*]

7. I see the counting numbers as the usual pictures, but set on a counting pattern. This is probably a throw-back to my elementary school days. For example:

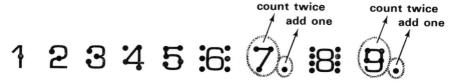

8. I think of numbers as numerical symbols.

9. The counting of numbers gives a quantity or a goal to reach.

10. I think numbers are fascinating and interesting once you have learned how to use all the different theories. I think of all kinds of numbers thrown together in the form of a problem, and some place, somehow, there is a solution.

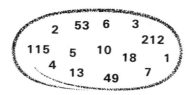

11. My feeling of numbers is a jumbled page with an enormous amount of number symbols.

12. . . . and going on in circles as

```
           33  34  35  36
      32                  37
   31    16   17  18    19   38
  30   15   6  7   8      20   39
  29   14   5  1         21   40
           4  3  2  9
  28   13        10    22   41
      27   12  11      23   42
        26  25  24        43
    · ·                44
      48  47  46  45
```

Numbers such as this are great fun for playing with my new calculator.

13. When someone speaks of numbers, I begin to think of wholeness. Sometimes the number *eight* gives me a picture of my children. The number *one* makes me think of my adopted daughter. The number *four* distinguishes my girls from my boys. When I am driving, I think of my car license or my house number.

14. That string of numbers refuses to become much of anything except a group of symbols used by the large pageful to find the right answers. They go off into space on a long string, perhaps like the string on a kite. I like to count and usually know the number of students in a class, and various subgroups. But that's objects first and numbers afterwards.

15. . . . apples and bananas that we had to count in first grade.

16. I think of: house numbers, zip codes, telephone numbers, distances to places, prices, mileage on a car, time of day, and bus schedules. It is an infinite field of thought and vision even within a few minutes.

17. When I think of numbers, I see the arabic numerals. I also think 2 + 2, for some reason.

18. I picture all consecutive numbers (principally 1–10, say) as a series of independent, finite units, each consecutive one a greater cumulative larger unit. As the numbers grow larger, I tend to "visualize" them less as identities; they become abstractions, or mass numbers without any tangible definition. I have long tended to visualize numbers with almost specific color property; i.e., 1 = black, 2 = blue with dark stripes, 3 = white, 4 = brown, 5 = whitish-yellow, 6 = blue, 7 = dark red, 8 = yellow, 9 = orange, 10 = bright red.

   0–even numbers = "good", with evenness, symmetry.

   –odd = "bad", discordant, uneven, improper.

19. A number is a very abstract thing—a word given to call certain elements. One thing that I find very interesting is that in all countries they look the same. They are easy to form their shape and don't change regardless of the languages or the country, even though in each language they are called differently. The use of the computer, if you are familiar with one, will be very helpful. It will be far easier for people to deal with arithmetic, since sometimes our mind can slip and make easy mistakes. With my experience in elementary school, I had a psychological problem or mental block. I always thought I would never get the right answer; i.e.,

$$
\begin{array}{r}
1\ 2\ 5 \\
9\ 3\ 7 \\
\hline
1\ 0\ 6\ 2
\end{array}
$$

My answer would have been either 1152, or 962, or 162, or 173. It was difficult to make my mind convince itself that I should have more numbers or less at the time of giving the answer.

When I see 2, I think of $\frac{3}{4}$ of a bird. 3 makes me think of two $\frac{1}{2}$'s of 0. 4 is a chair. 1 is a man. 0 as a playing ball, 8 as 2 balls.

20. Numbers are useful tools—necessary for dividing everything: my share, your share, our share. I still see them as figure people because that's the way I was taught to share numbers with children:

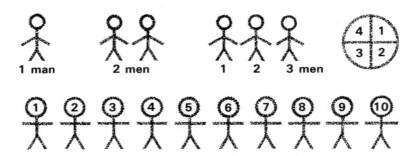

Numbers tell how to pay bills—amazing how carrying one can raise the total or lower the money you have to work with.

APPENDIX

B

California-Nevada Driving Distances

MILES: *210*   AVERAGE TIME (EXCLUDING STOPS): 4:10

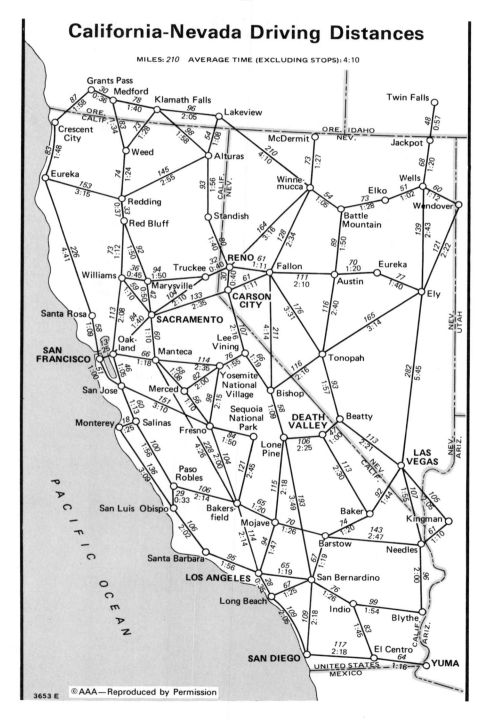

APPENDIX
C

Answers to Exercises Marked with • in Text

Exercises 1.3 (p. 11)

1. a. *26.* b. *306.* c. *780.* d. *0.9341* f. *6.032*
2. Most machines simply stop registering further input after *12345678.*
4. a. *1.5* or *1.50* b. *.55* or *0.55* e. *1000.01*
5. b. $987,654.32

Exercises 1.4 (p. 14)

4. *2.1763668*
8. *121.*
9. *1331.*
10. *14641.*
11. *1030301.*
12. *104060401.*

Exercises 1.5 (p. 16)

1. d. *117.* f. *51.* g. *72956.*
2. a. $1 + 2 = 3$ b. $1 + 2 + 3 = 6$ c. $1 + 2 + 3 + 4 = 10$
   d. $1 + 2 + 3 + 4 + 5 = 15$

pattern:	3	6	10	15
differences	3	4	5	

3. a. 3 b. 6 c. 10 d. 15
   Compare with corresponding parts in Exercise 2.

Exercises 2.2 (p. 21)

1. a.

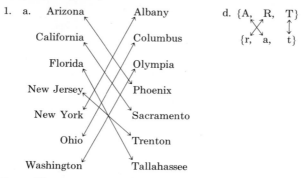

3. a. There are more students than possible grades.
   b. There may be more students than major subject areas.
   c. There may be more than one student sharing a particular birthdate.
6. For example, the sets of states of the United States and their capital cities (Exercise 1), or the states and their governors. Many other examples are possible.

Exercises 2.3 (p. 27)

1. b. 4027. c. 153.
2.

3. 004 013 022 031 040 103 112 121 130 202 221 220 301 310 400
6. 00 01 02 03 10 11 12 20 21 30

Exercises 2.4 (p. 33)

1. Sample answers (many others are possible):
   b. 78 < 79 < 80 c. 100 < 101 < 395 f. 1234 < 3099 < 4321
2. Compare the second digits, where the numerals begin to differ:

$$5 < 6; \text{ so } 1563 < 1673.$$

3. a. "1" is in the thousands place, but "9" is in the tens place.
   b. "0" is compared with the "1".
4. b. > d. < f. >
5. 479, 497, 749, 794, 947, 974
7. This problem corresponds to the total number of beans in all the pots.
12. a. 19 b. 109

2. 38, 80, 102, 200, 2947, 3100
3. a. 0, 3, 6, 9, 12, 15, 18, 21
   These are multiples of 3.
   b. By repeated addition of 3.
   c. 5, 10, 15, 20, 25, 30, 35, 40
7.

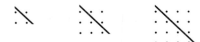

9. 1, 4, 9, 16, 25, 36, . . .
   These are squares of counting numbers.

Exercises 2.6 (p. 41)

1. a. *10.* b. *100.* c. *10000.*
2. a. *○10000000* or *○1.0000000*, etc.
   b. 100000000
6. 99999999 + 1 (see Exercise 2a). The shepherd needs another pot.

Exercises 2.7 (p. 47)

1. A: 3.2   B: 2.5   C: 1.1   D: 0.3 or .3
2.

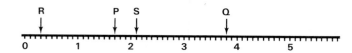

3. A. *2.5*   C. *1.25*   F. *0.75*

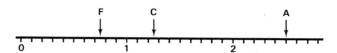

5. a. 163.753 > 61.3753   (The integer part 163 > 61)
   b. 69.4214 > 69.4124   (The integer parts are equal; the first place the
   digits differ is in the hundredths: 2 > 1.)
6. .07532468, 2.5347, 253.47, 253.74
8. b. Same size.
9. c. 9.00010 < 9.0010; the numerals differ in the thousandths place, 0 < 1.
13. a. *0.1111111*   c. *0.3333333*   f. *0.6666666*   i. *1.* = 1.0000000
14. b. $71925\frac{3}{9}$ d. $9504\frac{3}{9}$ f. 1167   h. $817557\frac{6}{9}$
16. a. $58738\frac{1}{9}$ ($28 \div 9 = 3\frac{1}{9}$)   c. $19295\frac{3}{9}$ ($30 \div 9 = 3\frac{3}{9}$)   f. $1173\frac{6}{9}$ ($15 \div 9 = 1\frac{6}{9}$)
    h. $1039845\frac{8}{9}$ ($35 \div 9 = 3\frac{8}{9}$)
    Remainders for each problem are equal.
18. a. *0.0769230* = 0.0769230   b. *0.1538461*   c. *0.2307692*
    d. *0.3076923*   e. *0.3846153*   f. *0.4615384*

Exercises 3.2 (p. 57)

1. a. 158 c. 989 e. 1,433 h. 2,088 j. 465,264
2. 131,213 The second and third digits: $3 + 1 = 4$
   The fourth and fifth digits: $2 + 1 = 3$
3. 200718; compare with Exercise 1h.
4.   285 ones added to ones: $5 + 9 = 14$; (carry "1" ten).
   +579 tens added to tens: $8 + 7 + (1) = 16$; [carry "1" hundred;
   notice (1) ten carried from above].
   hundreds added to hundreds: $2 + 5 + (1) = 8$
   $800 + 60 + 4 = 864$
   Numerals indicated by (1) correspond to beans in upper shelf for the "carries."
6. a. 5 b. 8 f. 18
7. a. 5 c. 1 (in "12"), 6 (in "36")
8. 44, 296, 264
10. San Diego to Long Beach   109
    Long Beach to Los Angeles   28
    Los Angeles to Bakersfield   114
    Bakersfield to Los Angeles   114
    Los Angeles to San Bernardino   65
    San Bernardino to San Diego   109
    539 miles
12. Bonnie $11.85; Cindy $12.03
    Bonnie spent less.
    Items costing more than $1.00: cheese, bacon, coffee.
    Items costing less than 50¢: paper towels (Bonnie).

Exercises 3.3 (p. 62)

1. $22.70
4. 812.56
6. For example,   7.0
                   0.7
                   0.07
                   0.007
                  +0.0007
                   7.7777
7. a. A: 5.7 cm B: 9.7 cm C: 8.8 cm D: 7.7 cm
   b. triangle C

Exercises 3.4 (p. 65)

1. a. 165391189
2. a. 103984921395149692
3. a. 12,643,364.64 acres
   b. Everglades   1,258,361.00
      Mt. McKinley   1,939,319.04
      Yellowstone   2,213,206.55
      Glacier   999,015.15
      Olympic   887,986.91
      7,297,888.65 acres

Exercises 3.5 (p. 69)

1. b. *8.* d. *2395.* f. *176004.*
2. a. *4.* b. *15 ⊟ 11 ⊜ 4.*
3. a. $35 profit
5. b. $538.90   c. $269.45
6. The total spent was $486, so each person's share should be $97.20. A procedure to settle the accounts might be as follows, where the one who paid out the least gives as much as he owes to the one who paid out the most and so forth: Frankie pays $84.20 to Tony; Mary pays $56.03 to Tony; and Billie pays $7.82 to Jim and $9.56 to Tony.

Exercises 3.6 (p. 74)

1. b. 105   c. 16.9   e. 3.519
3. a. 18¢   b. coffee, 21¢

Exercises 3.7 (p. 77)

1. a.
```
   65 74
 − 31 23
 ─────────
   34 51
```
   74 ⊟ 23 ⊜ 51.
   65 ⊟ 31 ⊜ 34.
   6574 ⊟ 3123 ⊜ 3451.

   d. "borrow" a "1" from the "5".
```
      4 1
   65̸ 12
 −  7 58
 ─────────
   57 54
```
   112 ⊟ 58 ⊜ 54.
   64 ⊟ 7 ⊜ 57.
   6512 ⊟ 758 ⊜ 5754.

2. a.
```
   9876 54367
 − 2531 42055
 ───────────────
   7345 12312
```
   c.
```
           2
   3463̸ 175482
 − 1789  96344
 ───────────────
   1673  79138
```

3. a.
```
            1
   86432̸ 189756
 − 54329  57889
 ────────────────
   32103  31867
```

4.
```
         1
   17932̸ 13175
 − 15132  5798
 ───────────────
    2799  7377
```

Exercises 4.1 (p. 83)

1.

x	1	2	3	4	5	6	7	8	9
1	1	2	3	4	5	6	7	8	9
2	2	4	6	8	10	12	14	16	18
3	3	6	9	12	15	18	21	24	27
4	4	8	12	16	20	24	28	32	36
5	5	10	15	20	25	30	35	40	45
6	6	12	18	24	30	36	42	48	54
7	7	14	21	28	35	42	49	56	63
8	8	16	24	32	40	48	56	64	72
9	9	18	27	36	45	54	63	72	81

2. b. $16 \boxed{\times} 3 \boxed{=} 48.$ d. $8 \times (16 \boxed{\times} 14) = 8 \boxed{\times} 224 \boxed{=} 1792.$
    e. $(8 \boxed{\times} 16) \times 14 = 128 \boxed{\times} 14 \boxed{=} 1792.$ g. $87654321 \boxed{\times} 0 \boxed{=} 0.$
3. Overflow!
4. $L = 9,999$: $9999 \boxed{\times} 9999 \boxed{=} 99980001.$
7. One day: 43,750 fish. One week: 306,250 fish. One year: 15,968,750 fish.
9. a. The two sides are symmetrical, so there is an even number in the two sides.
       The central section has an even number (10) of rows, so there is an even
       number of seats in the central section. Altogether, an even number plus an
       even number gives an even number of seats.
    b. $13 \times 10 = 130$ seats   c. 32   d. $130 + 32 + 32 = 194$
    e. Even, as predicted
11. a. $1760 \times 3 = 5280$ feet   b. $5280 \times 12 = 63,360$ inches
12. a. $100 \times 1000 = 100,000$ centimetres
    b. $1000 \times 1000 = 1,000,000$ millimetres
15. $135 \times 0.45359 = 61.23465$ kilograms
17. $1250 \times 50 = 62,500$ cubic feet. This would be a room, for example, 25 feet high
    with floor space of 50 feet by 50 feet.

Exercises 4.2 (p. 86)

1. a. 8, 8 b. 84, 8.4 c. 128, 12.8 d. 396, 39.6 e. 852, 8.52 f. 4952, 49.52
    g. 39996, 399.96 h. 8548, 8.548
2. a. 86, 8.6 b. 903, 9.03 c. 1376, 13.76 d. 4257, 42.57 e. 9159, 9.159
    f. 53234, 53.234 g. 429957, 429.957
    Answers in both parts are the same except for location of the decimal point.
3. a. 912, 9.12 c. 14592, 14.592 e. 97128, 9.7128 g. 4559544, 455.9544
    i. 3653472, 36.53472 k. 9626616, 9.626616 m. 199,178,064, 199.178064
4. a. 5 places c. 3 places

Exercises 4.3 (p. 92)

1. a. $6 \boxed{+} 8 \boxed{=} 14;$ $2 \boxed{\times} 7 \boxed{=} 14.$
    c. $54272 \boxed{+} 37888 \boxed{=} 92160;$ $1024 \boxed{\times} 90 \boxed{=} 92160.$
    e. $40 \boxed{+} 48 \boxed{+} 64 \boxed{=} 152;$ $8 \boxed{\times} 19 \boxed{=} 152.$

3.

Lodge	Number of Trees Sold	Receipts
Red Lodge	120	$420
Blue Lodge	150	525
Gold Lodge	80	280
Silver Lodge	230	805
Total	580	$2030

Exercises 4.4 (p. 98)

1. a. $12 \boxed{\times} 4 \boxed{=} 48.$ b. $4 \boxed{\times} 12 \boxed{=} 48.$ e. $38 \boxed{\times} 13 \boxed{=} 494.$
2. a. 60
4. 322 miles

Exercises 4.5 (p. 104)

1. a. 192,931,893 d. 9,876,543,210
2. a. $60 \times 60 \times 24 \times 365 = 31,536,000$
3. $31,536,000 \times 186,284 = 5,874,652,224,000$ miles

1. a. 12 e. 4 i. 3
    c. 9861 g. 57 k. 173
2. b. 3, 69
3. b. $11 \times 8 \neq 81$ d. $6.3 \times 4.7 = 29.61$
4. e. 200
5. c. 143, 1013
7. d. 3 e. 2149 g. 796
8. a. 6 c. 42 e. 111 g. 5332114
9. a. 137,500
10. Second car takes 1 hour more.
12. a. Alaska   0.3890979
       Arizona   11.602130
       California   99.955404
       Hawaii   99.441277
       Idaho   8.0380698
       Montana   4.6148786
       New Mexico   7.8344566
       Oregon   18.275136
       Utah   10.608861
       Washington   41.895868
       Wyoming   3.3859203
    c. Least dense: Alaska
14. a. Zero may be multiplied by any number and may be divided by any
       number other than zero.
    b. $(0 \; \boxed{\times} \; 1234) \div 1234 = 0 \; \boxed{\div} \; 1234 \; \boxed{=} \; 0.$

Exercises 4.7 (p. 111)

1. $\frac{1}{2} = 0.5$ $\frac{2}{2} = 1.$ $\frac{3}{2} = 1.5$ $\frac{4}{2} = 2.$ $\frac{5}{2} = 2.5$ $\frac{6}{2} = 3.$
    $\frac{1}{4} = 0.25$ $\frac{2}{4} = 0.5$ $\frac{3}{4} = 0.75$ $\frac{4}{4} = 1.$ $\frac{5}{4} = 1.25$ $\frac{6}{4} = 1.5$ $\frac{7}{4} = 1.75$ $\frac{8}{4} = 2.$
    $\frac{1}{8} = 0.125$ $\frac{2}{8} = 0.25$ $\frac{3}{8} = 0.375$ $\frac{4}{8} = 0.5$ $\frac{5}{8} = 0.625$ $\frac{6}{8} = 0.75$ $\frac{7}{8} = 0.875$
    $\frac{8}{8} = 1.$ $\frac{9}{8} = 1.125$ $\frac{10}{8} = 1.25$
2. $\frac{7}{2} = 3.5$ $\frac{9}{4} = 2.25$ $\frac{11}{8} = 1.375$ $\frac{8}{2} = 4.$ $\frac{10}{4} = 2.5$

Exercises 4.8 (p. 116)

1. a. $<$ d. $<$ e. $<$ f. $<$ h. $<$ j. $>$ l. $>$ o. $>$
    q. $(1234 \div 47) < (1240 \div 47)$
              $(1240 \div 47) < (1240 \div 36)$; so
    $(1234 \div 47) < (1240 \div 36)$
3. $(5 \times 5) < 28.7296 < (6 \times 6)$; so the number must be between 5 and 6. If
    there is an exact 2-place decimal, the last place must be either 4 or 6 because
    $4 \times 4 = 16$ and $6 \times 6 = 36$, the last digit agreeing with the last digit in
    28.7296. Using the calculator, we can then quickly try and notice
        $5.04 \; \boxed{\times} \; 5.04 \; \boxed{=} \; 25.4016 < 28.7296$
        $5.14 \; \boxed{\times} \; 5.14 \; \boxed{=} \; 26.4196 < 28.7296$
        $5.24 \; \boxed{\times} \; 5.24 \; \boxed{=} \; 27.4576 < 28.7296$
        $5.34 \; \boxed{\times} \; 5.34 \; \boxed{=} \; 28.5156 < 28.7296$
        $5.44 \; \boxed{\times} \; 5.44 \; \boxed{=} \; 29.5936 > 28.7296$
    Now try
        $5.06 \; \boxed{\times} \; 5.06 \; \boxed{=} \; 25.6036 < 28.7296$
        $5.16 \; \boxed{\times} \; 5.16 \; \boxed{=} \; 26.6256 < 28.7296$
        $5.26 \; \boxed{\times} \; 5.26 \; \boxed{=} \; 27.6676 < 28.7296$
        $5.36 \; \boxed{\times} \; 5.36 \; \boxed{=} \; 28.7296 = 28.7296$ BINGO!

Exercises 4.9 (p. 120)

1. a. 13.12 c. 27.02025 d. 278.775 f. 3.7221 g. 0.06 j. 8.37000
   k. 0.2697525332
   Notice that most calculators drop the three zeros in part j.

Exercises 4.10 (p. 122)

1. b.
   $$\begin{array}{r} 27 \\ -\ 9 \\ \hline 18 \ \checkmark \\ -\ 9 \\ \hline 9 \ \checkmark \\ -\ 9 \\ \hline 0 \ \checkmark \end{array}$$
   $27 \div 9 = 3$

   d.
   $$\begin{array}{r} 378 \\ -126 \\ \hline 252 \ \checkmark \\ -126 \\ \hline 126 \ \checkmark \\ -126 \\ \hline 0 \ \checkmark \end{array}$$
   $378 \div 126 = 3$

   f.
   $$\begin{array}{r} 1436889 \\ -\ 478963 \\ \hline 957926 \ \checkmark \\ -\ 478963 \\ \hline 478963 \ \checkmark \\ -\ 478963 \\ \hline 0 \ \checkmark \end{array}$$
   $1436889 \div 478963 = 3$

Exercises 4.11 (p. 124)

1. b. $95 \div 5 = (100 - 5) \div 5 = (100 \div 5) - (5 \div 5) = 20 - 1 = 19$
   d. $147 \div 7 = (140 + 7) \div 7 = (140 \div 7) + (7 \div 7) = 20 + 1 = 21$
   e. $117 \div 13 = (130 - 13) \div 13 = (130 \div 13) - (13 \div 13) = 10 - 1 = 9$
4. $120 \div (3 + 5) = 120 \div 8 = 15$; $(120 \div 3) + (120 \div 5) = 40 + 24 = 64$;
   So $120 \div (3 + 5) \neq (120 \div 3) + (120 \div 5)$
5. $6 \div (3 + 2) = 6 \div 5 = 1.2$; $(6 \div 3) + (6 \div 2) = 2 + 3 = 5$;
   So $6 \div (3 + 2) \neq (6 \div 3) + (6 \div 2)$

Exercises 4.12 (p. 134)

2. $23 = (7 \times 3) + 2$; Colonel gets two hotdogs.
4. a.
   $$\begin{array}{r} 30 \\ 6{\overline{)483}} \\ 480 \\ \hline \text{3 Remainder} \end{array}$$
   c.
   $$\begin{array}{r} 1 \\ 274{\overline{)483}} \\ 274 \\ \hline \text{209 Remainder} \end{array}$$
5. a. $23 = (4 \times 5) + 3$; $Q = 5$, $R = 3$ c. $94 = (17 \times 5) + 9$; $Q = 5$, $R = 9$
7. a. $9 \, 1 \, 3 \, 5 \, 1 \, 7 \, 8 \, 5 \ \boxplus \ 8 \, 1 \, 3 \, 1 \ \boxminus \ 1 \, 1 \, 2 \, 3 \, 5.$ c. $9 \, 1 \, 3 \, 5 \, 1 \, 7 \, 9 \, 0 \ \boxplus \ 8 \, 1 \, 3 \, 1 \ \boxminus \ 1 \, 1 \, 2 \, 3 \, 5.$
8. 91351786, 91351788, 91351789, 91351791, 91351792

Exercises 4.13 (p. 136)

1. a. 101,779,479; remainder 2 c. 100,046; remainder 25
2. a. 11.474956

Exercises 5.1 (p. 144)

1. Areas: $1.9 \times 7.7 = 14.63$; $2.9 \times 6.7 = 19.43$; $3.8 \times 5.8 = 22.04$;
   $4.8 \times 4.8 = 23.04$
   Perimeters: $2 \times (1.9 + 7.7) = 19.2$; $2 \times (2.9 + 6.7) = 19.2$;
   $2 \times (3.8 + 5.8) = 19.2$; $4 \times 4.8 = 19.2$
3. Illustrations are shown reduced here.

b.

11.72 cm   3.73 cm

c.

$5\frac{1}{8}$ in. $4\frac{7}{8}$ in.

5. Area of outer rectangle: $32 \times (30 + 12) = 32 \times 42 = 1344$
Area of inner rectangle: $(32 - 12) \times 30 = 20 \times 30 = 60$
Area of walk: $1344 - 600 = 744$ sq. ft.

Exercises 5.2 (p. 150)

2. a. $\frac{1}{2} \times 3 \times 4 = 6$ sq. in.   b. $\frac{1}{2} \times (4 + 9) \times 12 = 78$ sq. ft.
   d. $(2 \times \frac{1}{2} \times 3 \times 8) + (2 \times \frac{1}{2} \times 8 \times 6) = 24 + 48 = 72$ sq. ft.
   e. $7 \times 24 = 168$ sq. mi.
3. Figures are shown reduced here.

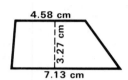

   b. 8.734375 sq. in.   c. 19.14585 sq. cm.
5. a. $1500 \div 40 = 37.5$ feet
   b. $40 \times 120 = 4800$; $1500 \div 4800 = 0.3125 < 0.65$;
      Mr. Anlife would not be violating the code.
6. $\frac{1}{2} \times (b + b) \times h = \frac{1}{2} \times 2 \times b \times h = b \times h$
   Because the two bases are equal, it is clear that the figure is simply a
   parallelogram whose area is equal to base $\times$ altitude.

Exercises 5.3 (p. 161)

3. a. Area of large semicircle: $\frac{1}{2} \times 4 \times 4 \times 355 \div 113 \doteq 25.132743$
      Area of small semicircles: $2 \times \frac{1}{2} \times 2 \times 2 \times 355 \div 113 \doteq 12.566371$
      Area of figure: $25.132743 - 12.566371 = 12.566372$
      Actually, as can be seen from the full expressions, the area of the small
      semicircles is half the area of the large semicircle.
   b. Large semicircle: $\frac{1}{2} \times 11 \times 11 \times 355 \div 113 \doteq 190.06637$
      Small semicircles: $2 \times \frac{1}{2} \times \frac{11}{2} \times \frac{11}{2} \times 355 \div 113 \doteq 95.033185$
      Figure: $190.06637 - 95.033185 = 95.033185$
   c. The area cut away is equal to the area that is left.
   d. The area of the "cashew nut" is equal to the area of the large
      semicircle (the convex portion can be cut off and fit into the concave
      portion). If the dimensions are those given in part a, the area of the
      "cashew nut" is about 25.132743 sq. cm.
4. a. The rope restricts the grazing to three-quarters of a circle:
      $\frac{3}{4} \times 15 \times 15 \times 355 \div 113 \doteq 530.1438$

Exercises 5.4 (p. 168)

1. $(2 \times 2.3 \times 3.4) + (2 \times 2.3 \times 4.5) + (2 \times 3.4 \times 4.5) = 66.94$ sq. cm.
2. a. $66.94 - (3.4 \times 4.5) = 51.64$ sq. cm.   c. $66.94 - (2.3 \times 3.4) = 59.12$ sq. cm.
5. See calculations in Example 5.16: $2200 + 960 + 216 = 3376$ sq. ft.
8. Area with 6 ft. radius: $4 \times 6 \times 6 \times 355 \div 113 \doteq 452.38938$ sq. ft.
   Area with 7 ft. radius: $4 \times 7 \times 7 \times 355 \div 113 \doteq 615.75221$ sq. ft.
   Increase in area: $615.75221 - 452.38938 = 163.36283$ sq. ft.

Exercises 5.5 (p. 176)

1. a. 2005.4408 cm³   b. 27 × 13.3 × 12 = 4309.2 cu. in.
   e. 2.32 × 2.32 × 6.51 × 355 ÷ 113 ≐ 110.0796 cu. in.
4. $\frac{4}{3}$ × 4000 × 4000 × 4000 × 355 ÷ 113
$$= \tfrac{4}{3} \times 4 \times 4 \times 4 \times 1000 \times 1000 \times 1000 \times 355 \div 113$$
$$≐ 268.08257 \times 1{,}000{,}000{,}000 = 268{,}082{,}570{,}000 \text{ cu. mi.}$$
5. a. $\frac{4}{3}$ × 432000 × 432000 × 432000 × 355 ÷ 113
$$≐ 337{,}706{,}863{,}000{,}000{,}000 \text{ cu. mi.}$$
7. a. Volume with 7 ft. radius: $\frac{4}{3}$ × 7 × 7 × 7 × 355 ÷ 113 ≐ 1436.7551 cu. ft.
   Volume with 6 ft. radius (from Example 5.24): 904.7787 cu. ft.
   Increase in volume: 1436.7551 − 904.7787 = 531.9764 cu. ft.
   About 531.98 cu. ft. of gas will be needed.
   b. 531.9764 ÷ 904.7787 ≐ 0.587963
11. a. 0.0012644   b. 0.0012644 ÷ 3.1415927 ≐ 0.0004024
12. Volume = $\frac{107}{16}$ × 2 × 2 × 355 ÷ 113 ≐ 84.03761 cu. in.
$$= 84.03761 \times 2.54 \times 2.54 \times 2.54 ≐ 1377.1926 \text{ cm}^3$$
   46 fluid ounces = 46 × 30 = 1380 cm³
   The can's capacity is a bit short of 46 fluid ounces (about 2.87 cm³ short).

Exercises 6.1 (p. 182)

1.

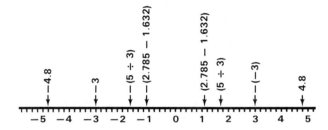

2. b. −87 < −78   c. −3 < 2   e. −(5 ÷ 2) < −(3 ÷ 2)
4. 65° + 30° = 95°

Exercises 6.2 (p. 185)

1. a. A = 6.54, −A = −6.54, A* = −6.54
   c. A = −0.37, −A = 0.37, A* = 0.37
2. a. A = 2.5, −A = −2.5
3. a. −3.05   b. 3.05   c. 9859.7461   d. −9859.7461
5. a. Time for first lap: 63 sec.
   Time for second lap: 129.5 − 63 = 66.5 sec.
   Time for third lap: 199.5 − 129.5 = 70. sec.
   Time for fourth lap: 264 − 199.5 = 64.5 sec.
   c. He slowed down on the second lap and still more on the third lap.

Exercises 6.3 (p. 190)

1. a. −360, −360   c. −427192, −427192   e. −42, 42   g. −56, 56
   j. −44, −44   k. −8.8, −8.8   n. −35.2, −35.2
2. a. −   c. +   e. −   g. −
3. a. 6 ÷ 3 = 2 = 4 − 2   c. 11 × 5 = 55 = 88 − 33
   e. 222.64 − 220.80 = 1.84 = 18.4 × 0.1
4. −3° F. or 3° F. below zero

5. Part a describes the result
7. a. Player 1: $[(175 - 147) \times 3] \div 4 = 21$
   Player 2: $[(175 - 119) \times 3] \div 4 = 42$
   Player 3: $[(175 - 171) \times 3] \div 4 = 3$
   Player 4: $[(175 - 175) \times 3] \div 4 = 0$
   Total handicap: $20 + 42 + 3 + 0 = 66$
   b. 678   c. 175

Exercises 7.1 (p. 196)

1. b. 97.65625
2. b. 18; 567; 37,116
3. $2^{19} = 524,288$

Exercises 7.2 (p. 198)

1. a. The number of zeros agrees exactly with the exponent.
3. In each case, 1.
4. $2^1 = 2$ $2^6 = 64$ $2^{11} = 2,048$ $2^{16} = 65,536$
   $2^2 = 4$ $2^7 = 128$ $2^{12} = 4,096$ $2^{17} = 131,072$
   $2^3 = 8$ $2^8 = 256$ $2^{13} = 8,192$ $2^{18} = 262,144$
   $2^4 = 16$ $2^9 = 512$ $2^{14} = 16,384$ $2^{19} = 524,288$
   $2^5 = 32$ $2^{10} = 1,024$ $2^{15} = 32,768$ $2^{20} = 1,048,576$
5. $3^1 = 3$ $3^5 = 243$ $3^9 = 19,683$ $3^{13} = 1,594,323$
   $3^2 = 9$ $3^6 = 729$ $3^{10} = 59,049$ $3^{14} = 4,782,969$
   $3^3 = 27$ $3^7 = 2,187$ $3^{11} = 177,147$ $3^{15} = 14,348,907$
   $3^4 = 81$ $3^8 = 6,561$ $3^{12} = 531,441$
7. $3^2 = 9 > 8 = 2^3$; $3^4 = 81 > 64 = 4^3$; $2^5 = 32 > 25 = 5^2$
8. a. base is 2, exponent is 10   b. base is 5, exponent is 3
   c. base is 7, exponent is 1

12.

N	N^2	N	N^2	N	N^2
1	1	12	144	23	529
2	4	13	169	24	576
3	9	14	196	25	625
4	16	15	225	26	676
5	25	16	256	27	729
6	36	17	289	28	784
7	49	18	324	29	841
8	64	19	361	30	900
9	81	20	400	31	961
10	100	21	441	32	1,024
11	121	22	484	33	1,089

13.

N	N^3	N	N^3	N	N^3
1	1	12	1,728	23	12,167
2	8	13	2,197	24	13,824
3	27	14	2,744	25	15,625
4	64	15	3,375	26	17,576
5	125	16	4,096	27	19,683
6	216	17	4,913	28	21,952
7	343	18	5,832	29	24,389
8	512	19	6,859	30	27,000
9	729	20	8,000	31	29,791
10	1,000	21	9,261	32	32,768
11	1,331	22	10,648	33	35,937

Exercises 7.3 (p. 208)

1. Same as Exercise 5 of Exercises 7.2, plus $3^{16} = 43,046,721$
2. $81 \times 243 = 3^4 \times 3^5 = 3^9 = 19,683$
   $27 \times 6561 = 3^3 \times 3^8 = 3^{11} = 177,147$
   $729^2 = 3^6 \times 3^6 = 3^{12} = 531,441$
6. a. $(2 \times 3)^2 = 6^2 = 36$; $2^2 \times 3^2 = 4 \times 9 = 36$ c. $(5 \times 5)^4 = 25^4 = 390,625$;
   $5^4 \times 5^4 = 625 \times 625 = 390,625$
7. For example, let $A = 2$, $B = 3$, $N = 3$, $M = 2$. Then
   $A^N \times B^M = 2^3 \times 3^2 = 8 \times 9 = 72$; $(A \times B)^{N+M} = (2 \times 3)^{3+2} = 6^5 = 7776 \neq 72$
9. b. $(6 \div 3)^4 = 2^4 = 16$; $6^4 \div 3^4 = 1296 \div 81 = 16$ d. $(12 \div 3)^3 = 4^3 = 64$;
   $12^3 \div 3^3 = 1728 \div 27 = 64$
10. a. $12^5 \div 6^5 = (12 \div 6)^5 = 2^5 = 32$ b. $8^7 \div 4^7 = (8 \div 4)^7 = 2^7 = 128$

Exercises 7.4 (p. 214)

1. a. $2^{63} = 2^{60} \times 2^3 = 8 \times (2^{10})^6 \doteq 8 \times (10^3)^6 = 8 \times 10^{18}$
   $= 8,000,000,000,000,000,000$
2. 1.609344×10^5

Exercises 7.5 (p. 218)

1. b. 2 d. 4 e. $\frac{1}{6}$
2. a. Yes c. Yes d. No
3. b. $<$ c. $<$ d. $>$ f. $=$ h. $=$
4. a. $350 \div 1087 \doteq 0.3219871$ second (less than $\frac{1}{3}$ second)
   b. $50 \div 1087 \doteq 0.0459981$ second (less than $\frac{1}{20}$ second)
   c. $20 \div 186,284 \doteq 0.0001073$ second
   $(50 + 10) \div 1087 \doteq \underline{0.0551977 \text{ second}}$
   0.0553050 second
   d. Joey hears the sound first. It takes a bit more than $\frac{1}{20}$ second to get to him,
   but about $\frac{1}{3}$ second to get to centerfield.

Exercises 8.1 (p. 224)

1. b. 1. c. 0.9999996 d. 1. h. 1. i. 0.9999999

   Of these, parts c and i did not turn out to be 1 because of round-off errors in $\dfrac{1}{A}$.

2. 91351786; 91351787; 91351788; 91351789; 91351790; 91351791; 91351792

4. a. One gear has $\frac{15}{53}$ as many cogs (teeth) as the other.

   c. Land area is $\frac{1}{5}$ as much as water area.

5. b. The price ratio of the two calculators is 2 to 3.

   c. The ratio of his father's age to Bob's age is 2 to 1.

6. b. 0.0000002 c. 0.0000005 f. No error

Exercises 8.2 (p. 232)

1. a. 3; 24 d. 0.5; 9 f. 0.8125; 13 g. 0.0234375; 3 i. 0; 0

2. b. $\frac{1}{20}$; $1 \div 20 \times 20 = 1$. d. $\frac{1}{125}$; $1 \div 125 \times 125 = 1$.

   g. $\frac{1}{50}$; $1 \div 50 \times 50 = 1$. i. $\frac{1}{10000}$; $1 \div 10000 \times 10000 = 1$.

3. b. 20 d. 125 e. 64 f. 80

5. a. $\dfrac{1}{1,000,000}$ b. 0.000001 c. 1×10^{-6}

6. a. $\frac{3}{8}$ e. $\frac{79}{160}$ f. $\frac{17}{32}$ g. $\frac{5}{5}$

7. b. $13 \times \frac{1}{4}$ d. $4 \times \frac{1}{8}$ f. $4 \times \dfrac{1}{1}$ g. $16 \times \frac{1}{16}$

8. Each is equal to 0.75 on the calculator.

10. a. $3 \times 8 = 24 = 4 \times 6$ d. $6 \times 24 = 144 = 8 \times 18$

    f. $15 \times 24 = 360 = 20 \times 18$

11. b. $\dfrac{\cancel{6} \times 4}{\cancel{6} \times 7} = \dfrac{4}{7}$

12. a. $\frac{2}{5} = \frac{4}{10} = \frac{6}{15} = \frac{8}{20} = \frac{10}{25} = \cdots$ d. $\frac{5}{16} = \frac{10}{32} = \frac{15}{48} = \frac{20}{64} = \frac{25}{80} = \cdots$

    f. $\frac{13}{4} = \frac{26}{8} = \frac{39}{12} = \frac{52}{16} = \frac{65}{20} = \cdots$

13. a. $\dfrac{3}{4} = \dfrac{3 \times 2}{4 \times 2} \quad = \quad \dfrac{3 \times 3}{4 \times 3} \quad = \quad \dfrac{3 \times 4}{4 \times 4} = \dfrac{3 \times 5}{4 \times 5} = \cdots$

    $\qquad = \quad \dfrac{6}{8} \quad = \quad \dfrac{9}{12} \quad = \quad \dfrac{12}{16} \quad = \quad \boxed{\dfrac{15}{20}} = \cdots$

    $\dfrac{2}{5} = \dfrac{2 \times 2}{5 \times 2} \quad = \quad \dfrac{2 \times 3}{5 \times 3} \quad = \quad \dfrac{2 \times 4}{5 \times 4} = \quad \cdots$

    $\qquad = \quad \dfrac{4}{10} \quad = \quad \dfrac{6}{15} \quad = \quad \boxed{\dfrac{8}{20}} = \quad \cdots$

    c. $\dfrac{3}{5} = \dfrac{3 \times 2}{5 \times 2} \quad = \quad \dfrac{3 \times 3}{5 \times 3} \quad = \quad \cdots$

    $\qquad = \quad \dfrac{6}{10} \quad = \quad \boxed{\dfrac{9}{15}} \quad = \quad \cdots$

    $\dfrac{2}{3} = \dfrac{2 \times 2}{3 \times 2} \quad = \quad \dfrac{2 \times 3}{3 \times 3} \quad = \quad \dfrac{2 \times 4}{3 \times 4} = \dfrac{2 \times 5}{3 \times 5} = \cdots$

    $\qquad = \quad \dfrac{4}{6} \quad = \quad \dfrac{6}{9} \quad = \quad \dfrac{8}{12} \quad = \quad \boxed{\dfrac{10}{15}} = \cdots$

Exercises 8.3 (p. 240)

1. a. $(5 \times \frac{1}{5}) \times 3 = 3$ b. $(8 \times \frac{1}{8}) \times 5 = 5$ e. $(25 \times \frac{1}{25}) \times 4 = 4$

2. a. $\dfrac{\cancel{5} \times 4}{1 \times \cancel{5}} = \dfrac{4}{1} = 4$ d. $\dfrac{234 \times \cancel{1000}}{\cancel{1000} \times 1} = 234$ f. $\dfrac{1 \times 3}{1 \times 1} = 3$

4. a. 9   d. 15   e. 15   g. 56

5. $5 \times 8 = 40$

6. a. $\dfrac{\overset{3}{\cancel{21}}}{\cancel{4}_1} \times \dfrac{\overset{3}{\cancel{12}}}{\cancel{7}_1} = 9$ c. $\dfrac{\overset{4}{\cancel{36}}}{\cancel{7}_1} \times \dfrac{\overset{11}{\cancel{77}}}{\cancel{9}_1} = 44$ d. $\dfrac{\overset{7}{\cancel{42}}}{\cancel{5}_1} \times \dfrac{\overset{13}{\cancel{65}}}{\cancel{6}_1} = 91$ h. $\dfrac{\overset{}{\cancel{13}}}{\cancel{4}_1} \times \dfrac{\overset{}{\cancel{14}}}{\cancel{3}_1} = 1$

7. b. $(3 \times 7) \times (\tfrac{1}{8} \times \tfrac{1}{4}) = \tfrac{21}{32} = 0.65625$ e. $(5 \times 1) \times (\tfrac{1}{16} \times \tfrac{1}{4}) = \tfrac{5}{64} = 0.078125$
   g. $(6 \times 14) \times (\tfrac{1}{7} \times \tfrac{1}{3}) = \tfrac{84}{21} = 4$

8. c. $0.25 \times 0.05 = 0.0125$ f. $0.04 \times 0.2 = 0.008$
   $\quad 1 \div 80 = 0.0125 \qquad\qquad 1 \div 125 = 0.008$
   i. $0.3333333 \times 0.1666666 = 0.0555555$
   $\qquad\qquad 1 \div 18 = 0.0555555$

11. b. $3\,\boxed{\times}\,7\,\boxed{\div}\,8\,\boxed{\div}\,4\,\boxed{=}\,0.65625$ e. $5\,\boxed{\times}\,1\,\boxed{\div}\,16\,\boxed{\div}\,4\,\boxed{=}\,0.078125$
   g. $6\,\boxed{\times}\,14\,\boxed{\div}\,7\,\boxed{\div}\,3\,\boxed{=}\,4.$

13. a. $\tfrac{4}{3}$ c. $\tfrac{8}{5}$ e. $\tfrac{16}{7}$ h. $\tfrac{23}{6}$

Exercises 8.4 (p. 250)

1. a. $23.45 = (2 \times 10) + (3 \times 1) + (4 \times \tfrac{1}{10}) + (5 \times \tfrac{1}{100})$
   d. $835.2007 = (8 \times 100) + (3 \times 10) + (5 \times 1) + (2 \times \tfrac{1}{10}) + (7 \times \tfrac{1}{10000})$
2. a. $(2 \times 10^1) + (3 \times 10^0) + (4 \times 10^{-1}) + (5 \times 10^{-2})$
   d. $(8 \times 10^2) + (3 \times 10^1) + (5 \times 10^0) + (2 \times 10^{-1}) + (7 \times 10^{-4})$
3. b. 20,000   c. 0.0002
4. a. 2 ten-thousands = 20 thousands
6. a. $2.537 \times 10^{-12} = 2.53700 \times 10^{-12}$ c. $7.284 \times 10^{-9} \;= 7.284000 \times 10^{-9}$
   $\;\,3.462 \times 10^{-14} = 0.03462 \times 10^{-12}$ $\;\,2.84\;\; \times 10^{-13} = 0.000284 \times 10^{-9}$
   $\;\;\,\overline{\phantom{3.462 \times 10^{-14} =}\; 2.57162 \times 10^{-12}}$ $\;\;\,\overline{\phantom{2.84 \times 10^{-13} =}\;\; 7.284284 \times 10^{-9}}$

   d. $7.284 \times 10^{-9}\;= 7.284000 \times 10^{-9}$
   $\;\,2.84\;\; \times 10^{-13} = 0.000284 \times 10^{-9}$
   $\;\;\,\overline{\phantom{2.84 \times 10^{-13} =}\; 7.283716 \times 10^{-9}}$

7. $\;\;4.78 \times 10^7 \;=\quad\; 47{,}800{,}000.0000000000$
   $\;\;2.83 \times 10^{-8} =\qquad\qquad\qquad 0.0000000283$
   $\phantom{7.\;\;2.83 \times 10^{-8} =}\;\overline{47{,}800{,}000.0000000283}$

8. a. $0.234\,\boxed{+}\,0.75\,\boxed{+}\,0.33\,\boxed{+}\,0.005\,\boxed{+}\,0.5612\,\boxed{=}\,1.8802$
   c. Split at the decimal point: $\begin{array}{r} 82921 \;|\, .976102 \\ -\,5431 \;|\, .653531 \\ \hline 77490 \;|\, .322571 \end{array}$

Exercises 8.5 (p. 255)

1. b. $8.$ c. 0.04263 d. $30.$
2. a. In each case, the last digit of the answer multiplied by the last digit in the
   divisor agrees with the digit in the appropriate decimal place in the
   dividend, so we suspect that each answer is exact. (NOTE: With some
   calculators, in part c, for example, the calculator algorithm may produce a
   result like 0.04299. Then, we might suspect that the correct answer is
   0.04263 and check the product accordingly.)
3. a. $1.40625\,\boxed{\div}\,0.6666666\,\boxed{=}\,2.1093752;$
   $\;\tfrac{45}{32} \div \tfrac{2}{3} = \tfrac{45}{32} \times \tfrac{3}{2} = \tfrac{135}{64} = 135\,\boxed{\div}\,64\,\boxed{=}\,2.109375$
   c. $4.5\,\boxed{\div}\,8\,\boxed{=}\,0.5625;\; \tfrac{9}{2} \div \tfrac{16}{2} = \tfrac{9}{2} \times \tfrac{2}{16} = 9\,\boxed{\div}\,16\,\boxed{=}\,0.5625$

   d. $4.5\,\boxed{\div}\,4.5\,\boxed{=}\,1.;\; 4.5 \div \dfrac{9}{2} = 4.5 \times \dfrac{2}{9} = \dfrac{4.5}{1} \times \dfrac{2}{9} = \dfrac{9}{9} = 9\,\boxed{\div}\,9\,\boxed{=}\,1.$

4. a. $(5.428 \times 10^5) \div (4.3987 \times 10^{-3}) = (5.428 \div 4.3987) \times (10^5 \div 10^{-3})$
   $\phantom{a. (5.428 \times 10^5) \div (4.3987 \times 10^{-3})}\; = 1.2340009 \times 10^8$
   c. $(7.425 \times 10^{-9}) \div (2.25 \times 10^{-7}) = (7.425 \div 2.25) \times (10^{-9} \div 10^{-7})$
   $\phantom{c. (7.425 \times 10^{-9}) \div (2.25 \times 10^{-7})}\; = 3.3 \times 10^{-2}$

5. $(1.65 \times 10^{-5}) \div (5.6 \times 10^3) = (1.65 \div 5.6) \times (10^{-5} \div 10^3) = 0.2946428 \times 10^{-8}$
$= 2.946428 \times 10^{-9}$ grams per millilitre

Exercises 9.1 (p. 261)

1. Parts b, c, and e have "7" in the thousandths place.
3. a. $2.40 \times 0.06 = \$0.144$ or 14¢ sales tax;
   purchase price $2.40 + \$0.14 = \$2.54$
   b. $2.43 \times 0.06 = \$0.1458$ or 15¢ sales tax;
   purchase price $2.43 + \$0.15 = \$2.58$
   c. $2.58 - \$2.54 = \$0.04$ or 4¢ difference
   d. $2.40 + \$0.16 = \$2.56$; $2.43 + \$0.16 = \$2.59$;
   $2.59 - \$2.56 = \$0.03$ or 3¢ difference
   e. Our "rounding off" procedure gave a break in this case to the tax that has not reached the half-cent ($0.005) mark. The official tax tables supposedly were designed "statistically" so that, in the long run, the overcharges are balanced by the undercharges. These statistical designs took into account the likelihoods for the total price in a purchase to reach various levels.

Exercises 9.2 (p. 265)

1.

	Percent	*Product Form*	*Fraction Form*	*Decimal Form*
a.	11%	$11 \times \dfrac{1}{100}$	$\dfrac{11}{100}$	0.11
b.	20%	$20 \times \dfrac{1}{100}$	$\dfrac{20}{100} = \dfrac{1}{5}$	0.20
c.	8%	$8 \times \dfrac{1}{100}$	$\dfrac{8}{100} = \dfrac{2}{25}$	0.08
d.	75%	$75 \times \dfrac{1}{100}$	$\dfrac{75}{100} = \dfrac{3}{4}$	0.75
e.	5%	$5 \times \dfrac{1}{100}$	$\dfrac{5}{100} = \dfrac{1}{20}$	0.05
f.	100%	$100 \times \dfrac{1}{100}$	$\dfrac{100}{100} = 1$	1.00
g.	820%	$820 \times \dfrac{1}{100}$	$\dfrac{820}{100} = 8\dfrac{1}{5}$	8.20
h.	7.5%	$7.5 \times \dfrac{1}{100}$	$\dfrac{7.5}{100} = \dfrac{75}{1000}$	0.075
i.	$\frac{7}{5}\%$	$\dfrac{7}{5} \times \dfrac{1}{100}$	$\dfrac{\frac{7}{5}}{100} = \dfrac{14}{1000}$	0.014
j.	$\frac{1}{4}\%$	$\dfrac{1}{4} \times \dfrac{1}{100}$	$\dfrac{\frac{1}{4}}{100} = \dfrac{25}{10000}$	0.0025
k.	12.5%	$12.5 \times \dfrac{1}{100}$	$\dfrac{12.5}{100} = \dfrac{125}{1000}$	0.125

	Percent	Product Form	Fraction Form	Decimal Form
l.	23%	$23 \times \dfrac{1}{100}$	$\dfrac{23}{100}$	0.23
m.	$2\frac{1}{2}\%$	$2\dfrac{1}{2} \times \dfrac{1}{100}$	$\dfrac{2\frac{1}{2}}{100} = \dfrac{25}{1000}$	0.025
n.	$6\frac{3}{4}\%$	$6\dfrac{3}{4} \times \dfrac{1}{100}$	$\dfrac{6\frac{3}{4}}{100} = \dfrac{675}{10000}$	0.0675
o.	0.5%	$0.5 \times \dfrac{1}{100}$	$\dfrac{0.5}{100} = \dfrac{5}{1000}$	0.005
p.	0.02%	$0.02 \times \dfrac{1}{100}$	$\dfrac{0.02}{100} = \dfrac{2}{10000}$	0.0002
q.	0.007%	$0.007 \times \dfrac{1}{100}$	$\dfrac{0.007}{100} = \dfrac{7}{100000}$	0.00007

4. a. $(\frac{1}{2} \times 100)\% = 50\%$; $(0.5 \times 100)\% = 50\%$
   b. $(\frac{1}{5} \times 100)\% = 20\%$; $(0.2 \times 100)\% = 20\%$
   c. $(\frac{1}{10} \times 100)\% = 10\%$; $(0.1 \times 100)\% = 10\%$
   d. $(\frac{1}{8} \times 100)\% = 12.5\%$; $(0.125 \times 100)\% = 12.5\%$
   e. $(\frac{3}{8} \times 100)\% = 37.5\%$; $(0.375 \times 100)\% = 37.5\%$
   f. $(\frac{3}{4} \times 100)\% = 75\%$; $(0.75 \times 100)\% = 75\%$
   g. $(1\frac{1}{2} \times 100)\% = 150\%$; $(1.5 \times 100)\% = 150\%$
   h. $(6\frac{1}{4} \times 100)\% = 625\%$; $(6.25 \times 100)\% = 625\%$
5. $(\frac{3}{20} \times 100)\% = 15\%$
7. $250\% = \frac{250}{100} = 2.5 = 2\frac{1}{2}$ times

Exercises 9.3 (p. 271)

1. a. 963.15 b. 16.00495 c. 17.35965 d. 0.00180647 e. 483 f. 4789 g. 76.5
   h. 15% i. 130% j. 0.270028%
3. $0.008 or 0.8¢
4. $(8 \times 10^{-3}) \times (10^1) \times (2 \times 10^7) = 16 \times 10^5 = 1.6 \times 10^6 = \$1,600,000.$
5. $0.04 \times 32 = 1.28$ ounces
8. 5.09138
10. $22,717.92
12. $17,180 − $16,000 = $1,180; $1,180 × 0.34 = $401.20;
    $3830.00 + $401.20 = $4231.20 tax
16. $5 \times 5280 \times 0.06 = 1584$ feet

By a 6% grade is meant 6 feet of drop for every 100 feet across in horizontal distance. Because we have not yet discussed the Pythagorean theorem (Chapter 12), for ease in computation we are using the hypotenuse as an estimate of the horizontal distance here. (Actually, this approximation is quite close. Using 5 miles for the hypotenuse, the drop should be about 1581.16 feet, or less than 3 feet off our answer.)

3. $87,500 \times 300\% = \$87,500 \times 3 = \$262,500$ increase needed;
   $262,500 + \$87,500 = \$350,000$ sales needed this year

4. a. $20\% \times E = 240$, so $E = 240 \div 20\%$
   $$= 240 \boxed{\div} 0.2 \boxed{=} 1200. \text{ total enrollment}$$
   b. $30\% \times 1200 = 0.3 \boxed{\times} 1200 \boxed{=} 360.$
   c. $40\% \times 1200 = 0.4 \boxed{\times} 1200 \boxed{=} 480.$
   d. $10\% \times 1200 = 0.1 \boxed{\times} 1200 \boxed{=} 120.$

6. a. $60\% \times N = 35$, so $N = 35 \div 60\% = 35 \boxed{\div} 0.6 \boxed{=} 58.333333$; there are
   probably 58 or 59 pizza parlors in Megalopolis (a quick check shows that
   35 is about 60.34% of 58 and about 59.32% of 59); let's assume there are
   58 pizza parlors.
   b. $40 \boxed{\div} 63 \boxed{=} 0.63492063 \doteq 63.5\%$
   c. $[(63.5 - 60.3) \div 60.3] \times 100\% \doteq 5.31\%$; note that we have computed the
   percent change in the *share* (that is, in the percent of total parlors
   owned); this is much smaller than the percent change in the *number* of
   parlors owned $[(40 - 35) \div 35] \times 100\% \doteq 14.29\%$

2. a. $4.84 c. $14.88 d. $2.38 f. $26.71
6. a. 95% b. $97\frac{1}{2}\%$ c. 85% d. $81\frac{1}{4}\%$ e. 10% f. 15% g. $66\frac{2}{3}\%$ h. 0%

1.

	Interest	*Amount Due*
a.	$324	$3,024
c.	$150.50	$13,000.50
d.	$973.95	$7,466.95
e.	$3.43	$109.43

2. 161.54%

4. $64.75 \div 1\frac{1}{4} = \$51.80$ interest per year; $51.80 \div 0.0925 = \$560$ principal

5. a. $2500 \times \left(1 + \dfrac{0.065}{12}\right)^{60} \doteq \$3457.04$; interest $957.04

   c. $750,220 \times \left(1 + \dfrac{0.105}{2}\right)^{30} \doteq \$3,482,184.50$; interest $2,731,964.50

   d. $750,220 \times \left(1 + \dfrac{0.105}{4}\right)^{60} \doteq \$3,551,230.40$; interest $2,801,010.40

   e. $750,220 \times \left(1 + \dfrac{0.105}{12}\right)^{180} \doteq \$3,599,376.10$; interest $2,849,156.10

1. a. 8.16% d. 10.38% f. $\left(1 - \dfrac{0.22}{12}\right)^{12} \doteq 24.36\%$

2. At 5.5% compounded semiannually, the effective rate is
   $$\left(1 + \frac{0.055}{2}\right)^{2} \doteq 5.58\%$$

   At 5.25% compounded daily, the effective rate is
   $$\left(1 + \frac{0.0525}{365}\right)^{365} \doteq 5.39\%$$

   So it is better to invest in 5.5% compounded semiannually.

1. In these schedules of payments, the last two columns are omitted. In each case, the *Installment Paid* is given above the table (with the last payment specified in the table). The final column, *Unpaid Balance at End of Period,* is the same as the *Unpaid Balance* (second column) for the following period.

b. Installment $124.95

Payment Period	Unpaid Balance	Interest on Balance	Amount Owed
1	$850.00	$31.88	$881.88
2	756.93	28.38	785.31
3	660.36	24.76	685.12
4	560.17	21.01	581.18
5	456.23	17.11	473.34
6	348.39	13.06	361.45
7	236.50	8.87	245.37
8	120.42	4.52	124.94

Last payment: $124.94

c. Installment $1,061.61

Payment Period	Unpaid Balance	Interest on Balance	Amount Owed
1	$10,450.00	$333.09	$10,783.09
2	9,721.48	309.87	10,031.35
3	8,969.74	285.91	9,255.65
4	8,194.04	261.19	8,455.23
5	7,393.62	235.67	7,629.29
6	6,567.68	209.34	6,777.02
7	5,715.41	182.18	5,897.59
8	4,835.98	154.15	4,990.13
9	3,928.52	125.22	4,053.74
10	2,992.13	95.37	3,087.50
11	2,025.89	64.58	2,090.47
12	1,028.86	32.80	1,061.66

Last payment: $1,061.66

3. $307.57

4. $300.51

6. a. The dealer's calculations were based on the assumption that the customer owes him $4000 each month throughout the loan period. In other words, in figuring the interest, the customer was not given credit for working off his debt.

   b. $232.92

   c. The amount overcharged per month was $242.22 − $232.92 = $9.30. For the entire period, the customer was overcharged $9.30 × 18 = $167.40.

8. Last payment: $1,413.21. Such a last payment (much much larger than the installment) is called a "balloon" payment. If Janese signed the loan agreement without realizing what the last payment would be, she may be startled to find the loaner threatening to take the security she put up for the loan if she can't immediately come up with $1,413.21, which turns out to be almost half the original loan.

1. For example: pounds per square inch (inflating tires), checkups per year (doctor's visits), yards gained per quarter (football), blooms per stem (cut flowers), or degrees per minute (cooling).
2. a. 20 miles per gallon or $\frac{1}{20}$ gallon per mile
   b. For miles per gallon, the larger the number, the more economical; for gallons per mile, the smaller the number, the more economical the fuel consumption.
4. The first may be compared with the second or the second with the first. That is, if the quantities are A and B, we may express the comparison as A to B or as B to A.

Exercises 10.2 (p. 308)

2. a.

No. Shots	Unit Price	No Shots	Unit Price
1	$2.50	11	$0.23
2	1.25	12	0.21
3	0.83	13	0.19
4	0.63	14	0.18
5	0.50	15	0.17
6	0.42	16	0.16
7	0.36	17	0.15
8	0.31	18	0.14
9	0.28	19	0.13
10	0.25	20	0.13

   b. $3.50 ÷ $.13 ≐ 26.9 or 27 shots. Actually, because the 13¢ ($.13) came from $2.50 ÷ 20, to be precise, we should solve the problem thus:

$$\$3.50 \div (\$2.50 \div 20) = \frac{\$3.50}{1} \div \frac{\$2.50}{20}$$

$$= \frac{\$3.50}{1} \times \frac{20}{\$2.50}$$

$$= 28 \text{ shots}$$

5. a. $2 \div \frac{1}{2} = 2 \boxplus 0.5 \boxminus 4.$
   b. Four $\frac{1}{2}$-lb. cups = 4 × $.35 = $1.40
      one 2-lb. cup   =   − 1.17
                        Saving: $0.23

Exercises 10.3 (p. 313)

1. a. 18 × 15.6 = 280.8 miles   b. 613.4 ÷ 21 ≐ 29.21 mpg
   c. 54.89 × 9.35 = 513.2215 km or about 31.89 miles
3. a. 449.8 miles   b. 17.1
   c.

Date	Odometer	Miles	Gallons	MPG
July 1	26539.8	—	18.8	—
July 1	26989.6	449.8	17.1	26.30
July 2	27521.3	531.7	19.4	27.41
July 3	27937.1	415.8	16.7	24.90
July 4	28470.7	533.6	18.3	29.16
July 5	28906.4	435.7	15.9	27.40
July 6	29401.5	495.1	17.6	28.13
July 7	29975.4	573.9	19.5	29.43

Exercises 10.4 (p. 320)

1. a. 573.5
   b. The total, 3441, has 3 remainder on division by 6, so we might conveniently interpret that in the average, every second class has an extra person beyond 573.
2. a. 6314.43   b. 45.02   c. 3.1898
4. a. 0.2666666; batting averages are usually given to 3-place decimals, which in this case would be 0.267.
   b. 26.7% or $26\frac{2}{3}\%$ c. About 27 hits
5. a. 64.3   b. 21.93   c. 490.57   d. 19.78
   NOTE: For part d, we have

$$\frac{(18.23 \times \frac{1}{7}) + (15.59 \times \frac{2}{7}) + (21.03 \times \frac{3}{7}) + (25.97 \times \frac{1}{7})}{\frac{1}{7} + \frac{2}{7} + \frac{3}{7} + \frac{1}{7}}$$

$$= [(18.23 \times 1) + (15.59 \times 2) + (21.03 \times 3) + (25.97 \times 1)] \times \frac{1}{7}$$
$$= [(18.23 \times 1) + (15.59 \times 2) + (21.03 \times 3) + (25.97 \times 1)] \div 7$$

Exercises 10.5 (p. 329)

1. a. weighted average: 52 mph   b. weighted harmonic average: 50 mph
   c. 80 miles at 40 mph; 180 miles at 60 mph
   d. 5 hours on each portion of trip
   e. Part b (one-half the total time compared with $\frac{2}{5}$ the total time)
3. To 2-place decimals, each replacement should be 400.00 ohms.
5. 500 gallons per hour.

Exercises 10.6 (p. 340)

1. a. weighted average: 52 mph   b. weighted harmonic average: 50 mph
   difference is approximately 0.0577
3. a. 2534.0134 miles   b. $4087.1 \div 116.28278 \doteq 35.507054$ km/hr
   c. about 21.791822 mph
4. a. $88 \times \frac{65}{60} \times \frac{3}{4} = 71.5$ feet   b. 302.5 feet, or almost $\frac{3}{5}$ of a city block
7. One imperial gallon $= \dfrac{277.418}{231}$ U.S. gallons $\doteq 1.2009437$ U.S. gallons

$$\$1.06 \div \frac{277.418}{231} = \$1.06 \times \frac{231}{277.418} \doteq \$0.8826391, \text{ or about } 88¢$$

8. a. $48 \times 90 = 4320$ ml $= 4\frac{1}{2}$ qt.
   b. $150 \times 0.4535924 \doteq 68.03886$ kg
   $68.03886 \times 90 \doteq 6123.4974$ ml

$$6123.4974 \div 960 \doteq 6 \text{ qt. } 363.4974 \text{ ml} = 6 \text{ qt. } \frac{363.4974}{480} \text{ pt.}$$

$$= 6 \text{ qt. } 0.7572862 \text{ pt., or about } 6 \text{ qt. } \frac{3}{4} \text{ pt.}$$

Exercises 10.7 (p. 350)

1. a. Similar; the ratio of corresponding sides is about 0.65
   b. Not similar. In ABCD the two shortest sides (AD and CD) are adjacent, but in EFGH the two shortest sides (EH and FG) are opposite. No alignment of the two figures will give a constant ratio for all corresponding sides.
   c. Not similar; for example, $\dfrac{AE}{HI} \doteq 0.8, \dfrac{DE}{IJ} \doteq 0.75$
   d. Similar; the ratio of corresponding sides is about 0.73

2. a. A ↔ D   d. A ↔ F   A ↔ F
       B ↔ F   B ↔ D or B ↔ E
       C ↔ E   C ↔ E   C ↔ D
4. a. A′B′ = 3 × AB   b. A′B′C′D′A′ = 3 × ABCDA   c. $h' = 3 \times h$
    d. Area of A′B′C′D′ = 9 × area of ABCD

6. a. $\dfrac{EF}{3.8} = \dfrac{3.51}{2.6}$ so EF $\doteq$ 5.13; $\dfrac{DF}{4.2} = \dfrac{3.51}{2.6}$ so DF $\doteq$ 5.67

    b. The ratio of areas is $\left(\dfrac{3.51}{2.6}\right)^2 = 1.8225$

9. 505.425 inches $\doteq$ 42 ft. 1.43 in., or about 42 ft. $1\frac{7}{16}$ in.
10. 34.782608 inches = 2 ft. 10.782608 in. $\doteq$ 2 ft. $10\frac{13}{16}$ in.

Exercises 10.8 (p. 354)

1. a. 4.125 in. = $4\frac{1}{8}$ in.
    b. 36 ÷ 4 = 9, so one dimension on the map must measure at least
       9 inches
3. 1 cm = 150 miles
5. a. $3 \div \frac{1}{45} = \frac{3}{1} \times \frac{45}{1} = 135$ cm   b. $135 \div 2.54 \doteq 53.149606$ in. $\doteq$ 4 ft. $5\frac{1}{8}$ in.

Exercises 11.1 (p. 357)

1. No; the only even prime number is 2.
2. a. 1 × 15, 3 × 5 b. 1 × 17 c. 1 × 21, 3 × 7 d. 1 × 22, 2 × 11 e. 1 × 23
    f. 1 × 24, 2 × 12, 3 × 8, 4 × 6, 2 × 2 × 6, 2 × 4 × 3, 2 × 2 × 2 × 3
    g. 1 × 25, 5 × 5
    h. 1 × 36, 2 × 18, 3 × 12, 4 × 9, 6 × 6, 2 × 2 × 9, 2 × 3 × 6, 3 × 3 × 4,
       2 × 2 × 3 × 3

3.

2	3	5	7	11
13	17	19	23	29
31	37	41	43	47
53	59	61	67	71
73	79	83	89	97

5. Three ways

Exercises 11.2 (p. 359)

1. b. $2^3 \times 3^2 \times 5^1$ e. $2^4 \times 3^1 \times 5^5$ f. $2^4 \times 5^6$ h. $2^{12} \times 5^{15}$
2. a. 13×29 c. 41×53 e. 37×11^2 f. $3^5 \times 5^1 \times 7^3$
    h. $3^7 \times 5^3 \times 13^1 \times 17^1$ j. $2^{12} \times 3^7 \times 5^1$

Exercises 11.3 (p. 364)

1. b. The next prime after 7 is 11. The smallest multiple of 11 not yet crossed out
       will be 11 × 11 = 121, which is beyond the end of the table. Therefore you
       can stop after crossing out multiples of 7. (Your list of primes should be the
       same as that given in the answer to Exercise 3 of Exercises 11.1, with the
       addition of 101 itself.)

2. 2, 3, 5, 7, 11, 13, 17, 19, 23, 29, 31, 37, 41, 43, 47, 53, 59, 61, 67, 71, 73, 79, 83, 89, 97, 101, 103, 107, 109, 113, 127, 131, 137, 139, 149, 151, 157, 163, 167, 173, 179, 181, 191, 193, 197, 199, 211, 223, 227, 229, 233, 239, 241, 251, 257, 263, 269, 271, 277, 281, 283, 293, 307, 311, 313, 317, 331, 337, 347, 349, 353, 359, 367, 373, 379, 383, 389, 397, 401, 409, 419, 421, 431, 433, 439, 443, 449, 457, 461, 463, 467, 479, 487, 491, 499, 503, 509, 521, 523, 541, 547, 557, 563, 569, 571, 577, 587, 593, 599, 601, 607, 613, 617, 619, 631, 641, 643, 647, 653, 659, 661, 673, 677, 683, 691, 701, 709, 719, 727, 733, 739, 743, 751, 757, 761, 769, 773, 787, 797, 809, 811, 821, 823, 827, 829, 839, 853, 857, 859, 863, 877, 881, 883, 887, 907, 911, 919, 929, 937, 941, 947, 953, 967, 971, 977, 983, 991, 997, 1009, 1013, 1019, 1021, 1031, 1033, 1039, 1049, 1051, 1061, 1063, 1069, 1087, 1091, 1093, 1097, 1103, 1109, 1117, 1123, 1129, 1151, 1153, 1163, 1171, 1181, 1187, 1193

Exercise 11.4 (p. 367)

1. a, b, f, g, and j are prime

Exercises 11.5 (p. 376)

1. a. 12 b. 6 c. 54 d. 108
2. a. 1 c. 18 d. 19 f. 432 h. 42 i. 1
3. a. 147 b. 12,096 d. 142,884
4. a. 42 c. 19

Exercises 11.6 (p. 382)

1. a. $\frac{13}{19}$ c. $\frac{31}{73}$ e. $\frac{15}{28}$ g. $\frac{7}{33}$
2. b. 455 d. 5355 e. 148,225
3. b. $\frac{166}{777}$ d. $\frac{5293}{11907}$
5. At 12:36 p.m.
7. Every 12 seconds

Exercises 11.7 (p. 384)

1. 1, 1, 2, 3, 5, 8, 13, 21, 34, 55, 89, 144, 233, 377, 610, 987, 1,597, 2,584, 4,181, 6,765, 10,946, 17,711, 28,657, 46,368
3. Each sum is one less than a corresponding term in the Fibonacci sequence, so we guess that the next sums will be 20 and 33.
   Check: $1 + 1 + 2 + 3 + 5 + 8 = 20$; $1 + 1 + 2 + 3 + 5 + 8 + 13 = 33$
4. The sums are equal to every other term in the Fibonacci sequence, so we guess that the next sums will be 55 and 144.
   Check: $1 + 2 + 5 + 13 + 34 = 55$; $1 + 2 + 5 + 13 + 34 + 89 = 144$
6. The sum always equals the product of two consecutive terms from the Fibonacci sequence, so the next sums should be $8 \times 13 = 104$ and $13 \times 21 = 273$.
   Check: $1^2 + 1^2 + 2^2 + 3^2 + 5^2 + 8^2 = 104$;
   $1^2 + 1^2 + 2^2 + 3^2 + 5^2 + 8^2 + 13^2 = 273$

Exercises 11.8 (p. 391)

1. a. $[7, 1, 4]$ c. $[1, 3, 1, 4]$ d. $[2, 2, 2]$ g. $[1, 5, 4, 5]$ h. $[0, 5]$
2. b. $\frac{30}{23}$ c. $\frac{194}{27}$ e. $\frac{129}{56}$ h. $\frac{11}{18}$

1. a. [7, 1, 4] c. [1, 3, 1, 4] d. [2, 2, 2] f. [4, 2, 1, 1, 2] g. [1, 5, 4, 5] h. [0, 5]

2. b. $(4 \times 10) - (3 \times 13) = 40 - 39 = 1$

    c. $(36 \times 11) - (5 \times 79) = 396 - 395 = 1$

    f. $(2 \times 4) - (3 \times 3) = 8 - 9 = 1$

4. a. $(8 \times 14) - (5 \times 21) = 7$

    $8 \times [14 + (5 \times t)] - 5 \times [21 + (8 \times t)] = 7$; let $t = -2$.

		Contents	
Action		8	5
Fill 8 gallon jug	→	8	0
Pour from 8 jug to 5 jug		3	5
Dump 5 gallon jug		3	0 →
Pour from 8 jug to 5 jug		0	3
Fill 8 gallon jug	→	8	3
Pour from 8 to 5		6	5
Dump 5 gallon jug		6	0 →
Pour from 8 to 5		1	5
Dump 5 gallon jug		1	0 →
Pour from 8 to 5		0	1
Fill 8 gallon jug	→	8	1
Pour from 8 to 5		4	5
Dump from 5 jug		4	0 →
Pour from 8 to 5		0	4
Fill 8 gallon jug	→	8	4
Pour from 8 to 5		7	5
Dump from 5 jug		7	0 →

For $t = -2$, we have $(8 \times 4) - (5 \times 5) = 7$; that is, fill the 8 gallon jug four times and dump out from the 5 gallon jug 5 times.

    b. Let $t = -1$; then $(8 \times 9) - (5 \times 13) = 7$

    Let $t = 0$; then $(8 \times 14) - (5 \times 21) = 7$

5. a. $\dfrac{27}{10}$ c. $\dfrac{178}{55}$ d. $\dfrac{972}{421}$ f. $\dfrac{79}{135}$ g. $\dfrac{1,807}{329,006,431}$ h. $\dfrac{11}{13}$

6. a. $\frac{3}{8}$ b. $\frac{27}{10}$

    The result in 6b is the same as that in 5a. With the addition of two zeros at the left, the $(n + 1)$st convergent then has the same value as the nth convergent without addition of the zeros.

7. $\frac{27}{10}$

8. b. $(73 \times a) + (31 \times b) = 25$

    $\frac{73}{31} = [2, 2, 1, 4, 2]$

    The double-column property gives $(31 \times 33) - (73 \times 14) = 1$, from which $(31 \times 825) - (73 \times 350) = 25$ and

    $31 \times [(825 + (73 \times t)] - 73 \times [350 + (31 \times t)] = 25$. Let $t = -11$; then $(31 \times 22) - (73 \times 9) = 25$.

9. $\frac{1247}{859} = [1, 2, 4, 1, 2, 13, 2]$.

    The convergents are:

		1	2	4	1	2	13	2
0	1	1	3	13	16	45	601	1247
1	0	1	2	9	11	31	414	859

$\frac{16}{11}$ is the fourth convergent and $\frac{45}{31}$ is the fifth convergent.

Exercises 11.10 (p. 406)

1. a. 1, 3, 5, 7, 9, 11, 13, 15, 17   b. the first nine odd numbers.
3. a. Yes   b. The square constructed in part 1a.

6. a.

11	16	15
18	14	10
13	12	17

b.

$\frac{1}{3}$	$\frac{7}{6}$	1
$\frac{3}{2}$	$\frac{5}{6}$	$\frac{1}{6}$
$\frac{2}{3}$	$\frac{1}{2}$	$\frac{4}{3}$

8. a. For example, $512 + 4 + 128 = 644 \neq 16 + 64 + 256 = 336$.
    b. The product of each row, column, or diagonal is 262,144.

c.

128	1	32
4	16	64
8	256	2

d.

3	729	243
6561	81	1
27	9	2187

e.

9	531441	59049
43046721	6561	1
729	81	4782969

Yes; by taking products. The product is 531,441.

This square is not "magic" by taking products.

Exercises 12.1 (p. 415)

1. a, c, g, and i are correct; h is not correct
2. b. $\sqrt{625} = \sqrt{25 \times 25} = \sqrt{25} \times \sqrt{25} = 5 \times 5 = 25$

    c. $\sqrt{196} = \sqrt{4 \times 49} = \sqrt{4} \times \sqrt{49} = 2 \times 7 = 14$ e. $\sqrt{\dfrac{25}{9}} = \dfrac{\sqrt{25}}{\sqrt{9}} = \dfrac{5}{3}$

    g. $\sqrt{1225} = \sqrt{25 \times 49} = 5 \times 7 = 35$
    i. $\sqrt{1296} = \sqrt{9 \times 9 \times 16} = 3 \times 3 \times 4 = 36$
    j. $\sqrt{1024} = \sqrt{4 \times 4 \times 4 \times 4 \times 4} = 2^5 = 32$

3. a. $2 \times \sqrt{2} \doteq 2.828$ b. $3 \times \sqrt{3} \doteq 5.196$ e. $\sqrt{3} \times \sqrt{5} \doteq 3.873$
    g. $\sqrt{4 \times 49 \times 3} = 2 \times 7 \times \sqrt{3} \doteq 24.248$

    h. $\sqrt{2 \times 9 \times 49} = 3 \times 7 \times \sqrt{2} \doteq 29.694$

    j. $\dfrac{3}{\sqrt{2} \times \sqrt{5}} \doteq \dfrac{3}{1.414 \times 2.236} \doteq 0.949$

4. a. $P = 2 \times \pi \times \sqrt{\frac{2}{32}} = 2 \times \pi \div 4 = \pi \div 2$
    $\doteq$ $355\ \boxplus\ 113\ \boxplus\ 2\ \boxed{=}\ 1.5707964$ seconds
    b. $60\ \boxplus\ 1.5707964\ \boxed{=}\ 38.197184$; about 38 ticks per minute

    c. $1 = 2 \times \pi \times \sqrt{\dfrac{\ell}{32}}$ means $4 \times \pi^2 \times \dfrac{\ell}{32} = 1$

    $\pi^2 \doteq (355 \div 113)^2 \doteq 9.8696053$; so
    $\ell \doteq 8\ \boxplus\ 9.8696053\ \boxed{=}\ 0.8105693$ feet, or about 9.7268316 inches. A more accurate estimate for π^2 is 9.8696044; using this estimate, the length is about 0.8105695 feet or 9.7268336 inches.

1. The number of steps will vary according to what one uses as the first guess. In the following table, we show the number of steps based on the first guess indicated.

A	First Guess	$\sqrt{A}$	Number of steps	A	First Guess	$\sqrt{A}$	Number of steps
0	0.0	0.0000000	1	2.8	1.7	1.6733200	3
0.2	0.4	0.4472135	5	3.0	1.7	1.7320508	4
0.4	0.6	0.6324555	5	3.5	1.9	1.8708286	5
0.6	0.8	0.7745966	5	4.0	2.0	2.0000000	1
0.8	0.9	0.8944271	4	4.5	2.1	2.1213203	3
0.9	1.0	0.9486832	5	5.0	2.2	2.2360679	4
1.0	1.0	1.0000000	1	5.5	2.4	2.3452079	4
1.2	1.1	1.0954451	4	6.0	2.5	2.4494897	4
1.4	1.2	1.1832159	4	6.5	2.6	2.5495098	4
1.6	1.3	1.2649110	5	7.0	2.6	2.6457513	4
1.8	1.3	1.3416407	5	7.5	2.7	2.7386128	4
1.9	1.4	1.3784048	4	8.0	2.8	2.8284271	4
2.0	1.4	1.4142135	4	8.5	2.9	2.9154759	4
2.2	1.5	1.4832396	5	9.0	3.0	3.0000000	1
2.4	1.6	1.5491933	5	9.5	3.1	3.0822070	4
2.6	1.6	1.6123415	4	10.0	3.3	3.1622777	4

3. $\sqrt{A} < A$ means $A < A^2$; dividing by A throughout, $1 < A$; that is, $A > 1$.

1. a.

A	$\sqrt{A}$	A	$\sqrt{A}$	A	$\sqrt{A}$
0.01	0.10	0.35	0.59	0.70	0.84
0.06	0.24	0.40	0.63	0.75	0.87
0.10	0.32	0.45	0.67	0.80	0.89
0.15	0.39	0.50	0.71	0.85	0.92
0.20	0.45	0.55	0.74	0.90	0.95
0.25	0.50	0.60	0.77	0.95	0.97
0.30	0.55	0.65	0.81	1.00	1.00

b.

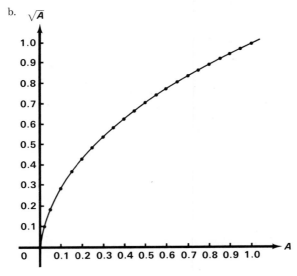

2.

A	$\sqrt{A}$	A	$\sqrt{A}$	A	$\sqrt{A}$
10	3.1622777	25	5.0000000	40	6.3245553
15	3.8729833	30	5.4772256	45	6.7082039
20	4.4721360	35	5.9160798	50	7.0710678

Exercises 12.4 (p. 434)

2. a. $\sqrt{17^2 + 33^2} \doteq 37.12$ c. 91.78
3. 746 feet
5. a, f, h, and j are right triangles
8. $BC = \sqrt{30^2 + 40^2}$
    $$\sqrt{CD^2 + BC^2} = \sqrt{20^2 + 30^2 + 40^2} = \sqrt{400 + 900 + 1600}$$
    $$= \sqrt{2900} \doteq 53.9 \text{ inches}$$
    The clerk should not accept the box.

Exercises 12.5 (p. 438)

3. a. 4   c. 3301.9272   f. 0.6406959
4.

A	$\sqrt[3]{A}$	A	$\sqrt[3]{A}$	A	$\sqrt[3]{A}$
0	0.00	4	1.59	8	2.00
1	1.00	5	1.71	9	2.08
2	1.26	6	1.82	10	2.15
3	1.44	7	1.91		

Exercises 12.6 (p. 441)

1. b. 2.2973967   d. 3.3302457   e. 2.5118864   f. 5.1186036   g. 13
2. a. $\sqrt{\sqrt{64}} = \sqrt{8} \doteq 2.8284271$ b. $\sqrt{\sqrt{100}} = \sqrt{10} \doteq 3.1622777$
    $\sqrt[4]{64} \doteq 2.8284271$ $\qquad\qquad\qquad$ $\sqrt[4]{100} \doteq 3.1622777$
3. a. $\sqrt[3]{\sqrt{64}} = \sqrt[3]{8} = 2$ b. $\sqrt{\sqrt[3]{64}} = \sqrt{4} = 2$
    $\sqrt[6]{64} = 2$ $\qquad\qquad\qquad$ $\sqrt[6]{64} = 2$

Exercises 12.7 (p. 446)

1.

a	$V = a \times [4 \times a^2 - (39 \times a) + 93.5]$
0	0
1	58.5
2	63.0
3	37.5
4	6.0

So $a \doteq 2$; by calculus, a more accurate value turns out to be approximately 1.6042985. Using the rough estimate 1.6,
$$V \doteq 1.6 \times [4 \times (1.6)^2 - (39 \times 1.6) + 93.5]$$
$$= 66.144$$

3.

r	$A = r \times \sqrt{(11.5)^2 - r^2}$	r	$A = r \times \sqrt{(11.5)^2 - r^2}$
0	0.000000	6	58.864251
1	11.456439	7	63.869007
2	22.649503	8	66.090847
3	33.305405	9	64.430195
4	43.127717	10	56.789083
5	51.780788	11	36.895122

$r \doteq 8$; again by calculus, it can be shown that a more accurate value for r is approximately 8.131728. Using 8.13 as a rough estimate,

$$V = 8.13 \times \sqrt{(11.5)^2 - (8.13)^2}$$
$$\doteq 66.124994$$

5.

r	$A = (2 \times \pi \times r^2) + \dfrac{1000}{r}$
1	1006.2832
2	525.13274
3	389.88201
3.5	362.68331
4.1	349.52279
4.2	348.93064
4.3	348.73425
4.4	348.91521
4.5	349.45674
5	357.07965
6	392.86136
8	527.12389
10	728.31858

Using the approximation 4.3, $A \doteq 348.73425$

$$H = 500 \div (\pi \times r^2) \doteq \frac{500}{(4.3)^2} \times \frac{113}{355} \doteq 8.6076219, \text{ or about 8.6 cm}$$

The least area is approximately 348.7 cm
Using a more accurate approximation for π, the area for $r = 4.3$ is about 348.73424, a difference of only 1 in the last decimal.

Index

Page numbers in **boldface** indicate definitions of terms.